Populations of Plant Pathogens

THEIR DYNAMICS AND GENETICS

Populations of Plant Pathogens

THEIR DYNAMICS AND GENETICS

Edited for the
British Society for Plant Pathology
by

M.S. WOLFE
Plant Breeding Institute,
Trumpington, Cambridge

and

C.E. CATEN
Department of Genetics,
University of Birmingham,
Birmingham B15 2TT

BLACKWELL SCIENTIFIC PUBLICATIONS
OXFORD LONDON EDINBURGH
BOSTON PALO ALTO MELBOURNE

Editorial offices:
Osney Mead, Oxford, OX2 0EL
8 John Street, London, WC1N 2ES
23 Ainslie Place, Edinburgh, EH3 6AJ
52 Beacon Street, Boston
Massachusetts 02108, USA
667 Lytton Avenue, Palo Alto
California 94301, USA
107 Barry Street, Carlton
Victoria 3053, Australia

First published 1987

Set by Setrite Typesetters Ltd, Hong Kong
Printed in Great Britain by
Butler & Tanner Ltd, Frome and London

DISTRIBUTORS

USA and Canada
Blackwell Scientific Publications Inc
P O Box 50009, Palo Alto
California 94303

Australia
Blackwell Scientific Publications
(Australia) Pty Ltd
107 Barry Street,
Carlton, Victoria 3053

British Library Cataloguing in Publication Data

Population of plant pathogens: their dynamics and genetics.
1. Micro-organisms, Phytopathogenic
2. Plant genetics
I. Wolfe, M.S. II. Caten, C.E.
III. British Society for Plant Pathology
581.2′3 SB731

ISBN 0-632-01433-4

Library of Congress Cataloging-in-Publication Data

Populations of plant pathogens.

Bibliography: p.
Includes index.
1. Micro-organisms, Phytopathogenic.
2. Microbial populations. 3. Plant diseases—Genetic aspects. 4. Microbial genetics.
5. Micro-organisms, Phyto-pathogenic—Host plants. I. Wolfe, M.S. II. Caten, C.E.
III. British Society for Plant Pathology.
SB731.P68 1987 632′.3 86-23229
ISBN 0-632-01433-4

Contents

Section 3
Genetic changes in pathogen populations

Preface

Pathogens are naturally present in most crops but become important only when their populations attain a size at which significant economic damage is produced. For centuries, and in a variety of conscious and unconscious ways, man has attempted to maintain pathogen populations at benign levels, generally with moderate success but with failures sometimes so dramatic as to change the course of history.

With the growth of scientific understanding of plant disease in the last few decades, the range of control strategies has increased but the essential problem remains that of manipulating a complex, dynamic living system, the pathogen population. The ability of this system to respond to change, either natural or introduced by man, was not fully appreciated at first leading to the false expectation that many diseases could be virtually eradicated by the breeding and widespread cultivation of resistant host genotypes. However, the evolutionary potential of pathogens soon became apparent from the rapid adaptive shifts in the genetic composition of their populations that followed and negated the changes in the host population introduced by man. More recently, similar evolutionary responses have occurred following the introduction of some of the new systemic fungicides, with equally unfortunate consequences for disease control.

Clearly, practical disease control is a problem in population biology and, for lasting crop protection, an understanding of the dynamics and genetics of pathogen populations must go alongside the rapidly expanding knowledge of the biochemistry, genetics and molecular biology of plants, of microorganisms and of plant-microorganism interactions.

Interest in pathogen populations is relatively new and there are many deficiencies in our knowledge of their size, potential growth rates, dispersal, genetic composition and evolutionary potential. Indeed in many instances both the theoretical framework and the methodology for the analysis of pathogen populations are still inadequate. Despite these limitations, strategies which aim to achieve stable disease control through the manipulation of pathogen populations are being developed and applied. While we may expect a growing interest in these strategies, their full exploitation will depend upon a better understanding of the dynamics and genetics of pathogen populations.

Two major aspects of pathogen populations are their size and their genetic composition. Unfortunately, in the past these have been largely examined separately; epidemiologists have followed the rise and fall in pathogen populations with little consideration of their genetic make-up, while geneticists have concentrated on the relative frequencies of different genotypes (races) with almost complete disregard for the actual numbers involved. This book attempts to bring these two aspects together and is divided into three sections. The first is concerned with the genetics of pathogenicity and with the genetic structure of pathogen populations. It aims to reveal the genetic potential for adaptation present in these populations and gives considerable attention to the development of appropriate practical methods of analysis. The second section examines the dynamics of pathogen populations and the development of epidemics. The impact of host and pathogen biology, sources of inoculum and methods of dispersal are considered and various models of pathogen dynamics discussed. These aspects of population genetics and population dynamics from the first two sections are drawn together, both theoretically and in relation to practical experience, in the final section. Several case histories are reviewed from the perspective of the impact of the control measures on the pathogen population. From the theory and the practical and empirical experience accumulated in this way, an attempt is made to look forward and to examine possibilities for the control of pathogen evolution so as to enhance and maintain the effectiveness of host resistance and fungicide activity.

This book arose from a meeting held by the British Society for Plant Pathology at the University of Leeds in December 1983. On behalf of the Society we thank the authors for their contributions to that meeting and for their cooperation and patience in the preparation of this volume. We also thank Margaret Howe for compiling the index. For our own part, we wish to thank the Society for providing the means to organize a stimulating and valuable meeting, and for encouraging us to produce this volume.

M.S. WOLFE
C.E. CATEN
British Society for Plant Pathology

Section 1
Basic concepts

1 The genetic basis of epidemics

P.R. DAY and M.S. WOLFE
Plant Breeding Institute, Trumpington,
Cambridge CB2 2LQ, UK

It is now common knowledge that the genetic make-up of crop plants is a key factor in the occurrence of plant disease epidemics. Unless a crop is genetically predisposed to be susceptible to a disease, an epidemic will not occur. Plant breeders strive to ensure that genetic predisposition is avoided and advisers try to ensure that varieties are protected from changes in such predisposition. For these efforts to be successful it is necessary to understand the nature of the problem and the way in which the genotype of the host, or the fungicide, affects the populations of plant pathogens.

In early agriculture, plant disease epidemics were probably common, particularly those caused by seed-borne pathogens. For inbreeding cereals like wheat and barley the increase in genetic uniformity that followed the primitive selection practised by farmers like Knight in the nineteenth century allowed epidemics to ebb and flow with changes in the environment. It was not until the discovery that plant pathogens were the cause rather than the result of disease that it became evident that epidemics could be associated with particular varieties. Crosses between wheat varieties resistant and susceptible to yellow rust (*Puccinia striiformis*) led Biffen (1905, 1912) to describe genetically inherited resistance. Farmers and agronomists came to expect that plant disease could be controlled completely and permanently by breeding resistant varieties. Indeed, Biffen's (1912) final comment was '... historical evidence ... [shows] that both the falling-off of immunity and the gradual advance of the parasitic properties of the rusts takes place too slowly to influence the work of the plant breeder.' However, hopes in this respect were soon shattered by the discovery of physiologic races of wheat stem rust (*P. graminis* f. sp. *tritici*) that were virulent on resistant wheat varieties, making them appear no less susceptible than the older varieties they had replaced (Levine & Stakman, 1918).

Breeders now faced the problem, apparently, of breeding for resistance to these new, common races. For many years it was assumed that if genes for resistance against all the common races of the pathogen could be assembled in one variety, then all would be well (Honecker, 1934). It took some years for breeders and pathologists generally to realize that the newly virulent races were being selected by the resistant varieties that they had developed. The more widely the varieties were grown, the greater was the selection and the shorter the time it took for a new race to develop (Johnson, this volume).

Wider recognition of this relationship required the demonstration by Flor (1946)

Wolfe M.S. & Caten C.E. (1987) *Populations of Plant Pathogens: their Dynamics and Genetics*. Blackwell Scientific Publications, Oxford.

that virulence in the pathogen was under similar genetic control to resistance in the host. The Dutch pathologist Oort (1944), working with loose smut of wheat (*Ustilago tritici*), had earlier reached similar conclusions to explain the specificity of host–parasite interactions, but it was Flor's studies that became associated with the gene-for-gene hypothesis, formalizing the genetic interdependence of host varieties and pathogen races. Person (1959) in a classic paper, clarified the implications of Flor's work for the genetic analysis of host–parasite interactions, showing how the patterns could be used to deduce genotypes. The dynamics of such systems were probably first considered seriously by Haldane (1948) who explored the negative and positive effects of parasites on the evolution of their host populations and vice versa. Mode (1958) was the first to include such considerations in a mathematical model of the evolution of a host–parasite system.

Despite these and other examples, the main thrust of plant pathology during the nineteen forties, fifties and sixties tended to ignore genetics, and particularly population genetics. On the one hand, there were those concerned with the practical matters of disease control, sprays, cultural methods, eradication and the search for new resistance genes in wild relatives. On the other hand, the laboratory pathologists tended to be immersed in biochemical and physiological studies to try to determine the nature of resistance.

A concern for population dynamics in crop protection developed first in the control of insect pests and arose from their response to the use of insecticides. Numerous authors have described the rapidity with which insecticides became outmoded by the development of insecticide resistance, and the consequences of increasing rates of application and the use of more potent chemicals (Georgiou, 1986). The destruction of beneficial insects and the emergence of new pests not controlled by the new agents conspired to force attention on the need for integrating the control methods and understanding the dynamics of insect populations.

Early fungicides evidently did not select for pathogen resistance (Georgopoulos & Zaracovitis, 1967). Revolutionary new and more effective compounds began to be introduced in the late nineteen sixties, which led to claims, analogous to that made fifty years earlier, that the problem of plant disease would be rapidly eliminated, this time by chemical control. Unfortunately, many of the new compounds were systemic and specific and they selected forms of the pathogens that varied in sensitivity to them (Georgopoulos, this volume; Skylakakis, this volume). Consequently, we now have for fungicides a situation that resembles that for race-specific resistance genes and insecticides.

Although various authors had earlier stressed the dangers of crop monoculture and genetic uniformity, the major corn crop losses caused by southern corn leaf blight in 1969 and 1970 (Leonard, this volume), focused attention on the vulnerability of modern agriculture to epidemic pests and diseases (National Academy of Sciences, 1972). This epidemic very clearly exposed the underlying genetic causes. These were the widespread use of Texas male sterile cytoplasm for hybrid seed production, coupled with the appearance of a race of the pathogen uniquely able to colonize and destroy plants with this cytoplasm. It is worth stressing that this feature of extraordinary uniformity in the corn crop initially appeared to be unrelated to disease or pest resistance.

Southern corn leaf blight prompted a variety of responses, ranging from concern over germplasm resources to the need for strategies to reduce the dangers not only in corn but in other crops where comparable genetic uniformity posed a similar threat. Even so, a strong, well-balanced interest in the population genetics of the host–parasite interaction has been slow to emerge. Attention has tended to be focused on simply describing the pathogen races that are present with no regard for the frequencies and dynamics of the virulence genes involved or for the formal genetics of the interaction (Caten, this volume). In trying to combine these and other approaches, Vanderplank (1968, 1978, 1982) played a central role and succeeded in stimulating widespread interest in the problem. This book brings together contributions from many who are now working towards a fuller understanding of the genetic complexities of host and parasite population dynamics.

There are many aspects left to explore. For each disease, the biology of the pathogen, the complexity of its life cycle, whether it is sexual, parasexual or asexual, its mode of overwintering or oversummering, the role of heterokaryosis and its control by vegetative incompatability all have profound effects and are often little understood. We still know little of the relative contributions of mutation, selection, migration and recombination. There are enormous practical problems in trying to obtain reliable estimates of such parameters, for example, population size, migration rates, outcrossing, and selection. Yet full understanding of their importance in population dynamics requires reliable quantitative estimates rather than qualitative judgements (Barrett, this volume; Gale, this volume). In a wider context, there is also the ecology of the pathogen, which may include competition with hyperparasites, grazing insects and fungal viruses, as well as the possible effects of fungicides on the whole community.

For the host, too, many problems remain unexplored, such as the effect of infection modifying the plant's ability to support subsequent pathogen development (Kuć, 1982). Differences in susceptibility between leaves and ears or flowers and between juvenile and adult forms of the host, which could affect pathogen selection, may or may not be paralleled by effects due to features of crop structure such as height, tiller density, drill-row orientation and the presence of other species. The frequency and distribution of varieties and fungicides are, to some extent, under the control of the agronomist, but the effects of changes in these parameters are only now beginning to be understood.

These factors are not enumerated merely to show that there are many, but to stress that their relative importance and interactions must also be assessed so that more rational and new, improved strategies for control may be developed. Clearly, the task is difficult and calls for considerable skill not only in assembling, but in making sense of complex data.

We must also bear in mind the rapid advances now being made in cell and tissue culture and in recombinant DNA technology and their implications for plant disease control (Wenzel, 1985). It seems likely that breeders will be able to isolate and, by transformation, introduce resistance genes at will. These genes may come from related species or unrelated genera, they may be synthetic oligonucleotides, or they may even be derived from the pathogen genome. One may also conceive of methods of attacking the pathogen directly, for example by distribution of a debilitating fungal virus, assuming the availability of a suitable means of delivery.

However, such heady possibilities should not allow us to fall into the trap of complacency previously observed with the advent first of resistant varieties and then of the new generation of fungicides. There is little reason to expect that pathogen populations will not be able to evolve to meet these new challenges. Indeed, whether we continue to use existing 'conventional' methods of control or take up new possibilities that may be offered in the future, the exploitation of all these weapons for disease control will depend heavily on the battlefield tactics and strategies designed by population biologists.

References

Biffen R.H. (1905) Mendel's laws of inheritance and wheat breeding. *Journal of Agricultural Science* **1**, 4–48.

Biffen R.H. (1912) Studies in the inheritance of disease resistance II. *Journal of Agricultural Science* **4**, 421–9.

Flor H.H. (1946) Genetics of pathogenicity in *Melampsora lini*. *Journal of Agricultural Research* **73**, 335–57.

Georgiou G.P. (1986) The magnitude of the resistance problem. In: *Pesticide Resistance: Strategies and Tactics for Management*, National Academy Press, Washington D.C. pp. 14–43.

Georgopoulos S.G. & Zaracovitis C. (1967) Tolerance of fungi to organic fungicides. *Annual Review of Phytopathology* **5**, 109–26.

Haldane J.B.S. (1948) Disease and evolution. *La Ricerca Scientifica*, supplement **19**, 68–76.

Honecker L. (1934) Über die Modifizierbarkeit des Befalles und das Auftreten verschiedener physiologischer Formen beim Mehltau der Gerste, *Erysiphe graminis* f. sp. *hordei* Marchal. *Zeitschrift für Züchtung* A **19**, 577–602.

Kuć J. (1982) Induced immunity to plant diseases. *BioScience* **32**, 854–60.

Levine M.N. & Stakman E.C. (1918) A third biologic form of *Puccinia graminis*. *Journal of Agricultural Research* **13**, 651–4.

Mode C.J. (1958) A mathematical model for the coevolution of obligate parasites and their hosts. *Evolution* **12**, 158–65.

National Academy of Sciences (1972) *Genetic Vulnerability of Major Crops*. National Academy of Sciences, Washington.

Oort A.J.P. (1944) Onderzoekingen over stuifbrand II. Overgevoeligheid van Tarwe voor stuifbrand (*Ustilago tritici*). *Tijdschrift voor Planzenziektenkundig* **1**, 73–106.

Person C.O. (1959) Gene-for-gene relationships in host: parasite systems. *Canadian Journal of Botany* **37**, 1101–30.

Vanderplank J.E. (1968) *Disease Resistance in Plants*. Academic Press, London.

Vanderplank J.E. (1978) *Genetic and Molecular Basis of Plant Pathogenesis*. Springer-Verlag, Berlin.

Vanderplank J.E. (1982) *Host-pathogen Interactions in Plant Disease*. Academic Press, New York.

Wenzel G. (1985) Strategies in unconventional breeding for disease resistance. *Annual Review of Phytopathology* **23**, 149–72.

2 The genetic determination of variation in pathogenicity

B.J. CHRIST*, C.O. PERSON and D.D. POPE
Department of Botany, University of British Columbia, Vancouver, BC V6T2Bl, Canada

Introduction

The concept that pathogenicity is determined through gene-for-gene interactions has been applied to a wide variety of parasites. From these reports, inheritance patterns for gene-for-gene systems have been proposed. The characteristics of these inheritance patterns are generally recognized as inherent in gene-for-gene relations. Exceptions to these characteristics can be found but do not necessarily negate the gene-for-gene concept.

Interest has focused recently on polygenes in host-parasite interactions and their role in the gene-for-gene framework. The scarcity of data has led to speculative discussion and to several viewpoints which will be discussed in a later section. This paper is concerned with observed exceptions to the normal pattern of inheritance in gene-for-gene systems and with the role and properties of polygenic systems in the determination of pathogenicity.

Gene-for-gene systems

Since Flor's parallel crossing experiments with flax and flax rust, many host–parasite systems have been shown to have gene-for-gene interactions (Sidhu, 1980). Based on theoretical considerations, Person (1959) proposed that gene-for-gene interactions would occur widely in parasitic systems, and this has been verified. These interactions have been found not only in plant-fungal interactions but also in plant interactions with viruses, bacteria, nematodes and insects (Sidhu, 1980).

Several characteristics are common to all plant-parasite systems where gene-for-gene relations have been demonstrated. First, resistance is usually dominant and controlled by single genes which may be linked, unlinked or in a multiple allelic series. Second, virulence is usually recessive and inherited by single independent genes. Third, a single combination for incompatibility is epistatic to all other interactions, and fourth, resistance and avirulence alleles function as conditional genes. Deviations have never been observed in the last two characteristics. It is generally agreed that the specificity of gene-for-gene interactions stems from the specificity of interaction between the conditional resistance and avirulence alleles.

* Present address: Department of Plant Pathology, The Pennsylvania State University, University Park, PA 16802, USA

Wolfe M.S. & Caten C.E. (1987) *Populations of Plant Pathogens: their Dynamics and Genetics.* Blackwell Scientific Publications, Oxford.

Dominance, at the biochemical level, is thought to be associated with a functional gene product. It is conceivable that a product from both the resistance allele and the avirulence allele may be required to form a hybrid molecule that acts as a 'stop signal'. It is difficult to distinguish the underlying genetic and biochemical mechanisms of gene-for-gene interactions, because the genetic interpretation in host–parasite systems is based on the disease reaction which is the phenotype expressed by two dissimilar interacting organisms. It is possible that the avirulence allele may provide some sort of recognition for the host which then may turn on certain subsequent events.

Exceptions to the typical characteristics of gene-for-gene systems

Exceptions to the typical pattern of inheritance of resistance in gene-for-gene systems (Table 2.1) include: (i) recessive resistance genes, (ii) two host genes controlling resistance to a particular pathogen genotype (it is not always clear whether or not the two genes in the host are interacting with only one gene in the pathogen), (iii) modifiers which can enhance the effects of a major resistance gene (there is speculation that 'matched' resistance genes may provide some residual resistance and may act as modifiers), (iv) occasional suppressors of resistance genes (especially when genes from a distant relative are incorporated into the host). The major exception to gene-for-gene characteristics for resistance is resistance controlled by several genes that have small additive affects (Table 2.2). This type of resistance is considered to be completely separate from major gene resistance.

The five types of exceptions to the inheritance of resistance listed above are also found to the typical inheritance patterns of virulence in gene-for-gene systems. The first exception is that dominant virulence has been reported for a number of pathogens (Table 2.3). Person & Mayo (1974) pointed out that the overall pattern of interaction of a host and parasite is not altered by changes in dominance relations of a host and/or parasite. An example of the complex nature of dominance relations is found in *Puccinia graminis* f. sp. *tritici*, where Williams *et al.* (1966) reported an avirulence gene to be recessive on one cultivar but dominant on another. Kao & Knott (1969) found a case where there were two alleles for virulence, one which was dominant and the other recessive.

The second exception to the general virulence characteristics is the case of two virulence genes matched to one resistance gene (Table 2.3). There are two possibilities for the evolution of two virulence genes matched to one resistance gene. One possibility is a stepwise gain in virulence where there is a single resistance gene in the host for which virulence in the pathogen progressively increases by changes at two loci such that both virulence genes become necessary for full pathogenicity. The other possibility is where two pathogen populations have separately gained virulence for a specific resistance gene through mutation of different genes. When these virulence genes are brought together in a single pathogen population, either will provide virulence to the specific resistance gene.

Table 2.1. Host–parasite systems where exceptions to the typical inheritance patterns of resistance in gene-for-gene relations have been reported

Nature of exception	Systems	Reference
1. Recessive inheritance	*Triticum/Puccinia striiformis*	Biffen, 1905 Lupton & Macer, 1962
	Triticum/P. graminis f. sp. *tritici*	Knott & Anderson, 1956 Sawhney *et al.*, 1981
	Triticum/P. recondita	Dyck & Samborski, 1968 Bartos *et al.*, 1969
	Avena/P. graminis f. sp. *avenae*	McKenzie & Green, 1965 McKenzie & Martens, 1968
	Zea/P. sorghi	Malm & Hooker, 1962 Hooker & Saxena, 1967
	Hordeum/Ustilago hordei	Wells, 1958
	Oryza sativa/Xanthomonas campestris pv. *oryzae*	Olufowote *et al.*, 1977 Yoshimura *et al.*, 1984
2. Control by two genes	*Hordeum/Ustilago nuda*	Konzak, 1953
	Triticum/P. graminis f. sp. *tritici*	Knott & Anderson, 1956 Knott, 1957
	Triticum/P. recondita	Dyck & Samborski, 1982
	Avena/P. graminis f. sp. *avenae*	Martens *et al.*, 1981
	Avena/P. coronata	Baker, 1966
3. Modifiers	*Triticum/P. recondita*	Dyck *et al.*, 1966 Samborski & Dyck, 1982
	Avena/P. graminis f. sp. *avenae*	McKenzie & Martens, 1968
4. Suppressors	*Triticum/P. graminis* f. sp. *tritici*	Kerber & Green, 1980

Haggag *et al.* (1973) presented evidence for two genes governing virulence where the second gene was possibly acting as a suppressor of the other avirulent allele. Christ & Groth (1982) found two genes in *Uromyces phaseoli* matched to one resistance gene. The initial segregation patterns indicated one gene with avirulence dominant and the second gene was discovered only when an isolate, which was postulated to be heterozygous for the first gene, did not segregate 3 avirulent: 1 virulent on selfing but gave all avirulent progeny. In another cross, a virulent F1 isolate that was homozygous for the first gene segregated 3 virulent: 1 avirulent when selfed, indicating that virulence was dominant for the second gene (Table 2.4b). Because of the interaction of these two genes, Christ & Groth (1982) alluded to the possibility of the second gene being an inhibitor. In *Melampsora lini*, Lawrence *et al.* (1981a) discovered that several avirulence alleles interacted with inhibitor genes.

Table 2.2. Host–parasite systems where polygenically determined resistance has been demonstrated

System	Reference
Triticum/Puccinia striiformis	Lewellen & Sharp, 1968 Lupton & Johnson, 1970
Triticum/P. graminis f. sp. *tritici*	Skovmand *et al.*, 1978
Triticum/P. recondita	Gavinlertvatana & Wilcoxson, 1978
Avena/P. coronata	Luke *et al.*, 1975
Hordeum/P. hordei	Parlevliet, 1978 Johnson & Wilcoxson, 1979
Solanum/Phytophthora infestans	Thurston, 1971
Zea/Cochliobolus carbonum	Hamid *et al.*, 1982
Brassica/Leptosphaeria maculans	Cargeeg & Thurling, 1980

From extensive analysis with *M. lini*, a model can be proposed where the virulence gene and inhibitor gene are matched to one resistance gene, with the dominant inhibitor allele suppressing the expression of the dominant avirulence allele (Lawrence *et al.*, 1981a) (Table 2.4a). The results from *U. phaseoli* (Christ & Groth, 1982) do not fit this model (Table 2.4b). In *U. phaseoli* it appears that the expression of the virulence allele is inhibited. Lawrence *et al's* (1981a) use of the term 'inhibitor' was not intended to imply any particular biochemical mode of action for these genes, but merely to describe the phenotype observed when the two genes interact. Interactions between genes determining virulence have been observed for several fungi (Table 2.3 and Crute, this volume), suggesting that caution should be exercized in inferring the genotype of the pathogen from its phenotype.

Other exceptions to the typical virulence inheritance characteristics are linkage of virulence genes, and modifiers of virulence genes (Table 2.3). Linkage has been reported only recently (Lawrence *et al.*, 1981b), probably because of the paucity of detailed genetic analysis of pathogens. Modifiers of virulence genes will be discussed in further detail in a later section. So far, these exceptions do not alter the overall pattern of disease reactions observed within the classic 'quadratic check' (i.e., the specificity of the interaction has not been altered).

The 'quadratic check' simply portrays the disease reaction that occurs between a host and pathogen involved in gene-for-gene relations hiding the underlying complexity. The antigen–antibody interactions in animal–parasite systems are also portrayed in a disease matrix, and can be used as an analogy to gene-for-gene interactions. Person & Christ (1983) compared the disease pattern of the 'quadratic check' with the disease matrix from antigen–antibody interactions and observed that the specific interaction in both leads to the failure of disease development. In both the gene-for-gene interactions of plant–parasite systems and the antigen–antibody interactions, the phenotype potentially hides the complex interactions

Table 2.3. Host–parasite systems where exceptions to the typical inheritance patterns of virulence in gene-for-gene relations have been reported

Nature of exception	Systems	Reference
1. Dominant inheritance	*Puccinia graminis* f. sp. *tritici/Triticum*	Johnson, 1954 Luig & Watson, 1961 Green, 1964, 1966 Williams *et al.*, 1966 Kao & Knott, 1969
	P. graminis f. sp. *avenae/Avena*	Green, 1965 Green & McKenzie, 1967
	P. recondita/Triticum	Samborski & Dyck, 1968, 1976 Haggag *et al.*, 1973 Statler, 1977, 1982 Statler & Jones, 1981
	P. coronata/Avena	Zimmer *et al.*, 1965
	Melampsora lini/Linum	Flor, 1946
	Uromyces phaseoli/Phaseolus	Christ & Groth, 1982
	Ustilago hordei/Hordeum	Ebba & Person (unpublished)
2. Control by two genes: additive and duplicate	*P. graminis* f. sp. *tritici/Triticum*	Kao & Knott, 1969
	P. recondita/Triticum	Statler, 1977, 1979a, 1982 Statler & Jones, 1981
	P. graminis f. sp. *avenae/Avena*	Green, 1965
	M. lini/Linum	Statler & Zimmer, 1976 Statler, 1979b
3. Control by two genes: inhibitors	*U. phaseoli/Phaseolus*	Christ & Groth, 1982
	M. lini/Linum	Lawrence *et al.*, 1981a
	Bremia lactucae/Lactuca	Michelmore *et al.*, 1984 Crute, this volume
4. Linkage	*P. recondita/Triticum*	Samborski & Dyck, 1976 Statler & Jones, 1981 Statler, 1982
	M. lini/Linum	Statler, 1979b Lawrence *et al.*, 1981b
	U. Phaseoli/Phaseolus	Christ & Groth, 1982
	U. hordei/Hordeum	Pedersen & Kiesling, 1979
5. Modifiers	*P. graminis* f. sp. *tritici/Triticum*	Green, 1965 Green & McKenzie, 1967
	P. recondita/Triticum	Samborski & Dyck, 1968 Dyck & Samborski, 1974
	U. hordei/Hordeum	Pedersen & Kiesling, 1979

Table 2.4. Models for the interaction of virulence and inhibitor genes in *Melampsora lini* (a) and *Uromyces phaseoli* (b)

(a) *M. lini* model based on Lawrence *et al.* (1981a)

	Parents		F1	F2			
genotype[a]	$aa\ II \times$	$AA\ ii$	$Aa\ Ii$	$9\ A-\ I-$:	$3\ A-\ ii$:	$3\ aa\ I-$:	$1\ aa\ ii$
reaction[b]	+	−	+	+	−	+	+

(b) *U. phaseoli* model based on Christ & Groth (1982)

	Parents		F1	
genotype	$aa\ II \times$	$Aa\ ii$	$1\ Aa\ Ii$:	$1\ aa\ Ii$
reaction	+	−	−	+
	F1 (selfed)		**F2**	
genotype	$aa\ Ii$		$3\ aa\ I-$:	$1\ aa\ ii$
reaction	+		+	−

[a] A/a = dominant avirulence/recessive virulence alleles at virulence gene.
I/i = dominant and recessive alleles at inhibitor gene.
[b] +/− = compatible/incompatible reaction.

involved. The complex basis underlying the antibodies was reviewed by Leder (1982).

For antigen–antibody interactions the underlying genetics can be complicated. To take one example, African trypanosomes evade the host immune response by varying their surface antigens. The antigen properties are due to a single glycoprotein. A single trypanosome can make more than a hundred types of glycoprotein that are so different that the host antibodies against one glycoprotein are not effective against other types. Recent studies, reviewed in Borst & Cross (1982), show that each trypanosome seems to contain the entire set of antigen genes and that the activation of some of these genes involves their transposition into an expression site. The antibodies of the host need only exert a selective effect in a heterogeneous trypanosome population because a low degree of switching of the antigen is an intrinsic genetic property.

For plant–parasite systems it appears that each resistance gene of an allelic series interacts with a separate and independent genetic locus of the pathogen that is capable of providing the 'matching' virulence allele (Flor, 1946). Transposition of virulence from one site to another is not known. For both systems the host exerts selection for a specific virulence that is advantageous to the parasite, regardless of the mechanism by which the virulence is expressed.

Polygenic determination of host–parasite interactions

One of the major deviations of the gene-for-gene determination of host–parasite interactions is when virulence or resistance is controlled or modified by several genes. Zimmer *et al.* (1965) identified a major virulence gene in *Puccinia coronata*, but observed some variation within the proposed discrete classes. They theorized that minor genes may be modifying the effects of the virulence gene and that the minor genes would be hard to distinguish because their effects could be of the same magnitude as those of the environment. The first definitive report on the control of pathogenicity by polygenes was by Emara (1972) working with *Ustilago hordei*. Since then polygenically determined pathogenicity has been demonstrated in a few other pathogens (Table 2.5).

The major question concerning polygenes in host–parasite systems is whether the gene interaction is specific or non-specific (see also Leonard, this volume). In order to answer this question, more knowledge of the nature of the polygenes and their action is needed. Most of the discussion concerning polygenes in host–parasite systems has concerned polygenically determined resistance. One view is that apparent polygenic resistance results from resistance genes (of gene-for-gene interactions) that have been matched by virulence genes (Clifford, 1975). It is inferred that matched resistance genes provide some residual resistance and that when several such genes are combined in the host a quantitative effect occurs on the resistant phenotype (Nelson, 1978). Based on this inference, it is then argued that there is no such thing as polygenes affecting resistance (Ellingboe, 1981). Another argument is that polygenic resistance does exist and can modify major gene resistance (Röbbelen & Sharp, 1978).

The same reasoning can be applied to virulence genes. For the first argument we could say that polygenically determined virulence results from pathogenicity genes for which there is no matching resistance genes. These genes would act in an additive manner to modify virulence. This argument can be extended to imply that there is no such thing as polygenes affecting virulence. The other viewpoint is

Table 2.5. Pathogen species where polygenically determined pathogenicity has been demonstrated

Pathogen	Reference
Ustilago hordei	Emara, 1972 Emara & Sidhu, 1974 Person *et al.*, 1982 Person *et al.*, 1983 Caten *et al.*, 1984
Ustilago maydis	Bassi & Burnett, 1980
Ceratocystis ulmi	Brasier & Gibbs, 1976 Brasier, 1977
Gaeumannomyces graminis var. *tritici*	Blanch *et al.*, 1981
Trichometasphaeria turcica	Nelson *et al.*, 1970
Rhizoctonia solani	Garza-Chapa & Anderson, 1966

that polygenes affecting virulence do exist, and that they exert a non-specific modifying role in gene-for-gene interactions.

The parasite providing the most information at this time on the quantitative determination of pathogenicity is *Ustilago hordei*, which causes the covered smut disease of barley. Emara (1972) used biometrical analysis to examine the inheritance patterns of pathogenicity in *U. hordei*. More recently, a new experimental approach was taken to determine the effects of both major virulence genes and polygenes (Person *et al.*, 1982, 1983). This approach took advantage of the fact that with *U. hordei*, the haploid products of meiosis (sporidia) can be isolated, grown in culture, stored and, when individual meiotic tetrads were initially isolated, used in tetrad analysis. Tetrad analysis has been used to study polygenic inheritance in *Neurospora crassa* (Pateman, 1955; Pateman & Lee, 1960) and is not unique to *U. hordei*.

Person & Ebba (unpublished) selfed a number of F1 teliospores of *U. hordei* and observed 3:1 segregation patterns in single progenies, an indication of major gene segregation. A comparison of selfed progenies of many teliospores which were genetically identical for the major gene showed that the levels of virulence were not consistent but varied between teliospores. The variation was not accounted for by another major gene, and it was hypothesized that polygenes were enhancing the effects of the major virulence gene. It was expected that sporidia from a single meiosis might have a variable polygene content, and therefore the four different dikaryons produced from compatible matings of the products of a single meiosis could vary in virulence levels. To test the hypothesis of varying polygene content, the prediction was made that sporidia could be ranked on the basis of their performance from selfing, and that when sporidia of opposite mating types were both ranked from high to low and a diallel cross performed, constant ranking of the gametes should be exhibited (Person *et al.*, 1982). This was the first time the concept of constant ranking was applied to pathogenicity. This prediction assumed that the polygenes would exert primarily additive effects.

The results of a large experiment on varying polygene content showed constant ranking of gametes, both from a visual inspection of the results and from mathematical analysis (Person *et al.*, 1982, 1983). Pope (1982), working with some of the same F1 teliospores, set up a diallel cross using random sporidia, and performed a biometrical analysis on the results. He demonstrated that there was a major gene for virulence together with polygenes which enhanced virulence. However, an analysis of variance showed that, in addition to additive effects, there were significant non-additive effects, genotype by environment interactions, and interactions between the major gene and the polygenes. In addition to this biometrical analysis of pathogenicity, two different constant ranking tests (*sensu* Vanderplank) were performed (Pope, 1982). While one of these tests showed significant rank correlations, the second did not. The low correlations of the second test were attributed to masking by the non-additive effects and by the genotype by environment and major gene by polygene interactions.

The hypothesis of constant ranking (Vanderplank, 1968; Robinson, 1976) is based upon the contention that all polygenes have small, equal and additive effects. However, such effects are not a general property of polygenic systems (Mather & Jinks, 1971). Any alleles involved in genotype by environment, major gene by polygene, or non-additive interactions would be expected to confuse the rank order of either organism, thereby distorting the predicted constant ranking. For example, two pathogen isolates could have similar disease level phenotypes but differ genotypically. The first isolate might have many pathogenicity polygenes that function additively; the second isolate might have fewer such polygenes but also possess a gene that interacts with the environment to enhance pathogenicity. Both isolates might share the same rank when, in fact, they differ with respect to the contributions of additive polygene effects to their respective phenotypes.

The analyses of polygenic systems in *U. hordei* described above and those of Caten *et al.* (1984) indicate that additive and non-additive gene action, genotype by environment interactions, and interactions between major genes and polygenes may all be involved in the genetic determination of pathogenicity. The extent to which genes in these polygenic systems interact with host genotypes of different levels of resistance is not known. These complexities make the simple constant ranking hypothesis insufficient. The pathogenicity of individual parasite genotypes cannot be reliably assessed from the absolute disease level readings in tests involving only one or a few hosts or environments. Reliable assessment and ranking of pathogen or host genotypes requires testing against an array of host or pathogen genotypes, respectively. Such tests would enable the identification of additive (non-specific), non-additive (specific) effects and, if conducted in more than one environment, genotype by environment interactions, and they could provide useful information to the plant breeder. Ranking and selection on the basis of additive rather than non-additive effects should allow the accumulation of more durable resistance in crop plants.

Variability from specific and non-specific genetic interactions can play an important role in disease epidemics. New hypotheses, techniques and analytical methods must be developed in order to understand fully, at the population level, the effect of resistance and pathogen variability on agronomically important crops. These analytical methods will not be worthwhile without knowledge of the inheritance of resistance and virulence genes, of their molecular basis, and of their mode of action. We have some knowledge of the inheritance of these genes, but only a restricted understanding of the importance of the deviations observed in gene-for-gene relations and of the difference between these genes and polygenes. Little or no information is available concerning the molecular or biochemical basis of resistance or virulence. Just as with antigen–antibody interactions, we must go back and start with the molecular basis in order to understand the complexity of events occurring at the population level.

References

Baker E.P. (1966) Isolation of complementary genes conditioning crown rust resistance in the oat variety Bond. *Euphytica* **15**, 313–18.

Bartos P., Dyck P.L. & Samborski D.J. (1969) Adult-plant leaf rust resistance in Thatcher and Marquis wheat: a genetic analysis of the host–parasite interaction. *Canadian Journal of Botany* **47**, 267–9.

Bassi F.G.A. & Burnett J.H. (1980) The genetic architecture of aggressiveness in *Ustilago maydis*. *Annals of Applied Biology* **94**, 281.

Biffen R.H. (1905) Mendel's law of inheritance and wheat breeding. *Journal of Agricultural Science* **1**, 4–48.

Blanch P.A., Asher M.J.C. & Burnett J.H. (1981) Inheritance of pathogenicity and cultural characters in *Gaeumannomyces graminis* var. *tritici*. *Transactions of the British Mycological Society* **77**, 391–99.

Borst P. & Cross G.A.M. (1982) Molecular basis for trypanosome antigenic variation. *Cell* **29**, 291–303.

Brasier C.M. (1977) Inheritance of pathogenicity and cultural characters in *Ceratocystis ulmi*; hybridization of protoperithecial and non-aggressive strains. *Transactions of the British Mycological Society* **65**, 45–52.

Brasier C.M., & Gibbs J.N. (1976) Inheritance of pathogenicity and cultural characters in *Ceratocystis ulmi*: hybridization of aggressive and non-aggressive strains. *Annals of Applied Biology* **83**, 31–7.

Cargeeg L.A. & Thurling N. (1980) Seedling and adult resistance to Blackleg (*Leptosphaeria maculans* Desm. Ces. et. de Not.) in Spring Rape (*Brassica napus* L.). *Australian Journal of Agricultural Research* **31**, 37–46.

Caten C.E., Person C., Groth J.V. & Dhahi S.J. (1984) The genetics of pathogenic aggressiveness in three dikaryons of *Ustilago hordei*. *Canadian Journal of Botany* **62**, 1209–19.

Christ B.J. & Groth J.V. (1982) Inheritance of virulence to three bean cultivars in three isolates of the bean rust pathogen. *Phytopathology* **72**, 767–70.

Clifford B.C. (1975) Stable resistance to cereal diseases: Problems and progress. *Report of the Welsh Plant Breeding Station for 1974*, 107–13.

Dyck P.L. & Samborski D.J. (1968) Genetics of resistance to leaf rust in the common wheat varieties. Webster, Loros, Brevit, Carina, Malakof and Centenario. *Canadian Journal of Genetics and Cytology* **10**, 7–17.

Dyck P.L. & Samborski D.J. (1974) Inheritance of virulence in *Puccinia recondita* on alleles at the Lr2 Locus for resistance in wheat. *Canadian Journal of Genetics and Cytology* **16**, 323–32.

Dyck P.L. & Samborski D.J. (1982) The inheritance of resistance to *Puccinia recondita* in a group of common wheat cultivars. *Canadian Journal of Genetics and Cytology* **24**, 273–83.

Dyck P.L., Samborski D.J. & Anderson R.G. (1966) Inheritance of adult-plant leaf rust resistance derived from the common wheat varieties Exchange and Frontana. *Canadian Journal of Genetics and Cytology* **8**, 665–71.

Ellingboe A.H. (1981) Changing concepts in host–pathogen genetics. *Annual Review of Phytopathology* **19**, 125–43.

Emara Y.A. (1972) Genetic control of aggressiveness in *Ustilago hordei*. I. Natural variability among physiological races. *Canadian Journal of Genetics and Cytology* **14**, 919–24.

Emara Y.A. & Sidhu G. (1974) Polygenic inheritance of aggressiveness in *Ustilago hordei*. *Heredity* **32**, 219–24.

Flor H.H. (1946) Genetics of pathogenicity in *Melampsora lini*. *Journal of Agricultural Research* **73**, 335–57.

Garza-Chapa R. & Anderson N.A. (1966) Behavior of single basidiospore isolates and heterokaryons of *Rhizoctonia solani* from flax. *Phytopathology* **56**, 1260–8.

Gavinlertvatana S. & Wilcoxson R.D. (1978) Inheritance of slow rusting of spring wheat by *Puccinia recondita* f. sp. *tritici* and host–parasite relationships. *Transactions of the British Mycological Society* **71**, 413–18.

Green G.J. (1964) A color mutation, its inheritance, and the inheritance of pathogenicity in *Puccinia graminis* Pers. *Canadian Journal of Botany* **42**, 1643–64.

Green G.J. (1965) Inheritance of virulence in oat stem rust on the varieties Sevnothree, Richland and White Russian. *Canadian Journal of Genetics and Cytology* **7**, 641–50.

Green G.J. (1966) Selfing studies with races 10 and 11 of wheat stem rust. *Canadian Journal of Botany* **44**, 1255–60.

Green G.J. & McKenzie R.I.N. (1967) Mendelian and extrachromosomal inheritance of virulence in *Puccinia graminis* f. sp. *avenae*. *Canadian Journal of Genetics and Cytology* **9**, 785–93.

Haggag M.E.A., Samborski D.J. & Dyck P.J. (1973) Genetics of pathogenicity in three races of leaf rust on four wheat cultivars. *Canadian Journal of Genetics and Cytology* **15**, 73–82.

Hamid A.H., Ayers J.E. & Hill R.R. Jr (1982) The inheritance of resistance in corn to *Cochliobolus carbonum* Race 3. *Phytopathology* **72**, 1173–7.

Hooker A.L. & Saxena K.M.S. (1967) Apparent reversal of dominance of a gene in corn for resistance to *Puccinia sorghi*. *Phytopathology* **57**, 1372–4.

Johnson D.A. & Wilcoxson R.D. (1979) Inheritance of slow rusting of barley infected with *Puccinia hordei* and selection of latent period and number of uredia. *Phytopathology* **69**, 145–51.

Johnson T. (1954) Selfing studies with physiologic races of wheat stem rust, *Puccinia graminis* var. *tritici*. *Canadian Journal of Botany* **32**, 506–22.

Kao K.N. & Knott D.R. (1969) The inheritance of pathogenicity in races 111 and 29 of wheat stem rust. *Canadian Journal of Genetics and Cytology* **11**, 266–74.

Kerber E.R. & Green G.J. (1980) Suppression of stem rust resistance in the hexaploid wheat cv. Canthatch by chromosome 7DL. *Canadian Journal of Botany* **58**, 1347–50.

Knott D.R. (1957) The inheritance of rust resistance. II. The inheritance of stem rust resistance in 6 additional varieties of common wheat. *Canadian Journal of Plant Science* **37**, 177–92.

Knott D.R. & Anderson R.G. (1956) The inheritance of rust resistance. I. The inheritance of stem rust resistance in ten varieties of common wheat. *Canadian Journal of Agricultural Science* **36**, 174–95.

Konzak C.F. (1953) Inheritance of resistance in barley to physiologic races of *Ustilago nuda*. *Phytopathology* **43**, 369–75.

Lawrence G.J., Mayo G.M.E. & Shepherd K.W. (1981a) Interactions between genes controlling pathogenicity in the flax rust fungus. *Phytopathology* **71**, 12–19.

Lawrence G.J., Shepherd K.W. & Mayo G.M.E. (1981b) Fine structure of genes controlling pathogenicity in flax rust, *Melampsora lini*. *Heredity* **46**, 297–313.

Leder P. (1982) The genetics of antibody diversity. *Scientific American* **246**, 102–15.

Lewellen R.T. & Sharp E.L. (1968) Inheritance of minor reaction gene combinations in wheat to *Puccinia striiformis* at two temperature profiles. *Canadian Journal of Botany* **46**, 2155–72.

Luig N.H. & Watson I.A. (1961) A study of inheritance of pathogenicity in *Puccinia graminis* var. *tritici*. *Proceedings of the Linnean Society of New South Wales* **86**, 217–19.

Luke H.H., Barnett R.D. & Pfahler P.L. (1975) Inheritance of horizontal resistance to crown rust in oats. *Phytopathology* **65**, 631–2.

Lupton F.G.H. & Johnson R. (1970) Breeding for mature plant resistance to yellow rust in wheat. *Annals of Applied Biology* **66**, 113–43.

Lupton F.G.H. & Macer R.C.F. (1962) Inheritance of resistance to yellow rust (*Puccinia glumarum* Erikes and Henn.) in seven varieties of wheat. *Transactions of the British Mycological Society* **45**, 21–45.

McKenzie R.I.H. & Green G.J. (1965) Stem rust resistance in oats. I. The inheritance of resistance to race 6AF in 6 varieties of oats. *Canadian Journal of Genetics and Cytology* **7**, 268–74.

McKenzie R.I.H. & Martens J.W. (1968) Inheritance in the oat strain C. I. 3034 of adult plant resistance to race C 10 of stem rust. *Crop Science* **8**, 625–7.

Malm N.R. and Hooker A.L. (1962) Resistance to rust, *Puccinia sorghi* Schm., conditioned by recessive genes in two corn inbred lines. *Crop Science* **2**, 145–7.

Martens J.W., Rothman P.G., McKenzie R.I.H. & Brown P.D. (1981) Evidence for complementary gene action conferring resistance to *Puccinia graminis avenae* in *Avena sativa*. *Canadian Journal of Genetics and Cytology* **23**, 591–5.

Mather K. & Jinks J.L. (1971) *Biometrical Genetics* (2nd ed.). Chapman & Hall, London.

Michelmore R.W., Norwood J.M., Ingram D.S., Crute I.R. & Nicholson P. (1984) The inheritance of virulence in *Bremia lactucae* to match resistance factors 3, 4, 5, 6, 8, 9, 10 and 11 in lettuce (*Lactuca sativa*), *Plant Pathology* **33**, 301–15.

Nelson R.R. (1978) Genetics of horizontal resistance to plant diseases. *Annual Review of Phytopathology* **16**, 359–78.

Nelson R.R., MacKenzie D.R. & Scheifle G.L. (1970) Interaction of genes for pathogenicity and virulence in

Trichometasphaeria turcica with different numbers of genes for vertical resistance in *Zea mays. Phytopathology* **60**, 1250–4.

Olufowote J.Q., Kush G.S. & Kauffman H.E. (1977) Inheritance of bacterial blight resistance in rice. *Phytopathology* **67**, 772–5.

Parlevliet J.E. (1978) Further evidence of partiai resistance in barley to leaf rust, *Puccinia hordei. Euphytica* **27**, 369–79.

Pateman J.A. (1955) Polygenic inheritance in *Neurospora. Nature* **176**, 1274–5.

Pateman J.A. & Lee B.T.O. (1960) Segregation of polygenes in ordered tetrads. *Heredity* **15**, 351–61.

Pedersen W.L. & Kiesling R.L. (1979) Effect of inbreeding on pathogenicity in race 8 of *Ustilago hordei. Phytopathology* **69**, 1207–12.

Person C.O. (1959) Gene-for-gene relations in host: parasite systems. *Canadian Journal of Botany* **37**, 1101–30.

Person C.O. & Christ B.J. (1983) Population genetics and evolution of host-parasite interactions. In: *Challenging Problems in Plant Health* (Ed. by T. Kommedahl & P. Williams), pp. 379–86. American Phytopathological Society, St Paul, MN.

Person C., Fleming R. & Cargeeg L. (1982) Non-specific interaction based on polygenes. In: *Resistance to Diseases and Pests in Forest Trees* (Ed. by H.M. Heybroek, B.R. Stephan & K. von Weissenberg), pp. 318–25. Pudoc, Wageningen.

Person C., Fleming R., Cargeeg L. & Christ B. (1983) Present knowledge and theories concerning durable resistance. In: *Durable Resistance in Crops* (Ed. by F. Lambert, J.M. Waller & N.A. Van der Graff), pp. 27–37. Plenum Press, New York.

Person C.O. & Mayo G.M.E. (1974) Genetic limitations of specific interactions between a host and its parasite. *Canadian Journal of Botany* **52**, 1339–47.

Pope D.D. (1982) Biometrical analysis of pathogenicity in the *Ustilago hordei–Hordeum vulgare* host–parasite system. *M.Sc. Thesis*, University of British Columbia.

Röbbelen G. & Sharp E.L. (1978) *Mode of Inheritance, Interaction and Application of Genes Conditioning Resistance to Yellow Rust.* Verlag Paul Parey, Berlin.

Robinson R.A. (1976) *Plant Pathosystems.* Springer-Verlag, New York.

Samborski D.J. & Dyck P.L. (1968) Inheritance of virulence in wheat leaf rust on the standard differential wheat varieties. *Canadian Journal of Genetics and Cytology* **10**, 24–32.

Samborski D.J. & Dyck P.L. (1976) Inheritance of virulence in *Puccinia recondita* on six backcross lines of wheat with single genes for resistance to leaf rust. *Canadian Journal of Botany* **54**, 1666–7.

Samborski D.J. & Dyck P.L. (1982) Enhancement of resistance to *Puccinia recondita* by interactions of resistance genes in wheat. *Canadian Journal of Plant Pathology* **4**, 152–6.

Sawhney R.N., Chopra V.R. & Swaminathan M.S. (1981) An analysis of genes for resistance against two Indian cultures of stem rust races of two bread wheats. *Theoretical Applied Genetics* **60**, 157–60.

Sidhu G.S. (1980) Genetic analysis of plant parasitic systems. *Proceeding XIV International Congress of Genetics* **1**, 391–408.

Skovmand B., Wilcoxson R.D., Shearer B.L. & Stucker R.E. (1978) Inheritance of slow rusting to stem rust in wheat. *Euphytica* **27**, 95–107.

Statler G.D. (1977) Inheritance of virulence of culture 73–47 *Puccinia recondita. Phytopathology* **67**, 906–8.

Statler G.D. (1979a) Inheritance of pathogenicity of culture 70–1, Race 1, of *Puccinia recondita tritici. Phytopathology* **69**, 661–3.

Statler G.D. (1979b) Inheritance of virulence of *Melampsora lini* race 218. *Phytopathology* **69**, 257–9.

Statler G.D. (1982) Inheritance of virulence of *Puccinia recondita* f. sp. *tritici* on Durum and spring wheat cultivars. *Phytopathology* **72**, 210–13.

Statler G.D. & Jones D.A. (1981) Inheritance of virulence and uredial color and size in *Puccinia recondita tritici. Phytopathology* **71**, 652–5.

Statler G.D. & Zimmer D.E. (1976) Inheritance of virulence of race 370, *Melampsora lini. Canadian Journal of Botany* **54**, 73–5.

Thurston H.D. (1971) Relationships of general resistance: late blight of potato. *Phytopathology* **61**, 620–6.

Vanderplank J.E. (1968) *Disease Resistance in Plants.* Academic Press, New York.

Wells S.A. (1958) Inheritance of reaction to *Ustilago hordei* in cultivated barley. *Canadian Journal of Plant Science* **38**, 45–60.

Williams N.D., Gough F.J. & Rondon M.R. (1966) Interaction of pathogenicity genes in *Puccinia graminis* f. sp. *tritici* and reaction of genes in *Triticum aestivum* sp. *vulgare* 'Marquis' and 'Reliance'. *Crop Science* **6**, 245–8.

Yoshimura A., Mew T.W., Khush G.S. & Omura T. (1984) Genetics of bacterial blight resistance in a breeding line of rice. *Phytopathology* **74**, 773–7.

Zimmer D.E., Shafer J.F. & Patterson F.L. (1965) Nature of fertilization and inheritance of virulence in *Puccinia coronata*. *Phytopathology* **55**, 1320–4.

discussion will be restricted to fungal pathogens, since it is here that the concept has been most fully developed and the term most widely used.

The concept of race

In biology, races are considered as 'genetically, and as a rule geographically, distinct mating groups within a species' (Darlington & Mather, 1949). The principal criterion for the separation of races is genetic difference, and the nature of the resultant phenotypic difference, morphological, physiological, pathological, etc., is not important. The maintenance of genetic difference between groups requires a barrier to gene flow, hence reproductive isolation by biological or geographical means is an essential feature of racial differentiation. Although the formal definitions of race give no indication of the magnitude of the genetic differences required before two populations are recognized as separate races, in practice these have usually been extensive, involving many genes and even variations in karyotype. Thus, biological races are characteristically a limited number of broad intraspecific groups.

Given this biological concept of race, it is not surprising that the term should have been applied to intraspecific pathological variants at a time when the genetic basis of the host specialization was unknown. Our current difficulties have arisen, I believe, through the application of the concept and the term over a wider range of genetic differences than in other branches of biology. At one extreme the broad biological concept of races as major subspecific groups is retained (e.g. Brasier, 1982, and this volume), while at the other, races are differentiated by a single gene (e.g. Yoder & Gracen, 1975). The tendency of plant pathologists to narrow the concept of race is clearly illustrated by Stakman & Harrar's (1957) definition: 'a biotype or group of biotypes within a species or variety that can be distinguished with reasonable certainty and facility by physiologic characters, including pathogenicity, and, in some fungi, by growth on artificial media'. Rigorously applied this definition would lead to virtually every genotype being designated a race. Narrowing of the race concept has been encouraged by our ability to clone many plant pathogens. Artificial cloning imposes a reproductive isolation which may or may not also occur in nature, and makes reliable discrimination possible on very fine grounds. In many cases the pathologists races are the geneticists genotypes.

Increased knowledge of the genetics of plant pathogenic fungi and, in particular, the wider appreciation of the gene-for-gene hypothesis (Flor, 1956) is now influencing our concept of race. In practice pathogen genomes have been almost entirely studied through their effects on the phenotype. Direct cytological examination is limited by the small size of fungal chromosomes, while the new molecular methodologies have yet to be applied to questions of racial differentiation.

What aspects of the phenotype have been used to define races? Morphology has been the cornerstone of fungal taxonomy and pragmatically there is much to recommend this. A consequence of this policy is that morphologically delineated groups have frequently been assigned to formally recognized taxa, such as subspecies, varieties or forms, and considered outside the race concept. Race has largely been reserved for those groups distinguished primarily by physiologic characters, as is apparent from the definition of Stakman and Harrar (1957) cited above. Physiology encompasses a wide spectrum of attributes and in practice most workers have concentrated on pathogenic properties as the sole or major criterion for the classification of races. Thus the Federation of British Plant Pathologists (1973) defined physiologic race or race as 'a taxon of parasites (particularly fungi) characterized by specialization to different cultivars of one host species'. In this chapter I will restrict myself to races defined by pathogenicity, although other criteria such as temperature optima have occasionally been used by plant pathologists to recognize ecological races.

Even where the diagnostic criteria are restricted to pathogenicity, race has been used in plant pathology to encompass a wide range of intraspecific categories with very different genetic structures and evolutionary possibilities. It has therefore become a heterogeneous concept necessitating the recognition of different types of race. While in reality there is a continuous range of levels of genetic differentiation of race, from single genes to ploidy differences, I have for convenience divided this continuum into four groups representing four types of race. In order of increasing complexity of the genetic determination of race these will be referred to as: (i) simple races, (ii) physiologic races, (iii) aggressive races, and (iv) biologic races.

Simple races

In certain necrotrophic fungi, the difference between the recognized pathogenic races is controlled by one or two genes determining toxin production (Scheffer *et al.*, 1967; Yoder & Gracen, 1975). With such simple genetic determination the number of races is limited, frequently to two. Race O and race T of *Helminthosporium maydis* (Leonard, this volume) show many features of such races. When inoculated on to the differential hosts (maize lines with normal and Texas male-sterile cytoplasms) the observed pattern of variation can be explained by race T possessing a property lacking in race O and there is no differential interaction (Table 3.1). This property is the ability to synthesize T-toxin conferred by the dominant allele at the *Tox 1* locus (Leach *et al.*, 1982).

Physiologic races

The recognition of physiologic races

The most widely recognized and familiar races are those of biotrophic fungi characterized by the reactions on a set of host cultivars carrying different sources of qualitative resistance (the differential hosts). The races originally described by Stakman (1914) in *Puccinia graminis* f. sp. *tritici* were of this type, and comparable races have since been identified in many rusts, powdery mildews and downy

Table 3.1. Reaction of races O and T of *Helminthosporium maydis* on maize with Normal and Texas male-sterile (Tms) cytoplasms

Race	Maize cytoplasm	
	Normal	Tms
O	+	+
T	+	+++

+ = pathogenic with low level of virulence.
+++ = pathogenic with high level of virulence.

mildews. The potential number of races that can be recognized is determined by the number of different sources of resistance (resistance factors) included in the differential hosts. Where each resistance factor/pathogen isolate combination is scored as compatible or incompatible, n resistance factors will permit the identification of 2^n races. Assuming each differential carries a unique resistance factor, then n will be equivalent to the number of host differentials. As the number of differentials is increased, so more and more races can be identified, and the limits to racial classification now appear to depend as much upon the diligence of the workers as upon the biology of the system.

One consequence of this system is that the precise grouping of isolates into races is completely dependent upon the particular differentials used; add a differential and two previously identical isolates may separate into different races and *vice versa*. It now appears that populations of biotrophic fungal pathogens exist as mixtures of a variable and potentially very large number of physiologic races. While most knowledge of this derives from agricultural systems (e.g. Wolfe & Schwarzbach, 1978; Groth & Roelfs, 1982; Crute, this volume), the few studies of natural pathosystems suggest a similar complexity (Harry & Clarke, 1986; Dinoor & Eshed, this volume).

The genetic determination of physiologic race

The multiplicity of physiologic races stems from the gene-for-gene control of host–pathogen specificity (Flor, 1956; Day, 1974). Since racial classification is based on the pattern of compatible and incompatible reactions on the differential hosts, the genetic determination of race will be the sum of the determinants of the reaction on each differential. In many instances the genetic basis of each resistance factor and of the corresponding avirulence factor in the pathogen are unknown, and one, two or more genes may be involved in each case. For simplicity, it will be assumed for the rest of this chapter that each host differential carries a different single resistance gene (R), which interacts with a corresponding avirulence gene (A) in the pathogen. It follows that with n differentials, the number of genes determining race in the pathosystem is also n and the potential number of racial genotypes in a haploid pathogen, or of phenotypes in a diploid or dikaryotic pathogen (assuming dominance), is 2^n, i.e. the same as the number of races that can be distinguished. Given that the number of resistance genes in the most extensively studied systems runs into tens, the number of possible races is very large; e.g. the 35 *R* genes conditioning low reaction to wheat leaf rust (Browder, 1980) imply the possible existence of 2^{35} (> a billion) races of *Puccinia recondita*. Any one race possesses a particular combination of alleles for avirulence or virulence at the 35 corresponding *A* genes, and hence is equivalent to an avirulence/virulence genotype (or more strictly a phenotype in diploid or dikaryotic pathogens). Two races may differ at only one gene, if their reactions differ on only one host, or by *n* genes if they differ on all *n* hosts. Hence, simple systems of nomenclature may obscure the genetic relationships between races. As Ellingboe (1981) has pointed out, this view of a physiologic race as a particular combination of genes in the pathogen has been slow to develop and 'a race seems to have been considered as a genetic unit'.

Because more than a few genes determine physiologic race it seems inappropriate to describe the control as 'monogenic' or 'oligogenic'. On the other hand, 'polygenic' is potentially confusing since this term implies not only the involvement of many genes but also that these genes are of individually small effect, leading to the production of a phenotypic continuum (Darlington & Mather, 1949). Discontinuous variation with many genes is unusual and there is something to be said for Vanderplank's (1984) proposal to adopt the new term 'pseudomonogenic' for this situation. Despite the number of genes involved, the variation remains discontinuous because of the strong duplicate epistasis between different corresponding gene pairs which ensures that an incompatible reaction is, at least at the macroscopic level, virtually the same whether it is determined by one, two or *n* interacting A/*R* allele combinations. Consequently there are only two basic phenotypes, compatible and incompatible, and it is the independent occurrence of these two phenotypes on the different tester hosts that leads to the multiplicity of races.

Exceptions to the simple gene-for-gene relationship (Christ *et al.*, this volume) will complicate the genetic determination of physiologic race. Thus, ambiguity in the genotype corresponding to a particular racial phenotype may arise through interactions of the inhibitor or duplicate gene types, as well as through dominance. Consequently, in both haploid and diploid pathogens the same race may be produced by more than one genotype (Table 3.2). Despite these complexities, the genes involved in the differentiation of physiologic races represent only a small part of the pathogen's genome.

Physiologic races differ from simple races in that appropriate combinations of host and pathogen genotypes show a large differential interaction (the diagonal check), and hence the pattern of pathogen variation can not be explained by the stepwise acquisition of a series of pathogenicity determinants, such as the ability to synthesize a toxin. The interactions between the phenotypes of a single corresponding gene pair produce the familiar quadratic check of one incompatible and three compatible reactions, the reverse of the situation with simple races. Ellingboe

Table 3.2. The same physiologic race may be produced by several different genotypes

	Differential resistance genes		
	R_1	R_2	R_3
Phenotype[a]	+	−	+
Possible genotypes[b]			
Haploid	a_1	A_2	a_3
	A_1I_1	A_2	a_3
	a_1	A_2	A_3I_3
		etc	
Diploid	a_1a_1	A_2A_2	a_3a_3
	a_1a_1	A_2a_2	a_3a_3
	$A_1A_1I_1I_1$	A_2A_2	a_3a_3
	$A_1A_1I_1i_1$	A_2a_2	a_3a_3
		etc	

[a] + = compatible = virulent = susceptible,
− = incompatible = avirulent = resistant.

[b] A_1/a_1, alleles for avirulence/virulence to resistance gene R_1. The allele for avirulence is dominant.
I_1/i_1, dominant/recessive alleles at an inhibitor gene interacting with A_1 (see Christ *et al.*, this volume).

(1976) has discussed the implication of this distinction for the underlying physiology of the host–pathogen interaction.

Aggressive races

Vanderplank (1968) recognized two sorts of race: virulent races, which are equivalent to physiologic races, and aggressive races. He defined the latter as pathogenic variants which 'do not interact differentially with varieties of the host plant', and cited isolates of *Fusarium oxysporum* f. sp. *lycopersici* studied by Wellman & Blaisdell (1940). This example reflects the variation in degree of pathogenicity (aggressiveness) present in most populations of plant pathogens but only revealed when quantitative assays are carried out (e.g. Brierly, 1931; Jeffrey *et al.*, 1962; Habgood, 1973). The underlying distribution of variation in these cases is continuous, although apparent discontinuities may arise through sampling and suggest the existence of discrete groups or races. However, as more isolates are examined the continuous nature of the variation becomes clear, and either virtually every isolate has to be recognized as a race or the attempt to recognize races has to be abandoned. For example, Allingham & Jackson (1981) compared 282 isolates of *Septoria nodorum* for three quantitative components of pathogenicity and found that nearly every isolate was unique and that the pattern of variation did not 'conform to conventional race differentiation'.

The continuous nature of the variation with aggressive races suggests that the relevant aspects of the phenotype are affected by many genes of individually small effect, i.e. polygenic control. This is borne out by crosses which produce progenies with a continuous range of levels of aggressiveness (Blanch *et al.*, 1981; Caten *et al.*, 1984). Recognition of aggressiveness as a polygenically-determined, continuously varying character makes racial classification on this basis not only impractical but also meaningless, since the existence of discrete classes is central to any concept of race. In my view aggressive races are generally an artefact of sampling and the use of this term should be discontinued; I will not consider them further in this chapter. This is not to say that aggressiveness is unimportant, but rather that it is better handled by the statistical methods of quantitative genetics (Caten, 1979) than by division into necessarily arbitrary classes.

Biologic races

Biologic races are major intraspecific groups differentiated by morphological features as well as other characters. Of the four types of race proposed here, they correspond most closely to the broad races recognized by geneticists and evolutionary biologists as genetically distinct mating groups. The number of biologic races recognized within a species is limited, and individuals in different races differ by many associated characters and are frequently geographically or reproductively isolated.

The three forms of the dutch elm disease fungus, *Ceratocystis ulmi*, provide a good example of biologic races in plant pathology. These forms, which are referred to as races or strains, rank consistently in pathogenicity from low to high in the order non-aggressive strain, EAN aggressive race, NAN aggressive race, irrespective of the species or genotype of elm on which they are tested. In addition they differ for several morphological and physiological characters expressed in culture, and do not readily hybridize (Brasier, 1982; and this volume). These complexes of associated characters indicate that the strains or races are major subgroups within the species and clearly different biological categories from simple, aggressive or physiologic races.

Crosses between the non-aggressive and the NAN aggressive race of *C. ulmi* confirmed their extensive genetic differentiation. The progenies of these crosses exhibited a wide range of continuous variation in cultural morphology, growth rate and pathogenicity, most offspring being distinct from both parents (Brasier & Gibbs, 1976; Brasier, 1977). In particular, many of the offspring were less vigorous than their parents, as shown by low growth and pathogenicity scores. This loss of vigour on crossing indicates a lack of co-adaptation between the genotypes of the different forms, and provides evidence of their genetic isolation (Mather, 1943).

The genetic composition of races

Theoretical considerations

To be classified as the same race, two isolates must express the same phenotype in respect of the diagnostic criteria. While in certain cases a particular phenotype may be produced by more than one genotype, this phenotypic identity reflects near genotypic identity, and for the following discussion we can make the approximation that isolates of the same race carry the same alleles at loci involved in the determination of race. However, as discussed above, the number of race-determining genes varies greatly between systems and in many cases (especially simple and physiologic races) such genes constitute only a very small part of the pathogen's genome. It is important, therefore, to consider to what extent the distribution of the variation in the rest of the genome reflects the racial classification, i.e. is the total genetic and phenotypic variation within the species distributed at random over races, or is it associated with races? In the former case, races will be heterogeneous populations with regard to all characteristics except their reactions on the differentials, while in the latter case, racial classification will provide a reliable guide to overall genetic relatedness and two isolates of the same race will be generally similar, if not identical.

A priori we may expect the relative amount of variation within, as opposed to between, races to be inversely related to the complexity of the genetic control of race. Where racial classification is based on one or a few genes, two isolates may carry identical alleles at these genes, and hence be classified as the same race, but differ for all other genes in the genome; while two isolates identical at all genes except one may be placed in different races if that gene happens to control the reaction with a differential. Thus, groupings into simple races need not reflect overall genetic similarity, and such races are potentially highly heterogeneous. The same argument applies to physiologic races but becomes progressively weaker as the number of A genes increases (i.e. as n increases). Although when n is large (>20), A genes still represent only a very small proportion of total genes (<0.001), linkage will ensure genetic identity over a significant proportion of the genotype unless the A genes are clustered. Thus the value of physiologic race as an indicator of overall genetic relatedness will improve as n increases. Nevertheless, we should still expect to find genetic and phenotypic variation within physiologic races.

In the extreme case of biologic races, where a significant part of the genome is directly involved in race determination, the groupings will reflect general genetic similarity and the variation between races will be greater than that within. This is not to say that biologic races are genetically uniform entities, only that two isolates of the same race will, in most respects, be more alike than two of different race.

Variation within biologic races is generally accepted, probably because this race concept has been developed predominantly in species which can be cultured. The potential genetic heterogeneity of physiologic races, however, has not been widely recognized, despite the fact that the simple genetic control of avirulence/virulence was established over 40 years ago (Johnson, 1946). Thus, physiologic races have often been treated as genetically uniform entities, as evidenced by the many physiological and epidemiological studies which drew general conclusions about differences between races from comparisons of single isolates of each race (e.g. French, 1953; Watson & Singh, 1952).

Experimental data

In view of this discrepancy between the theory and the common perception of a physiologic race as a homogeneous population, it is of value to examine available data on the genetic composition of such races. This requires characterization of a population of isolates firstly with regard to their reaction on a set of differentials, permitting racial classification, and secondly for one or more independent traits. Where pathogens can be cultured, isolates of the same race have been shown to differ for characters such as colony morphology, growth rate and sporulation, e.g. race 4 isolates of *Phytophthora infestans* (Upshall, 1969; Caten, 1974). Isozymes provide another set of secondary characters and have the added advantage of being readily relatable to the genotype. They have been widely used as characters and markers in higher organisms and their recent introduction into studies of rust populations has yielded very interesting results. Populations of *Puccinia graminis* f. sp. *tritici*, where the sexual stage is a vital part of winter survival each year, are very diverse with respect to both virulence and isozyme phenotype. Furthermore, these two aspects of the phenotype vary independently, and isolates with the same virulence phenotype (i.e. of the same race) are generally quite distinct electrophoretically (Burdon & Roelfs, 1985b). These results are clearly consistent with the theoretical expectation of a physiologic race as a genetically heterogeneous entity.

A different picture, however, has emerged from similar studies of asexually-reproducing rust populations (Burdon *et al.*, 1983; Newton *et al.*, 1985; Burdon & Roelfs, 1985a). These populations vary in virulence but are relatively uniform for isozymes, containing just one (*P. graminis* f. sp. *tritici* in Australia, *P. recondita* f. sp. *tritici* in Australia, *P. striiformis* f. sp. *tritici* in the UK), two (*P. recondita* f. sp. *tritici* in eastern and central North America) or nine (*P. graminis* f. sp. *tritici* in eastern and central North America) isozyme phenotypes. In these populations a common isozyme phenotype is shared not only by all isolates of the same race but also by isolates of different races. It appears that these asexually reproducing populations consist of a very limited number of clones, in some cases just one, which are evolving and differentiating into races by mutation at genes governing virulence (Burdon & Roelfs, 1985a; Newton *et al.*, 1985). Under these circumstances a physiologic race does correspond to the idea of a homogeneous genetic unit implicit in much of the early literature, but this arises as an accident of the biology of the species and not from the genetic determination of race.

These isozyme studies of rust populations emphasize the importance of the reproductive biology of the pathogen in racial classification. Recognition of races requires that the same virulence phenotypes are reisolated on a significant number of occasions. Where sexual reproduction occurs regularly gene flow and recombination will generate a vast array of virulence genotypes and phenotypes; consequently, the probability of repeat isolation of the same phenotype will be low and races are unlikely to be recognised. This would be the case in the sexual population of *P. graminis* f. sp. *tritici*, where 100 virulence phenotypes occurred among 426 isolates sampled (Roelfs & Groth, 1980). With fewer opportunities for sex, successful clones can become dominant, and the population comes to consist of a limited and non-random set of virulence phenotypes. For example, only 17 were found among 2377 isolates of *P. graminis* f. sp. *tritici* collected from eastern and central North America, where sexual reproduction does not occur (Roelfs & Groth, 1980). Under such conditions, reisolation of the same phenotype is likely and races can be recognized. Thus with the advantage of hindsight, we can see that predominantly asexual reproduction makes the recognition of races possible by perpetuating what Vanderplank (1984) calls 'non random assemblages of virulences and avirulences'. These effects of reproductive biology upon population structure and dynamics are also seen in plants, where different terms are used to distinguish individuals derived from different zygotes (genets) from individuals derived by clonal propagation from the same zygote (ramets) (Harper, 1977).

In the light of the argument developed above the situation in the smuts is particularly interesting. These fungi undergo an obligatory sexual cycle at least each season and typically possess no asexual multiplication of the pathogenic dikaryon, circumstances in which it is difficult to see how physiologic races could be defined in the first place, and then maintained as stable entities. Nevertheless, physiologic specialization occurs in several smuts (Halisky, 1965), including *Ustilago hordei* where the gene-for-gene system operates (Sidhu & Person, 1972). The answer to this apparent paradox may be that smut species, although heterothallic, naturally inbreed by sib mating, so that infective dikaryons are homozygous and breed true for race. Evidence for this natural inbreeding comes from progenies derived from single teliospores collected on the Canadian prairies, which resemble progenies obtained from artificially inbred dikaryons, and contrast with the progeny of a synthetic dikaryon (Table 3.3). In smuts therefore, as in the rusts, racial classification is made possible by the biology of the organism, and we may best view a smut race as an inbred population established and maintained in a particular seed lot.

The value of the race concept

The genetic nature of races varies greatly from system to system, making it desirable to recognize different types of race and hence have not one but several concepts of race. We should now ask whether each of these concepts is still of value as a means of describing the variation in populations of phytopathogenic fungi. The

Table 3.3. Genetic and environmental components of variation in populations derived from natural and artificial dikaryons of *Ustilago hordei*

Parent dikaryon	Origin	Component		% Genetic
		Genetic	Environmental	
Natural[b]				
N1	Single teliospore	3.2[a]	31.8	9
N2	Single teliospore	4.1	36.3	10
N3	Single teliospore	1.6[a]	43.9	4
N4	Single teliospore	0.0[a]	40.1	0
Artificial[c]				
E3*a* + I4*A*	Synthetic	26.0	27.6	49
3*a* + 1*A*	Inbred	1.5[a]	26.8	5
BC3*a* + BC3*A*	Inbred	5.9	25.7	19

[a] Not significantly different from zero.
[b] Data of Person & Caten, unpublished.
[c] Data from Caten *et al.* (1984).

value of any concept may be judged by the amount and relevance of the information it conveys. Where pathogen races are distinct, stable, biologically important groups, the concept is of undoubted value in providing a simple, concise and informative description.

Simple, aggressive and biologic races

It has already been pointed out that the underlying quantitative nature of the variation makes the concept of aggressive races impracticable and potentially misleading, and that in such cases racial classification should be abandoned in favour of statistical procedures appropriate to continuous variables. Exceptions to this general rule might arise when, through massive clonal propagation, one or a few particular genotypes become exceptionally frequent, thereby introducing discontinuities into the population.

The concept of simple races also carries little or no information about an isolate other than its phenotype with respect to one or two loci. In these cases, it is just as simple, and potentially less misleading, to describe isolates by the relevant phenotype and drop the race prefix. Once the genetic basis of the phenotypic difference has been established, the phenotypic designation can be replaced by the genotype. For example, race T and race O isolates of *Helminthosporium maydis* would be designated as phenotypically Tox+ and Tox− and genotypically as *TOX 1* and *tox 1*, respectively. This approach is consistent with the recent proposal to standardize genetic nomenclature for plant pathogenic fungi (Yoder *et al.*, 1986).

In contrast to simple races and aggressive races, assignment to a biologic race conveys much useful information and the only question that might be asked is whether the differences involved are not such as to merit a formal taxonomic rank, such as subspecies or variety. Taxonomic recognition in this way forces compliance with the International Code of Botanical Nomenclature (Hawksworth, 1974) and so serves as a brake on unnecessary proliferation of subspecific groups.

Physiologic races

Most concern has been expressed over the concept of physiologic races. Vanderplank (1982, 1983) has raised two objections: firstly that the way races are identified leads to their potential number increasing geometrically as new differentials are added, which is at variance with normal taxonomic procedures, and secondly that physiologic races are fixed taxa whereas the evidence demands gene flow, not fixity. Although the validity of these objections may be disputed, each nevertheless makes a point.

The first objection confuses taxa with taxonomic criteria. Thus Vanderplank (1982, 1983) compares the consequences of recognizing a new species or variety (a taxon) with those of adding a new differential resistance gene (a taxonomic criterion). If standard taxonomy continually added criteria and each criterion was independent of all others then taxa would increase geometrically in the same way as races. Thus the difference between taxonomy and the physiologic race concept is not as fundamental as Vanderplank implies, but arises from the fact that taxonomic criteria generally remain fixed and are not independent. Nevertheless, he has a point in highlighting the practical absurdity posed by the enormous number of races that can, in theory, be distinguished by ten or twenty resistance genes. There must come a point where it is no longer of practical value to continue classifying by reactions on an ever increasing number of standard differentials. The difference between taxonomy and the physiologic race concept is that the former does not attempt to recognize individual genotypes.

The second objection corresponds to a point considered above, namely that a physiologic race is not a fixed genetically homogeneous entity, but a population characterized by common possession of a particular combination of alleles for avirulence and virulence. Theoretically this is a very important point and some populations do show the fluid associations of avirulence and virulence that Vanderplank envisages, e.g. *Puccinia graminis* f. sp. *tritici* in north western USA (Roelfs & Groth, 1980; Groth & Roelfs, this volume). However, it is clear that in other populations the associations are not fluid and that the alleles are fixed into a limited number of combinations, e.g. *P. graminis* f. sp. *tritici* in eastern and central North America (Roelfs & Groth, 1980; Groth & Roelfs, this volume). The critical factor is the amount of gene flow in the population, something about which little is known. Where gene flow is high the classical concept of physiologic races will be misleading in the way Vanderplank (1982, 1983) suggests and should be abandoned. On the other hand, where gene flow is limited the pathogen population may correspond to the classical picture of a simple mixture of fixed, genetic units and the race concept remains valid.

Another criticism of the concept of physiologic race is that by concentrating on the whole phenotype it obscures information regarding the frequencies of particular, important virulences in the pathogen population. What concerns the plant breeder is the frequency of individual virulences or virulence combinations matching the resistance genes in his breeding lines, not the complete spectrum of reactions to a large set of resistance genes, many of which are no longer of agricultural importance. Yet to extract the required information from race survey data is at best laborious and may be impossible if keys relating the race code to the reactions on the differentials are not readily available. This latter problem may be overcome by adopting race nomenclature systems which directly reflect the reactions with the differentials. The first such system was the international nomenclature proposed for races of *Phytophthora infestans* (Black *et al.*, 1953). This system was extended by Green (1965) to include specification of those differentials to which the isolate was avirulent as well as those to which it was virulent. Such virulence formulae are phenotypic specifications in respect to the tested differentials and are used in several pathosystems (e.g. Green, 1971; Loegering & Browder, 1971). They allow any subset of reactions to be readily extracted from the data, but are cumbersome if many differentials are involved. Habgood (1970) proposed a method of coding a set of differential reactions into a more compact format and this has been adopted for *Puccinia striiformis* f. sp. *tritici* in Europe (Johnson *et al.*, 1972).

However, any system of coding carries the risk of obscuring the basic data to all but the initiated. There is no simple solution to this problem, but any system of nomenclature should permit ready extraction of the frequencies of the individual virulences. Virulence formulae allow this and are consistent with the standard practice in fungal genetics of designating cultures by their complete genotype. Each virulence formula can be assigned a shorthand code number, as proposed by Green (1965), but such codes are secondary to the full formula.

The physiologic race concept not only obscures information but may also direct resources away from analysis of those aspects of the pathogen population of real importance. Standard host differentials may be of no practical relevance because they do not reflect the resistance factors in agricultural usage at that time or place. But classical race classification depends upon a fixed set of differentials and any change makes comparisons over time and locations difficult. Thus formal race designation has a built-in inertia which mitigates against the flexible change of differentials required to monitor a pathogen population that is evolving rapidly in response to the breeders efforts. The temptation is simply to keep on adding differentials but this entails expending resources on tests of doubtful practical value and can not continue indefinitely.

Because of all these problems and criticisms, several workers have proposed that races should be abandoned as the basic unit for describing populations of fungi which show extensive physiological specialization (Wolfe & Schwarzbach, 1975; Browder *et al.*, 1980). Instead, populations of these pathogens should be categorized by the frequencies of currently important virulences and secondly by the frequencies of selected combinations of such virulences. This approach, which has been termed virulence analysis (Wolfe & Schwarzbach, 1975), has much to recommend it; the basic information is readily comprehensible, the differentials can be changed in a flexible manner to incorporate currently important resistances, and it conforms with the standard practice in population genetics of using gene rather than genotype frequencies to describe populations. Given the frequencies of individual virulences, the frequencies of particular combinations may be predicted and these predictions compared with the actual frequencies of the same combinations (Wolfe & Schwarzbach, 1975; Wolfe *et al.*, 1976; Wolfe, this volume). Agreement between the predicted and observed frequencies indicates that the population is in linkage equilibrium. However, deviations between observed and predicted may arise from several causes and should be interpreted with caution (Wolfe & Knott, 1982). Where gene flow is restricted, as in the asexual populations of *Puccinia graminis* f. sp. *tritici*, populations are unlikely to be in linkage equilibrium and data on virulence frequencies alone can give a misleading picture of the genotypes actually present. In this circumstance, race frequency data have the advantage over virulence analysis since they concisely convey information on not only the individual virulence frequencies but on all possible combinations.

Conclusions

The term race has been used to classify very different aspects of pathogenic variation and hence there is not one but several concepts of race in plant pathology. This heterogeneity is reflected in the genetic composition of races which vary from near isogenic clones to highly polymorphic populations. Where races are major subspecific groups differing in many respects classification into races is a useful way of documenting the composition of pathogen populations. However, the classical system of identifying physiologic races using fixed sets of differential hosts no longer serves a useful purpose and is being replaced by small, flexible groups of host genotypes which carry resistances of contemporary and local importance. These allow the pathogen populations to be characterized primarily by the frequencies of the individual virulences rather than by the frequencies of complex combinations of avirulence and virulence, i.e. by gene frequencies rather than genotype frequencies. Not only does this approach make the information of relevance to the resistance breeder more accessible but it also emphasizes the potential dynamic nature of the genetic make-up of pathogen populations. In addition, by bringing plant pathology in line with population genetics it facilities application of the theory and analytical methods of the latter discipline. This is not to say that physiologic races should be totally abandoned; they can still provide useful shorthand codes for particularly common pathogen genotypes.

References

Allingham E.A. & Jackson L.F. (1981) Variation in pathogenicity, virulence and aggressiveness of *Septoria nodorum* in Florida. *Phytopathology* **71**, 1080–5.

Barrus M.F. (1911) Variation of varieties of beans in their susceptibility to anthracnose. *Phytopathology* **1**, 190–5.

Black W., Mastenbroek C., Mills W.R. & Peterson L.C. (1953) A proposal for an international nomenclature of races of *Phytophthora infestans* and of genes controlling immunity in *Solanum demissum* derivatives. *Euphytica* **2**, 173–9.

Blanch P.A., Asher M.J.C. & Burnett J.H. (1981) Inheritance of pathogenicity and cultural characters in *Gaeumannomyces graminis* var. *tritici*. *Transactions of the British Mycological Society* **77**, 391–9.

Brasier C.M. (1977) Inheritance of pathogenicity and cultural characters in *Ceratocystis ulmi*; hybridization of protoperithecial and non-aggressive strains. *Transactions of the British Mycological Society* **68**, 45–52.

Brasier C.M. (1982) Occurrence of three sub-groups within *Ceratocystis ulmi*. In: *Proceedings of the Dutch Elm Disease Symposium and Workshop* (Ed. by E.S. Kondo, Y. Hiratsuka & W.B.G. Denyer), pp. 298–321. Manitoba Department of Natural Resources, Manitoba, Canada.

Brasier C.M. & Gibbs J.N. (1976) Inheritance of pathogenicity and cultural characters in *Ceratocystis ulmi*: hybridization of aggressive and non-aggressive strains. *Annals of Applied Biology* **83**, 31–7.

Brierly W.B. (1931) Biological races in fungi and their significance in evolution. *Annals of Applied Biology* **18**, 420–34.

Browder L.E. (1980) A compendium of information about named genes for low reaction to *Puccinia recondita* in wheat. *Crop Science* **20**, 775–9.

Browder L.E., Lyon F.L. & Eversmeyer M.G. (1980) Races, pathogenicity phenotypes and type cultures of plant pathogens. *Phytopathology* **70**, 581–3.

Burdon J.J., Luig N.H. & Marshall D.R. (1983) Isozyme uniformity and virulence variation in *Puccinia graminis* f. sp. *tritici* and *P. recondita* f. sp. *tritici* in Australia. *Australian Journal of Biological Science* **36**, 403–10.

Burdon J.J. & Roelfs A.P. (1985a) Isozyme and virulence variation in asexually reproducing populations of *Puccinia graminis* and *P. recondita* on wheat. *Phytopathology* **75**, 907–13.

Burdon J.J. & Roelfs A.P. (1985b) The effect of sexual and asexual reproduction on the isozyme structure of populations of *Puccinia graminis*. *Phytopathology* **75**, 1068–73.

Butler E.J. & Jones S.G. (1949) *Plant Pathology*. Macmillan, London.

Caten C.E. (1974) Intra-racial variation in *Phytophthora infestans* and adaptation to field resistance for potato blight. *Annals of Applied Biology* **77**, 259–70.

Caten C.E. (1979) Quantitative genetic variation in fungi. In: *Quantitative Genetic Variation* (Ed. by J.N. Thompson, Jr & J.M. Thoday), pp. 35–59. Academic Press, New York.

Caten C.E., Person C., Groth J.V. & Dhahi S.J. (1984) The genetics of pathogenic aggressiveness in three dikaryons of *Ustilago hordei*. Canadian Journal of Botany **62**, 1209–19.

Darlington C.D. & Mather K. (1949) *The Elements of Genetics*. Allen & Unwin, London.

Day P.R. (1974) *Genetics of Host–Parasite Interaction*. W.H. Freeman & Co., San Francisco.

Ellingboe A.H. (1976) Genetics of host–parasite interactions. In: *Encyclopedia of Plant Physiology, Vol. 4: Physiological Plant Pathology* (Ed. by R. Heitfuss & P.H. Williams), pp. 761–78. Springer Verlag, Heidelberg & New York.

Ellingboe A.H. (1981) Changing concepts in host-pathogen genetics. *Annual Review of Phytopathology* **19**, 125–43.

Eriksson J. & Henning E. (1896) *Die Getreideroste*. Norstedt, Stockholm.

Federation of British Plant Pathologists (1973) *A Guide to the Use of Terms in Plant Pathology*. Phytopathological Papers No. 17. Commonwealth Mycological Institute, Kew.

Flor H.H. (1956) The complementary genic systems in flax and flax rust. *Advances in Genetics* **8**, 29–54.

French A.M. (1953) Physiologic differences between two physiologic races of *Phytophthora infestans*. *Phytopathology* **43**, 513–16.

Green G.J. (1965) Stem rust of wheat, rye and barley in Canada in 1964. *Canadian Plant Disease Survey* **45**, 23–9.

Green G.J. (1971) Physiologic races of wheat stem rust in Canada from 1919 to 1969. *Canadian Journal of Botany* **49**, 1575–88.

Groth J.V. & Roelfs A.P. (1982) Effect of sexual and asexual reproduction on race abundance in cereal rust fungus populations. *Phytopathology* **72**, 1503–7.

Habgood R.M. (1970) Designation of physiological races of plant pathogens. *Nature* **227**, 1268–9.

Habgood R.M. (1973) Variation in *Rhynchosporium secalis*. *Transactions of the British Mycological Society* **61**, 41–7.

Halisky P.M. (1965) Physiologic specialization and genetics of the smut fungi. *Botanical Review* **31**, 114–50.

Harper J.L. (1977) *Population Biology of Plants*. Academic Press, London.

Harry I.B. & Clarke D.D. (1986) Race-specific resistance in groundsel (*Senecio vulgaris*) to the powdery mildew *Erysiphe fischeri*. *New Phytologist* **103**, 167–75.

Hawksworth D.L. (1974) *Mycologists Handbook*. Commonwealth Mycological Institute, Kew.

Hawksworth D.L., Sutton B.C. & Ainsworth G.C. (1983) *Dictionary of the Fungi (7th ed)*. Commonwealth Mycological Institute, Kew.

Jeffrey S.I.B., Jinks J.L. & Grindle M. (1962) Intra-racial variation in *Phytophthora infestans* and field resistance to potato blight. *Genetica* **32**, 323–38.

Johnson R., Stubbs R.W., Fuchs E. & Chamberlain N.H. (1972) Nomenclature for physiologic races of *Puccinia striiformis* infecting wheat. *Transactions of the British Mycological Society* **58**, 475–80.

Johnson T. (1946) Variation and the inheritance of certain characters in rust fungi. *Cold Spring Harbor Symposia on Quantitative Biology* **11**, 85–93.

Leach J., Tegtmeier K.J., Daly J.M. & Yoder O.C. (1982) Dominance at the *Tox 1* locus controlling T-toxin production by *Cochliobolus heterostrophus*. *Physiological Plant Pathology* **21**, 327–33.

Loegering W.Q. & Browder L.E. (1971) A system of nomenclature for physiologic races of *Puccinia recondita tritici*. *Plant Disease Reporter* **55**, 718–22.

Mather K. (1943) Polygenic inheritance and natural selection. *Biological Reviews* **18**, 32–64.

Newton A.C., Caten C.E. & Johnson R. (1985) Variation for isozymes and double-stranded RNA among isolates of *Puccinia striiformis* and two other cereal rusts. *Plant Pathology* **34**, 235–47.

Robinson R.A. (1969) Disease resistance terminology. *Review of Applied Mycology* **48**, 593–606.

Roelfs A.P. & Groth J.V. (1980) A comparison of virulence phenotypes in wheat stem rust populations reproducing sexually and asexually. *Phytopathology* **70**, 855–62.

Scheffer R.P., Nelson R.R. & Ullstrup A.J. (1967) Inheritance of toxin production and pathogenicity in *Cochliobolus carbonum* and *Cochliobolus victoriae*. *Phytopathology* **57**, 1288–91.

Sidhu G. & Person C. (1972) Genetic control of virulence in *Ustilago hordei*. III. Identification of genes for host resistance and demonstration of gene-for-gene relations. *Canadian Journal of Genetics and Cytology* **14**, 209–13.

Stakman E.C. (1914) A study in cereal rusts. Physiological races. *Minnesota Agricultural Experiment Station Bulletin* **138**.

Stakman E.C. & Harrar J.G. (1957) *Principles of Plant Pathology*. Ronald Press, New York.

Stakman E.C. & Levine M.N. (1922) The determination of biologic forms of *Puccinia graminis* on *Triticum* spp. *Technical Bulletin of the Minnesota Agricultural Experimental Station* **8**, 1–10.

Upshall A. (1969) Intraracial variation in *Phytophthora infestans*, and its relationship to different host varieties. *Canadian Journal of Botany* **47**, 863–7.

Vanderplank J.E. (1968) *Disease Resistance in Plants*. Academic Press, New York.

Vanderplank J.E. (1982) *Host–Pathogen Interactions in Plant Disease*. Academic Press, New York.

Vanderplank J.E. (1983) Durable resistance in crops: should the concept of physiological races die? In: *Durable Resistance in Crops* (Ed. by F. Lamberti, J.M. Waller & N.A. Van der Graaff), pp. 41–4. Plenum Press, New York.

Vanderplank J.E. (1984) *Disease Resistance in Plants* (2nd ed.). Academic Press, Orlando.

Walker J.C. (1969) *Plant Pathology* (3rd ed). McGraw-Hill, New York.

Watson I.A. & Singh D. (1952) The future of rust resistant wheat in Australia. *Journal of the Australian Institute of Agricultural Science* **18**, 190–7.

Wellman F.L. & Blaisdell D.J. (1940) Differences in growth characters and pathogenicity of Fusarium wilt isolations tested on three tomato varieties. *United States Department of Agriculture Technical Bulletin* **705**, 28 pp.

Wolfe M.S., Barrett J.A., Shattock R.C., Shaw D.S. & Whitbread R. (1976) Phenotype–phenotype analysis: field application of the gene-for-gene hypothesis in host–pathogen relations. *Annals of Applied Biology* **82**, 369–74.

Wolfe M.S. & Knott D.R. (1982) Populations of plant pathogens: some constraints on analysis of variation in pathogenicity. *Plant Pathology* **31**, 79–90.

Wolfe M.S. & Schwarzbach E. (1975) The use of virulence analysis in cereal mildews. *Phytopathologische Zeitschrift* **82**, 297–307.

Wolfe M.S. & Schwarzbach E. (1978) Patterns of race changes in powdery mildews. *Annual Review of Phytopathology* **16**, 159–80.

Yoder O.C. & Gracen V.E. (1975) Segregation of pathogenicity types and host-specific toxin production in progenies of crosses between races T and O of *Helminthosporium maydis* (*Cochliobolus heterostrophus*). *Phytopathology* **65**, 273–6.

Yoder O.C., Valent B. & Chumley F. (1986) Genetic nomenclature and practice for plant pathogenic fungi. *Phytopathology* **76**, 383–5.

4 The dynamics of genes in populations

JOHN A. BARRETT*
Department of Genetics, The University of Liverpool, Liverpool L69 3BX, UK

Introduction

When studying plant diseases, we are all too often confronted with the consequences of evolution. The use of resistant varieties and pesticides often results in genetic changes in pathogen populations which render the control measure ineffective. During recent years there has been a growing appreciation of the importance of understanding the processes which bring about these changes. The study of genetic change in populations requires both a knowledge of the mode of inheritance of the character and a method of following changes in the frequencies of the genes involved. The first part of this chapter is a brief review of the extent of our knowledge of the genetic basis of characters which are important in disease development and of the data available on gene frequency change. In the second part of the chapter, I shall review the major factors effecting changes in gene frequencies and the influence of the biology of pathogens on them.

Genetic change in plant pathogen populations

Genetic variation

For many purposes in plant pathology it is often sufficient to show that different strains of a plant pathogen vary for some character and that the character difference is stable over time. From such observations the inference can be made that at least part of the difference between strains is genetic in origin. However, if changes in gene frequency are to be studied, the formal genetics of the character must be determined. To some extent, plant pathologists are in a difficult position because classical genetic analysis requires that the strains showing the different states of a character can be crossed. This immediately restricts formal analyses to those pathogens with a sexual or parasexual cycle. To obtain accurate analyses, progenies should be large or large numbers of replicate crosses should be carried out; this may be difficult with obligate biotrophs. Also, because many plant pathogens cannot be cultured on defined media, conventional genetic methods using auxotrophic marker stocks cannot be used for mapping. Consequently, the range of plant pathogens which have been investigated genetically is restricted, and often the choice of

* Present Address: Department of Genetics, University of Cambridge, Downing Street, Cambridge CB2 3EH, UK

Wolfe M.S. & Caten C.E. (1987) *Populations of Plant Pathogens: their Dynamics and Genetics*. Blackwell Scientific Publications, Oxford.

pathogen is more a matter of its experimental tractability than its economic importance. Notwithstanding these problems, genetic analyses of some plant pathogens have been carried out and the following list gives a brief indication of the range of characters which have been investigated.

(a) Specific pathogenicity
Knowledge of the genetics of specificity is almost exclusively confined to gene-for-gene interactions, in the loosest sense, because if the genetics of the pathogen is known it is almost certain that the genetics of resistance in the host has been worked out first. Gene-for-gene interactions have been demonstrated or suggested for about two dozen pathogens (Vanderplank, 1982), a very small number in comparison to the number of diseases which can attack even a single host species. Although the gene-for-gene hypothesis is a convenient way of describing reciprocal variation in host and parasite, the simple two-allele locus-for-locus relationship described by Flor (1956) does not seem to hold in many cases (Christ *et al.*, this volume; Christ, 1984).

(b) Non-specific pathogenicity
This has received most attention in prokaryotic pathogens, e.g. fluidal and non-fluidal phenotypes of *Pseudomonas solanacearum* (Staskawicz *et al.*, 1983), and the ability of *Rhizobium* and *Xanthomouas* to establish infections (Beringer, 1983; Daniels *et al.*, 1983). A number of studies have examined non-specific pathogenicity in fungi, but the character seems to be largely quantitative and the analyses are difficult (Emara & Sidhu, 1974; Christ, 1984; Christ *et al.*, this volume).

(c) Insensitivity to chemicals
Despite the widespread observation of insensitivity and its economic importance, there has been relatively little work to determine the genetics of insensitivity. In only about half a dozen plant pathogens has there been any success (Georgopoulos, 1982, and this volume; Hollomon, 1981).

(d) Non-pathogenic characters
In some plant pathogens which can be cultured on defined media there have been investigations using auxotrophic mutants (e.g. Garber *et al.*, 1980; Fincham *et al.*, 1979; Day & Jones, 1969). Following on the success of the use of electrophoretic methods in conventional population genetic studies, a few studies have been carried out using these techniques on plant pathogens (e.g. Burdon *et al.*, 1982). Because of its importance in the pathology of frost damage, the ice-nucleating properties of *Pseudomonas syringae* have been subjected to intense genetic analysis (Lindow, 1983).

Although a casual perusal of the literature might leave one with the impression that a lot is known about the genetic basis of characters of pathological interest, the truth is that very few pathogens have been investigated relative to the number of diseases of economic importance. Even among those systems that have been examined, only a small number have been subjected to any detailed analyses, e.g. the flax/flax rust system, and, unfortunately, these few analyses have led to generalizations about the inheritance of pathological characters which may not be true in any general sense (Christ *et al.*, this volume). What little is known points towards the conclusion that virtually all types of inheritance can be involved in pathogenic and insensitivity characters. To insist that any particular mode of inheritance is characteristic of pathogenicity or insensitivity is just not possible with the data available at the moment.

Gene frequencies

Until quite recently, little attention was paid to gene or even phenotype frequencies in plant pathogen populations. Two approaches predominated: race surveys, and the assay of bulk samples for the presence or absence of certain phenotypes. During the 1960s a number of plant pathologists realized independently that the actual composition of pathogen populations could be important in understanding epidemics. Two basic approaches have since been used to estimate phenotype frequencies in populations.

(a) Single isolate tests

A large number of individual isolates are tested on a standard range of host differentials in the same way that races were originally identified, only now importance is placed on the actual frequencies (e.g. Dinoor & Eshed, this volume). This technique tends to be very time-consuming and limits the number of populations which can be tested and the number of isolates per population, but the technique can yield very accurate results within the constraints of restricted sample sizes.

(b) Bulk sample tests

If mobile nurseries consisting of a range of host plants with different resistance characters or chemical treatments are exposed in the field along with a universally susceptible or untreated control, the number of infections on each plant relative to the control can give an estimate of the relative frequency of the corresponding pathogen phenotype in the population (Eyal *et al.*, 1973; Wolfe & Minchin, 1976; Wolfe & Schwarzbach, 1975). A laboratory version of this can be used on bulk samples using spray techniques or settling towers. There are a number of problems with this approach: (i) the definition of a universally susceptible control, (ii) a pathogen phenotype (genotype) on its compatible host may not give the same number of infections as on the control plants, (iii) host differentials with the right combinations of resistance factors may not be available, e.g. when there are multiple alleles for resistance.

The mobile nursery technique does not yield precise estimates of phenotype frequencies, but may be accurate to within an order of magnitude. On the other hand, if the method is used for successive sampling of a population, the biases should be the same on each occasion and the method is quite sufficient to determine relative increases and decreases in phenotype frequencies.

It is essential to note that both single isolate and bulk sample methods estimate phenotype frequencies. In haploid pathogens this may give a direct estimate of genotype freqencies if the genetics of the character is known. In diploid pathogens it may be possible to determine the frequencies of recessive homozygotes, but not to distinguish between heterozygotes and dominant homozygotes. Thus, even in cases where the frequency changes of genotypes cannot be obtained directly, the underlying crude gene frequency changes can be determined by inference from the data.

Recent work using these approaches and the reanalysis of race survey data has been useful in throwing light on frequency changes in populations, e.g. stem rust of wheat in the USA (Roelfs & Groth, 1980; Groth & Roelfs, this volume) and in Australia (Burdon *et al.*, 1982; Watson, 1958), powdery mildew of barley in the UK and Western Europe (Wolfe & Schwarzbach, 1978; Wolfe *et al.*, 1982; Wolfe, this volume). However, a close look at the literature reveals a dearth of good data; the same results appear time and time again: Waterhouse (1952), Watson & Luig (1963) and Luig & Watson (1970) in Australia, Flor (1953) in the USA, and Green (1971) in Canada. These extensive sets of data were not collected for many of the purposes to which they are being put today, and consequently they must be treated very cautiously. The following two examples illustrate this need for caution.

1. The data of Watson & Luig (1963) and Luig & Watson (1970) were collected from eight areas throughout Australasia. During the period 1954–1969 a total of 120 samples could have been obtained; however, 60% of all samples taken were smaller than 50 isolates (Table 4.1). With a sample size of 50 the minimum frequency which can be detected with a probability greater than 5% is about 0.06; if the expected phenotype frequency is 0.5, then the 95% confidence intervals on the phenotype frequency estimates are 0.64–0.36. Consequently, it would be difficult to detect either relatively rare phenotypes (less than 5%) or frequency changes of less than about 10% per annum in these samples. Furthermore, the sample sizes varied from year to year and place to place; in some years, sites were not sampled. If such data are pooled over different locations or different years this can have severe consequences for interpretation, especially if the composition of the population varies from one year to the next or from location to location (see Wolfe & Knott, 1982).

Table 4.1. Distribution of number of isolates in samples of *Puccinia graminis* f. sp. *tritici* collected in Australasia

	Number of isolates in sample						
	0	1–20	21–50	51–100	101–250	251–500	500
Frequency	7	37	27	21	20	7	1

Data from Watson & Luig (1963) and Luig & Watson (1970).

2. Vanderplank (1968; Table 4.3) used the data of Watson & Luig (1963) and Waterhouse (1952) to support his ideas about 'stabilising selection' and 'strong' and 'weak' resistance genes. He pooled data over all locations and then further pooled the data into three-year periods for his analysis (Table 4.2 and 4.3). He interpreted these data as showing a more rapid decline of race 126–1, 6 than race 126–6. However, if the data are broken down into the contributions from each area in each year, then it becomes immediately clear that the samples from New Zealand (Table 4.4) were absent from three of the six years in the interval 1955–1960, and of 72 race 126–6 isolates obtained during 1958–60 (Table 4.2), 54 were obtained in 1960 from New Zealand which had contributed virtually nothing during the preceding five years. When Watson (1958) pointed out the association between the area of Eureka planted and the frequency of 'races' pathogenic on Eureka, he commented: 'The figures for 1941–51 have been taken from those of Waterhouse (1952) and it is probable that there is some bias in the number of isolations of strains virulent on Eureka as many of his collections were made on this variety.' Consequently, any attempt to use these data for more elaborate analyses is fraught with difficulty, e.g. the attempt by Grant & Archer (1983) to estimate the relative fitnesses of the races (see also Johnson, this volume).

The empirical study of gene frequency change in plant pathogen populations requires both a knowledge of the genetics of the character and the availability of survey methods designed to obtain information on gene frequency change. These two requirements are satisfied in very few cases and it is only in recent years that any attempt has been made to grapple directly with these problems (e.g. Schwarzbach, 1978; Groth & Roelfs, this volume). Retrospective analysis of data obtained for other purposes is stricken with pitfalls and interpretations on anything other than a qualitative basis must be treated with extreme caution.

Causes of gene frequency changes

There are four major factors which contribute to changes in gene frequency in plant pathogen populations: (i) finite population size; (ii) mutation; (iii) migration; (iii) selection.

Finite population size

On those occasions when they are agriculturally important, plant pathogens exist in very large numbers, so for most purposes, the effects of finite population size can be ignored. However, there are two circumstances under which the size of pathogen populations becomes important (see also Gale, this volume).

(a) In very large populations, the frequencies of individual alleles may not be seriously affected by sampling effects, but if particular combinations of genotypes at different loci are considered, and these are expected at low frequencies, then even very large populations may not contain them. For example, if the frequencies of

Table 4.2. Number of isolates of races 126–6 and 126–1,6 of *Puccinia graminis* f. sp. *tritici* collected in each year during the period 1952–1960

Year	Number of isolates				
	Race 126–6		Race 126–1,6		All races
	annual total	3-year total	annual total	3-year total	annual total
1952	49		22		86
1953	14	95	7	30	44
1954	32		1		74
1955	116		14		317
1956	84	215	8	22	417
1957	15		0		317
1958	4		0		114
1959	12	72	5	5	584
1960	56		0		774

Data from Watson & Luig (1963).

Table 4.3. Total number of isolates of races 126–6 and 126–1,6 of *Puccinia graminis* f. sp. *tritici* in three-year periods between 1945 and 1963

Race	Period ending					
	1948	1951	1954	1957	1960	1963
126–6	326	132	95	215	72	14
126–1,6	502	244	30	22	5	4

Data from Vanderplank (1968).

Table 4.4. Numbers of isolates of races 126–6 and 126–1,6 of *Puccinia graminis* f. sp. *tritici* collected in New Zealand during the period 1955–1960

Race	Year					
	1955	1956	1957	1958	1959	1960
126–6	–	–	4	–	1	54
126–1,6	–	–	0	–	1	0
All races	–	–	11		4	95

– = no collections made in years shown.
Data from Watson & Luig (1963).

phenotypes *A* and *B* are 10^{-3} when estimated independently, then the expected frequency of the combination *AB* is 10^{-6}; and in populations of 10^6, about 36% of populations would not contain any *AB* phenotypes.
(b) Where a small number of individuals or propagules establish a new population, the gene frequencies in the founding population may differ significantly from those in the population from which they were derived (*Founder principle*). Perhaps a classic example of this may be the changes in the composition of the Australian stem rust population as demonstrated by Burdon *et al.*, (1982) using electrophoretic markers.

Mutation

The ultimate origin of variation is mutation, but the formal demonstration of mutation in plant pathogens is not easy in those cases where conventional genetic methods cannot be used. In these cases, two basic methods have been employed to demonstrate that mutation to specific phenotypes can occur. Firstly, selection of specific phenotypes from spores or propagules derived from a single isolate; this is an efficient method, if contamination can be controlled. Secondly, mutagenesis; although the use of mutagenic agents makes it easier to detect novel variants because of the increased frequency of mutation, the fact that specific variants can be obtained by mutagenesis does not necessarily mean that they will occur in the field, e.g. *Sr26* virulence (Luig, 1979).

The drawback with plant pathogens is that although the occurrence of mutation can be demonstrated, it is difficult to estimate the mutation rate (e.g. Flor, 1958; Vanderplank, 1982). Whilst it is relatively easy to estimate the number of mutants surviving selection, it is more difficult to estimate the number of non-mutants which did not survive. Consequently, mutation rates have not been estimated in many plant pathogens.

Migration

Migration is perhaps the most neglected area of the dynamics of gene frequencies, and yet it is known from experimental epidemiology that migration can be a significant factor in the biology of many pathogens. The epidemiological consequ-ences of long-distance migration have been well documented in some cases where disease is seasonal and episodic, e.g. the Puccinia pathway in North America. But where diseases are persistent, the importance of migration may not be so readily apparent. However, genetic differences between subpopulations can allow cryptic migration to be detected, for example in the migration of powdery mildew spores between Germany and Denmark (Hermansen, 1968; Hermansen & Wiberg, 1972; Hermansen *et al* 1978), and between Austria and Germany (Limpert & Schwarzbach, 1981). Migration from genetically different sources can have profound effects on the efficacy of control measures. For example, insecticidal control of green leafhoppers in Japan has been maintained despite intensive use of insecticides because a large proportion of the population is composed of immigrants from the Chinese mainland where the insecticides are not used (Nagata, 1983).

Selection

Natural selection is the necessary consequence of three observable facts: (i) characters vary within populations; (ii) this variation is at least in part inherited; and (iii) the variation has an effect on the probability of leaving offspring. It is not necessary to catalogue the properties of plant pathogens which satisfy these conditions, as they are probably self-evident. However, a problem arises when it is necessary to quantify the process. When different phenotypes vary in their probability of leaving offspring, they are said to vary in *fitness*. The magnitude of the differences in fitness determines how rapidly evolution can proceed. The problem then arises of how differences in fitness can be measured.

Field survey data which show systematic changes in phenotype, genotype or gene frequencies can be analysed using the techniques of population genetics to estimate fitness differences retrospectively. But as I have shown already, much of the available data are not suitable for such analyses (see also Leonard, this volume; Johnson, this volume). Alternatively, known phenotypes can be taken, and various characters believed to be important in determining survival and reproduction measured, e.g. growth rates, sporulation period, i.e. probable components of fitness. The problem here is whether the characters can be measured with sufficient accuracy; a fitness difference of 5% will cause rapid genetic change, but in many cases characters of a plant pathogen cannot be measured to within this level of accuracy. There is also the problem that the characters chosen may not be important in determining fitness and that measurements in the laboratory may have little relevance to performance in the field, e.g. attempts to equate aggressiveness with fitness. One way around the problems inherent in the two methods outlined above is to duplicate experimentally the process of selection under field conditions using populations of known composition (e.g. Watson & Singh, 1952; Dovas *et al.*, 1976; McGee & Zuck, 1981).

Mutation and selection

Although the four factors, finite population size, mutation, migration, and selection, have been treated separately for convenience, in the real world all of them can be acting simultaneously, giving rise to interactions. For example, even a cursory examination of the plant pathology literature will reveal a common assumption that novel phenotypes are always present at the mutation rate. For this to be true, the variant must be lethal in the absence of the factor which permits its selection. If it is not lethal it will increase in the population until the rate of increase due to mutation is exactly balanced by the rate of loss due to selection. For example in a haploid, if allele A mutates to a at rate u, and the fitness of a relative to A is $(1-s)$, then:

Genotype	A	a
frequency	p	q
fitness	1.0	$1-s$

at equilibrium $\hat{q} \cong u/s$, assuming $s \gg u$.
If $u = 10^{-6}$ and $s = 10^{-2}$, then $\hat{q} \cong 10^{-4}$. In other words the variants will be present at a frequency one hundred times higher than the mutation rate (see also Gale, this volume).

The importance of pathogen and host biology

These, then, are the factors which produce gene frequency change; but they must have a stage on which to act. This is provided by the breeding biology and ecology (epidemiology) of the pathogen and its host(s).

Breeding biology

The breeding system, in its broadest sense, determines the relationship between genotype and phenotype, and between the frequencies of genes, genotypes and phenotypes, their rates of change and the generation of variation by recombination.

(a) Modes of inheritance
Whether a pathogen species is haploid, diploid or polyploid determines whether the effects of a particular allele are expressed phenotypically and the segregation ratios following a sexual cycle. Hence ploidy may modify the effectiveness of any factors causing selective differences. On the other hand, the transmission of cytoplasmically determined characters is quite different from nuclear genes, and this will modify the rates of response to selection (e.g. Marquis virulence (Johnson, 1946)).

(b) Sex
In a pathogen with a sexual cycle, novel combinations of genes can be produced by recombination. Whether the sexual stage is an obligate stage of the life cycle and how frequently it occurs can have profound effects on both the type and rate of genetic response to selection. In an infinitely large population, there is no advantage of sexual reproduction over asexual reproduction (Maynard Smith, 1974). Sexual processes only generate novel combinations more easily when there is linkage disequilibrium. But as has been demonstrated earlier, linkage disequilibrium can be generated in very large populations if the expected frequency of combinations of genes is small enough. A second effect of sexual reproduction in diploid and polyploid organisms is that the genotype frequencies of the offspring will be diffe-

rent from those in their parents, and this too can have significant effects on rates of evolution.

If a sexual stage occurs on a different host or in a different place from the asexual cycle in those pathogens which show a strict alternation of generations, the effective breeding structure of the population can be altered, e.g. the contrast between powdery mildew on barley and stem rust on barberry. The precise form of the genetic control of sex or mating type will also have an effect on the rates of gene frequency change because of possible linkage with the controlling loci, for example, in the fungi, whether a pathogen is heterothallic or homothallic and whether it has a bipolar or tetrapolar mating system.

Even in the absence of sexual processes, parasexual processes may occur and these too will affect the rate of gene frequency change. In the fungi, the existence of hyphal incompatibility systems may break up populations into reproductively isolated sub-populations with a consequent effect on the possibility of forming novel gene combinations or heterokaryons.

Ecology and epidemiology

Of critical importance to the evolutionary dynamics of plant pathogens is the environment in which they exist. This is composed of both biotic factors (other organisms and especially the host plants), and physical factors, for example, the distribution of host plants, the period of conditions conducive to growth and reproduction and the temperature-sensitivity of some traits. Where pathogens have a strict alternation of generations, the number of asexual generations per sexual cycle will be critical in determining rates of genetic change; for example, the epidemiology and evolutionary dynamics of stem rust show remarkable differences between populations east of the Rockies, where barberry has been eradicated, and west of the Rockies where barberry has persisted (Roelfs & Groth, 1980). Perhaps the most important difference between treatments of evolutionary dynamics in plant pathogens and more conventional population genetic studies is that population size must be an inherent consideration in studies of plant pathogens, because an understanding of epidemic development is, after all, the reason for studying them (see also Gale, this volume).

The interaction of factors in the evolution of pathogen populations

The rate and direction of evolution depend on how all of these factors interact. The following examples illustrate the consequences of the joint action of some of these factors when assumptions specific to plant pathogens are incorporated.

Example 1. Mutation–selection equilibrium

In a random mating population, the dynamics of mutation can be considered at the gamete level because random mating is the same as random fusion of gametes (see above). If allele A mutates to allele a at rate u, and the mutant character is recessive in both its expression and fitness, and the frequency of A is p and the frequency of a is q, then:

Genotype	AA	Aa	aa	
frequency	p^2	$2pq$	q^2	where $p + q = 1$
fitness	1	1	$1-s$	

At equilibrium the rate of gain of a by mutation is exactly balanced by its rate of loss by selection and it can easily be shown that, approximately:

$$\hat{q} \cong \sqrt{u/s} \text{ if } s \gg u$$

and the frequency of the mutant phenotype is $q^2 \cong u/s$.

On the other hand, mutation in an asexual population cannot be considered at the gamete level. In an asexual diploid population exhibiting the same character, the genotypes will not be in Hardy–Weinberg equilibrium, and the frequencies and fitnesses of each genotype are:

Genotype	AA	Aa	aa	
frequency	x	y	z	$x + y + z = 1$
fitness	1	1	$1-s$	

If A mutates to a at rate u, then, the equilibrium frequency of aa can easily be obtained as $\hat{z} = u/s$, as for a random mating population. But under these conditions the equilibrium frequencies of the other genotypes are $\hat{x} = 0$ and $\hat{y} \cong 1$ and consequently $\hat{q} = x + \tfrac{1}{2}y \cong 0.5$. In other words at mutation–selection equilibrium for a deleterious recessive in an asexual diploid population, the population will consist almost entirely of heterozygotes. This further implies that, if the oft-quoted statement that virulence is recessive to avirulence in diploid pathogens is true (Vanderplank, 1978; see also Christ *et al.*, this volume), then homozygous avirulent genotypes should be very rare. Conversely, if it is found that avirulent homozygotes are common, then this would imply that although virulence may be phenotypically recessive, its effects on fitness are not recessive. Unfortunately, there appear to be no data available which would allow these predictions to be tested. For the more general case:

Genotype	AA	Aa	aa
frequency	x	y	z
fitness	1	$1-hs$	$1-s$

Where h is a constant between 0 and 1 which allows the fitness of the heterozygote to be expressed in terms of the fitness of the two homozygotes. It can be shown that if $h < u/s$, the AA genotype cannot be maintained in the population (Barrett, in preparation).

Example 2. Recombination and breeding system

Consider a fungus with a bipolar mating system. If a strain heterozygous for a marker unlinked to the mating-type locus is selfed, then the expected ratio of the diploid progeny is 1*AA*:2*Aa*:1*aa*. Under sustained selfing of all the progeny from this cross, the heterozygotes would eventually be lost. But if the fungus is secondarily homothallic with intra-tetrad selfing (e.g. some of the smuts) and the marker locus is tightly linked to its centromere, but unlinked to the mating-type locus, which is also tightly linked to its centromere, then all of the progeny from a strain heterozygous for the marker will be heterozygotes and the heterozygosity will be maintained under sustained selfing. Unless this was realized, difficulties could be created in any attempt to interpret gene and genotype frequencies in secondarily homothallic fungus populations (Zakharov, 1972; Barrett & Kirby, in preparation).

Example 3. Estimation of selection coefficients

Haldane (1924) showed that for an asexual population consisting of two phenotypes A and a with frequencies p and q respectively, the relative fitnesses can be estimated from the slope of the plot of $1n(p/q)$ against time if selection is not too strong. Although this method can, and has, been used with plant pathogens (e.g. MacKenzie, 1978), it contains the implicit assumption that the population size remains constant. Under the conditions in which this type of experiment would be carried out with a plant pathogen, the population size would almost certainly not remain constant. Under epidemic conditions in which a population is subjected to density-dependent regulation, the relative fitnesses will also be density dependent and consequently estimates of relative fitness made at different stages of the epidemic are likely to be different (Barrett, 1983).

Example 4. Ecology and gene frequency dynamics

Simple population genetic models can be built to investigate the dynamics of genetic change of plant pathogens on a mixture of different host genotypes. If a discrete generation model is used, it can be shown that very strong selection is required to prevent the spread of the 'super-race' (Person *et al.*, 1977; Marshall & Pryor, 1978). This approach assumes completely random distribution of pathogen genotypes over all host genotypes in each generation. But if non-random distribution of propagules is built into the model, it can be shown that simple or complex races can predominate, depending on the degree of non-random distribution (Barrett, 1980).

Discussion

In putting together this chapter, I have followed the normal practice in plant pathology of lumping together everything about genetic changes in populations. But if the work which has been carried out is examined closely, two quite distinct and almost mutally exclusive objectives, namely tactical and strategic, can be discerned. Tactical objectives aim to provide a guide to what is happening in the field, e.g. whether insensitivity to a pesticide is increasing. It is a purely empirical approach and consequently there is little need for a detailed knowledge of the genetics of the character in question. Most of the data available come from this source. Strategic objectives on the other hand, entail a more theoretical approach in which the possible evolutionary consequences of different control methods are examined, or an attempt is made to understand why a certain train of events gave rise to a particular outcome. More details of the system being studied are required. Whilst it may be possible to use data originally obtained for tactical objectives, the fact that the data were collected for other purposes may not make them suitable except in a crude qualitative way. At the practical level, experiments with strategic objectives are usually concerned with testing specific hypotheses.

The idea that problems of plant pathogen adaptation are problems in evolution and amenable to analysis using the principles of population genetics is relatively recent. However, the fact that the same basic principles can be applied, glosses over the differences which exist between classical (or academic) population genetics and evolutionary plant pathology. Most studies in population genetics are based on sexually reproducing diploid species, but relatively few plant pathogens conform to this model, while the range of breeding and genetic systems exhibited by plant pathogens has hardly been touched by population geneticists. Consequently, the theoretical background which is a prime necessity for the successful study of gene frequency dynamics is lacking in many problems in evolutionary plant pathology. Studies in population genetics tend to choose organisms which are amenable to genetic analysis. In plant pathology there is rarely a choice of organism, and consequently the genetic basis of the character to be studied may be poorly understood and inappropriate models may be applied. The study of the dynamics of genes in plant pathogen populations requires that we understand the genetic basis of the characters, that we have good data on genotypic or phenotypic changes, and that we have appropriate models with which to analyse and interpret the data. Unfortunately, these criteria are not fulfilled in the majority of problems which plant pathologists have to face.

References

Barrett J.A. (1980) Pathogen evolution in multilines and variety mixtures. *Zeitschrift für Pflanzenkrankheiten und Pflanzenschutz* **87**(7), 383–96.

Barrett J.A. (1983) Estimating relative fitness in plant parasites: some general problems. *Phytopathology* **73** , 510–12.

Barrett J.A. Mutation–selection equilibria in asexual diploids. (In preparation).

Barrett J.A. & Kirby G.C. Population genetics of secondarily homothallic fungi. (In preparation)

Beringer J.E. (1983) The *Rhizobium*–plant interaction. In: *Molecular Genetics of the Bacteria–Plant Interaction* (Ed. by A.Pühler), pp. 9–13. Springer-Verlag, Berlin.

Burdon J.J., Marshall D.R., Luig N.H. & Gow D.J.S. (1982) Isozyme studies on the origin and evolution of *Puccinia graminis* f.sp. *tritici* in Australia. *Australian Journal of Biological Sciences* **35**, 231–8.

Christ B.J. (1984) Effects of selection on populations of *Ustilago hordei*. *Ph.D. Thesis*. University of British Columbia.

Daniels M.J., Turner P.C., Barber C.E., Cleary W.G. & Reed G. (1983) Towards the genetical analysis of pathogenicity of *Xanthomonas campestris*. In: *Molecular Genetics of the Bacteria–Plant Interaction* (Ed. by A. Pühler), pp. 340–4. Springer-Verlag, Berlin.

Day A.W. & Jones J.K. (1969) Sexual and parasexual analysis of *Ustilago violacea*. *Genetical Research, Cambridge* **14**, 195–221.

Dovas C., Skylakakis G. & Georgopoulos S.G. (1976) The adaptability of the benomyl-resistant population of *Cercospora beticola* in northern Greece. *Phytopathology* **66**, 1452–6.

Emara Y.A. & Sidhu G. (1974) Polygenic inheritance of aggressiveness in *Ustilago hordei*. *Heredity* **32**, 219–24.

Eyal Z., Yurman R., Moseman J.G. & Wahl I. (1973) Use of mobile nurseries in pathogenicity studies of *Erysiphe graminis hordei* on *Hordeum spontaneum*. *Phytopathology* **63**, 1330–4.

Fincham J.R.S., Day P.R. & Radford A. (1979) *Fungal Genetics* (4th edn). Blackwell Scientific Publications, Oxford.

Flor H.H. (1953) Epidemiology of flax rust in the North Central States. *Phytopathology* **43**, 624–8.

Flor H.H. (1956) The complementary genic systems in flax and flax rust. *Advances in Genetics* **8**, 29–54.

Flor H.H. (1958) Mutation to wider virulence in *Melampsora lini*. *Phytopathology* **48**, 297–301.

Garber E.D., Ruddat M. & Merza A.P. (1980) Genetics of *Ustilago violacea* VII the *pumpkin* locus. *Botanical Gazette* **141**, 210–12.

Georgopoulos S.G. (1982) Genetical and biochemical background of fungicide resistance. In: *Fungicide Resistance in Crop Protection* (Ed. by J. Dekker & S.G. Georgopoulos), pp. 46–52. Centre for Agricultural Publishing and Documentation (PUDOC), Wageningen.

Grant M.W. & Archer S.A. (1983) Calculation of selection coefficients against unnecessary genes for virulence from field data. *Phytopathology* **73**, 547–51.

Green G.J. (1971) Physiologic races of wheat stem rust in Canada from 1919 to 1969. *Canadian Journal of Botany* **49**, 1575–88.

Haldane J.B.S. (1924) A mathematical theory of natural and artificial selection. *Transactions of the Cambridge Philosophical Society* **23**, 19–41.

Hermansen J.E. (1968) Studies on the spread and survival of cereal rust and mildew diseases in Denmark. *Friesia* **8**, 161–359.

Hermansen J.E., Torp U. & Prahm L.P. (1978) Studies of transport of live spores of cereal mildew across the North Sea. *Grana* **17**, 41–6.

Hermansen J.E. & Wiberg A. (1972) On the appearance of *Erysiphe graminis* f. sp. *hordei* and *Puccinia hordei* in the Faröes and possible primary sources of inoculum. *Friesia* **10**, 30–4.

Hollomon D.W. (1981) Genetic control of ethirimol resistance in a natural population of *Erysiphe graminis* f. sp. *hordei*. *Phytopathology* **71**, 536–40.

Johnson T. (1946) Variation and the inheritance of certain characters in rust fungi. *Cold Spring Harbor Symposia on Quantitative Biology* **11**, 85–93.

Limpert E. & Schwarzbach E. (1981) Virulence analysis of powdery mildew of barley in different European regions in 1979 and 1980. *Barley Genetics. IV.* (*Proceedings of the 4th International Barley Genetics Symposium*), *Edinburgh*, pp. 458–65.

Lindow S.E. (1983) The role of bacterial ice nucleation in frost injury to plants. *Annual Review of Phytopathology* **21**, 363–84.

Luig N.H. (1979) Mutation studies in *Puccinia graminis tritici*. *Proceedings of the 5th International Wheat Symposium*, pp. 533–39.

Luig N.H. & Watson I.A. (1970) The effect of complex genetic resistance in wheat on the variability of *Puccinia graminis* f. sp. *tritici*. *Proceedings of the Linnaean Society of New South Wales* **95**, 22–45.

McGee D.C. & Zuck M.G. (1981) Competition between benomyl-resistant and sensitive strains of *Venturia inaequalis* on apple seedlings. *Phytopathology* **71**, 529–32.

MacKenzie D.R. (1978) Estimating parasitic fitness. *Phytopathology* **68**, 9–13.

Marshall D.R. & Pryor A.J. (1978) Multiline varieties and disease control. I. The 'dirty crop' approach with each component carrying a unique single resistance gene. *Theoretical and Applied Genetics* **51**, 177–84.

Maynard Smith J. (1974) Recombination and the rate of evolution. *Genetics* **78**, 299–304.

Nagata T. (1983) Insecticide resistance in rice pests, with special emphasis on the brown leafhopper, (*Nilaparvata lugens*, Stål). In: *Plant Protection for Human Welfare* (*Proceedings of the 10th International Congress of Plant Protection, Brighton, 1983*). British Crop Protection Council, Croydon, pp. 599–607.

Person C., Groth J.V. & Mylyk O.M. (1977) Genetic change in host–parasite populations. *Annual Review of Phytopathology* **14**, 177–84.

Roelfs A.P. & Groth J.V. (1980) A comparison of virulence phenotypes in wheat stem rust populations reproducing sexually and asexually. *Phytopathology* **70**, 855–62.

Schwarzbach E. (1978) Monitoring airborne populations of cereal mildew. In: *Plant Disease Epidemiology* (Ed. by P.R. Scott & A. Bainbridge), pp. 55–62. Blackwell Scientific Publications, Oxford.

Staskawicz B.J., Dahlbeck D., Miller J. & Damm D. (1983) Molecular analysis of virulence genes in *Pseudomonas solanacearum*. In: *Molecular Genetics of the Bacteria-Plant Interaction* (Ed. by A. Pühler), pp. 345–52. Springer-Verlag, Berlin.

Vanderplank J.E. (1968) *Disease Resistance in Plants*. Academic Press, New York.

Vanderplank J.E. (1978) *Genetic and Molecular Basis of Plant Pathogenesis*. Springer-Verlag, Berlin.

Vanderplank J.E. (1982) *Host–parasite Interactions in Plant Disease*. Academic Press, New York.

Waterhouse W.L. (1952) Australian rust studies. IX. Physiologic race determinations and surveys of cereal rusts. *Proceedings of the Linnaean Society of New South Wales* **77**, 209–58.

Watson I.A. (1958) The present status of breeding disease resistant wheats in Australia. *Agricultural Gazette of New South Wales* **69**, 630–60.

Watson I.A. & Luig N.H. (1963) The classification of *Puccinia graminis* var *tritici* in relation to breeding resistant varieties. *Proceedings of the Linnaean Society of New South Wales* **88**, 235–58.

Watson I.A. & Singh D. (1952) The future for rust resistant wheat in Australia. *Journal of the Australian Institute of Agricultural Sciences* **18**, 190–7.

Wolfe M.S., Barrett J.A. & Slater S.E. (1982) Pathogen fitness in cereal mildews. In: *Durable Resistance in Crops* (Ed. by F. Lamberti, J.M. Waller & N.A. Van der Graaff), pp. 81–100. Plenum Press, New York.

Wolfe M.S. & Knott D.R. (1982) Populations of plant pathogens: some constraints on analysis of variation in pathogenicity. *Plant Pathology* **31**, 79–90.

Wolfe M.S. & Minchin P.N. (1976) Quantitative assessment of variation in field populations of *Erysiphe graminis* f. sp. *hordei* using mobile nurseries. *Transactions of the British Mycological Society* **66**, 332–4.

Wolfe M.S. & Schwarzbach E. (1975) The use of virulence analysis in cereal mildews. *Phytopathologische Zeitschrift* **82**, 297–307.

Wolfe M.S. & Schwarzbach E. (1978) The recent history of the evolution of barley powdery mildew in Europe. In: *The Powdery Mildews* (Ed. by D.M. Spencer), pp. 129–57. Academic Press, London.

Zakharov I.A. (1972) Homozygosity in intratetrad and intraoctad fertilization in fungi. *Soviet Genetics* **4**, 636–42.

5 Factors delaying the spread of a virulent mutant of a fungal pathogen: some suggestions from population genetics

J.S. GALE
Department of Genetics, University of Birmingham B15 2TT, UK

The problem

While to the plant breeder it is a matter of great concern that the useful life of a new resistant variety may be limited to a few years, to the population geneticist it is at first sight puzzling that resistant varieties can be bred at all. According to a popular model, a non-virulent strain of pathogen produces a protein recognized by the host as foreign; the acquisition of virulence being a change leading either to the failure to produce the protein or to the production of an altered form that is no longer recognized (Ellingboe, 1982). From such information as we have on loss of antigenic activity in general in fungi, it seems unlikely that the rate of mutation to virulence would be smaller than 10^{-7} (Fincham, *et al.*, 1979). Further, *established* strains of pathogen spread very rapidly. It would appear, then, that by the time a new resistant variety had completed trials, a corresponding virulent strain of pathogen would already have become well established. Thus the problem is not that virulent strains arise and spread, but that frequently their spread is delayed long enough to allow a useful commercial life for a new resistant variety.

Factors delaying the spread of advantageous alleles have long been studied by population geneticists. I shall list these factors (some of which are probably familiar). Of course, a theoretical discussion of this kind can only suggest possibilities. However, it should be possible to decide, on the basis of empirical measurements, on the relative importance of the various factors described and thus resolve, at least in part, the problem.

'Primordial' versus 'established' virulent strains

The initial alteration of the recognized protein, although conferring virulence, could be very crippling to the pathogen in other ways, leading perhaps to a greatly reduced growth or sporulation rate. If so, the primordial virulent strain would spread very slowly. Only after a fairly lengthy evolution of the virulent strain, involving the acquisition of genes, perhaps at several loci, which together compensate for the loss or reduced efficiency of the crucial protein, would the virulent

Wolfe M.S. & Caten C.E. (1987) *Populations of Plant Pathogens: their Dynamics and Genetics.* Blackwell Scientific Publications, Oxford.

strain acquire the rapid rate of spread found in established virulents. This acquisition of compensatory or modifying genes corresponds to what Parlevliet (1981) termed the second stage in the adaptation of the pathogen population to a newly introduced resistance gene; analogous processes were first proposed, in a different context, by Fisher (1930).

Two points emerge. In crosses between established virulent strains and non-virulent strains, in which progeny were scored qualitatively for the presence or absence of virulence, virulence versus avirulence would still appear as a single locus difference. Only if other quantitative features of the phenotype were noted would the presence of the compensatory genes be revealed (see also Christ *et al.*, this volume).

Secondly, if the formerly resistant variety, R say, were withdrawn, the established virulent, being compensated for the deficiency of the crucial protein, might have only a slight selective disadvantage in comparison with other strains of the pathogen and thus decline in numbers rather slowly, as has been observed in practice (Parlevliet, 1981; see also Johnson, this volume). On the other hand, the primordial virulent would, in the absence of R, suffer extreme selective disadvantage. This point is crucial. As explained below, rapid establishment of the virulent is decidedly more likely if it is not too rare at the time when R is introduced. Of course, other things being equal, the greater the disadvantage before R is introduced, the lower the frequency of the primordial virulent.

Fitness of a genotype

To avoid vagueness, it is necessary at this stage to give some definitions. In traditional population genetics, we think of fitness of a particular genotype in terms of the viability and fecundity of individuals bearing that genotype. However, in the case of fungi, the concept of an individual is notoriously elusive. The following definition will serve in the present context. An individual is a mass of mycelium derived from a single spore; by a mature individual we mean an individual that has reached reproductive age. The fitness of a mature individual is defined as the number (possibly 0) of mature progeny individuals to which it gives rise. Consider, then, a large set of mature individuals, all of the same genotype. By the fitness of that genotype, we mean the average number of mature progeny individuals per member of the set.

Theoretical population geneticists have usually been concerned with sexually reproducing species; in the interest of simplicity, it is often supposed that such species are outbreeders, with much opportunity for recombination between different genomes. However, in species reproducing mainly asexually, many generations could pass with only a very limited number of occasions on which recombination could take place. While the details will vary with species, it seems best in the present context, as a first approximation, to ignore these occasions, even for pathogens in which sexual reproduction is not too infrequent. This being so, we may treat the complete genotype of a virulent strain as if it were a single locus, a decided simplification. In fact, even given occasional recombination, the genetic background in an individual in which a mutation to virulence occurs will be of crucial importance. Further, the rate of spread of virulent strains in different species of pathogen does not seem to be related to their mode of reproduction. We may, therefore, use our simplification with some confidence, although it will be exact only for strictly asexually reproducing species. When speaking of the fitness of a virulent strain, then, we mean the fitness of the complete genotype of that strain.

In traditional population genetics, we normally imagine the mutant and previous wild type as coexisting for some considerable time ('transient polymorphism'). In that case, it is the relative fitness of mutant and wild type that matters. In the present case, where the non-virulent fails completely on the resistant variety, the absolute fitness of the virulent strain will be the quantity relevant to its spread. It is quite straightforward to adapt standard formulae from population genetics to this situation.

For simplicity, we consider haploids, although the factors we shall mention also apply to diploids.

Effect of chance

We write $(1 + \alpha)$ for the fitness of a virulent genotype, v say. In the case of a primordial virulent, α could be quite small. We stress that $(1 + \alpha)$ is the *mean* number of mature progeny per mature parent of constitution v. Purely by chance, the actual number of mature progeny will vary from parent to parent. Quite possibly the number will be zero in any individual case, especially if α is small. Even if the number were not zero, the progeny in turn could fail to leave descendants. Similar considerations apply in later generations. Hence a whole line of descent, stemming from a single virulent individual, could die out. Under a wide range of circumstances, this random extinction is *very probable*; it is quite mistaken to suppose that all, or most, advantageous mutants leave descendants in the long run.

We ask, then: what is the probability that a mature individual of genotype v gives a line of descent which persists in the long run? Equivalently, we imagine a very large set of mature individuals of genotype v and ask: what fraction of these individuals gives a long-term line of descendants? For brevity, we call this fraction *the probability of survival.*

Naturally, this probability will be larger the larger the value of α, but there is another factor affecting our probability which, while less obvious, is of crucial importance, namely the ratio of the 'effective' population size N_e to the 'census' (i.e. actual) population size N. Suppose, in principle, that we count the total number of mature individuals of genotype v present at some time; this is the census size. As already explained, individuals will vary in respect of number of mature progeny they leave. In principle, this variation could be 'purely random' in nature (giving a Poisson distribution of mature progeny number). In practice, however, this does not seem to be so. For example, parental environment and background genotype can have a very large effect on fecundity. Thus the author's colleague, Dr I.J. Mackay, counted the number of seeds produced by every plant in a natural population, size 2316, of the long-headed poppy, *Papaver dubium*. Twenty-one per cent of these plants produced no seed at all, whereas four plants gave more than 20 000 seeds *each*. Such variation must be attributed to differences between parents in genes affecting fecundity and in environment, probably mainly the latter. While this may perhaps be an extreme case, all species investigated so far show the same phenomenon to some extent and there seems no reason why fungi should be an exception. I have tried to deal with genetic sources of variation in fecundity by treating the complete genotype of a virulent as a single unit. We must, however, allow for environmentally induced variation, which could, in the field, be very substantial.

It will be apparent that if, as in our poppy population, only a small proportion of parents contribute much to the next generation, the population will behave *as if* its size were substantially less than its census size N. In technical language, the effective population size N_e (Wright, 1931) would be very much less than N. The whole matter can be treated quantitatively and a numerical value assigned to the ratio N_e/N. In practice, factors other than variation in parental fecundity can also affect this value (see below). Ignoring these for the moment, we note that a quantitative formulation for random mating diploids was given by Wright (1938) (see also Nei & Murata (1966)) and for haploids by Kimura (1964). We suppose here that variation in fecundity is wholly environmental in origin. Let σ^2 be the variance of the number of mature progeny per parent. Then for haploids we have the simple formula

$$\left[\frac{N_e}{N} = \frac{1}{\sigma^2}\right]$$

provided α is small (or zero). Obviously, given large variation, σ^2 will be large and N_e/N quite small (for larger α, the mathematics is more complicated but the same conclusion applies).

Probability of survival

Generally, the smaller the effective population size in relation to the census size, the lower the probability of survival. In fact, if α is small, this probability is simply

$$\left[2\alpha\frac{N_e}{N}\right]$$

under a wide range of circumstances (Haldane, 1927; Fisher, 1930; Wright, 1931; Kimura, 1957; Moran, 1961; Maruyama, 1977). Clearly, a small value of N_e/N greatly reduces the probability of survival in cases where α is small (in fact, this reduction still occurs, although less marked in magnitude, in cases where α is too large for our simple formula to apply). In view of the importance of N_e/N, it is necessary to discuss briefly an additional factor which can lead to a further substantial reduction in this ratio.

Suppose that virulent strain v can grow and reproduce in a number of localities. Owing to natural or human agencies, conditions in any given locality L may change in such a way that all v individuals in L become extinct and leave no local descendants. If extinction occurs fairly rapidly after v first appeared in L, there may have been inadequate time for any spores produced by v individuals in L to have migrated and infected a host elsewhere. Then our v individuals in L contribute nothing to the long-term survival of v and *effectively* form no part of the overall population of v individuals. However, at some later stage, v individuals may again appear in L. On our assumptions, such individuals must necessarily have originated *elsewhere*. This is the process of subpopulation extinction and replacement, investigated theoretically by Maruyama & Kimura (1980). Suppose that, for localities in general, extinction and replacement occurs frequently. Then, if we count the world population of v individuals at a given time, our result will be very misleading, since, as we have seen, many of these individuals do not contribute to the effective population and should not therefore be counted. To allow for this, a substantial reduction in N_e/N must be made.

Finally, we note that, owing to seasonal and climatic factors, there will be periods in which conditions are particularly favourable for growth, reproduction and spread of v. During such periods, the fitness of mature v individuals will exceed the average value $(1 + \alpha)$ and the probability of survival will temporarily increase (Fisher, 1930). In fact, if a mutant first arises at the start of a favourable period lasting several generations, this increase in probability can be very marked (Kojima & Kelleher, 1962; Ewens, 1967). Similarly, a marked decrease is found if a mutant first arises at the start of an unfavourable period. In practice, such favourable and unfavourable periods (each covering several generations) will alternate. Now a new mutation is most likely to appear towards the end of a favourable period when the population size is largest, that is when conditions are *about to deteriorate*. Hence, as Ewens (1967) has shown, the increase and decrease in probability *do not cancel*. On the contrary, the probability of survival is lower than it would have been had the fitness of v remained at $(1 + \alpha)$ throughout.

The number of mutants required for success of the virulent

Although the probability of survival of a single virulent mutant may be very low, new virulent mutants will continually arise, so that sooner or later one (or more) mutant(s) will give a persistent line of descent. More strictly, a stage will be reached at which at least one line of descent will have become so large in number that its chance of dying out is altogether negligible. At this stage the future of the virulent is assured, although it may take some considerable time before the virulent becomes common. We shall, therefore, use the term 'successful' for a virulent that has attained this stage.

Now, although given enough separate mutations, success is certain, this may well not be so with a more restricted number of separate mutations, especially if the probability of survival is low. Even though mutations to virulence occur many times, every line of descent could fail to survive. We ask, then: how many separate mutations to virulence are required for success to be near-certain?

We count only those newly arisen mutants which appear in a mature individual (no doubt many additional mutants arise which never even reach this stage—these are irrelevant to the argument). Let n such mutants arise, *de novo*, over a period of time. Just for the moment, suppose all n have the same fitness $(1 + \alpha)$. Let y be the probability of survival; since the large majority of lines of descent which die out do so fairly soon, the fitness of the *primordial* virulent will be appropriate when calculating y. The probability that all n lines of descent die out is $(1-y)^n$. When y is small, this is well approximated by e^{-ny} (e = 2.71828 ...). Hence, to a close approximation, the probability of success is

$$1-e^{-ny}$$

which, for given ny, is easily found on almost any pocket calculator.

In practice, of course, α will vary from virulent mutant to virulent mutant, so that y will vary. It may be proved that, to allow for this variation, we must replace y in the above formulae by $\bar{y}$, the mean of the ys for a large sample of separate new mutants.

For a sporting chance (0.5) of success, $n\bar{y}$ will have to be about 0.69; for the chance of success to be 0.9, $n\bar{y}$ must be about 2.30. Thus if we suppose, purely for illustrative purposes, that $\bar{y}$ is as small as 1 in 100 000, the number of mutants (each represented in a mature adult) required for a 0.9 probability of success will be 230 000. Thus, random extinction of lines of descent could be a crucial factor in delaying the advance of the virulent. Eventually, however, as we have argued, these random effects will be overcome; in the later stages of spread of the virulent, any effects of chance will be very subsidiary.

Balance between mutation, selection and random effects

It is sometimes suggested that prior to the introduction of a resistant variety R, the corresponding virulent is already present at a frequency which, while low, well exceeds the rates of mutation to virulence (Editors' comment: the same suggestion is also made for mutants to increased resistance to fungicides; for examples in this volume see Barrett, Georgopoulos, Leonard, Skylakakis, Wolfe). The argument is as follows. Let the *relative* fitnesses of avirulent and virulent on the initial susceptible host be 1; $(1-s)$. If μ is the mutation rate per generation to virulence, the frequency of virulents among all juvenile individuals on the susceptible host, calculated *on the assumption that chance effects are absent*, eventually reaches an equilibrium value μ/s.

This formula is normally derived on the supposition, unrealistic in the present context, that different generations do not overlap. Given overlapping generations, we can be almost certain that no equilibrium is possible (Charlesworth, 1980). Suppose, as must almost surely be the case, that the intensity of selection against the virulent varies with age of the individual. Then the relative fitnesses of virulent and avirulent, appropriately defined, will depend on the age composition of the population. Unless, as seems out of the question, the latter remains stable, the frequency of the virulent will never reach an equilibrium value, but will vary as the demographic composition of the population varies with changing season and climate.

Even if we suppose that the theory of non-overlapping generations will provide a *rough guide* to the frequency of the virulent, it does not necessarily follow that this frequency will much exceed the mutation rate. Firstly, $(1-s)$ is here the relative fitness, on the suceptible host, of the primordial, not of the established, virulent. Thus $(1-s)$ might be very close to zero and thus s very close to unity. Secondly, the assumption that chance effects may be neglected is very questionable. As Wright (1937) has shown, the formula μ/s can be quite inappropriate in cases where chance effects are important. Let $2N_e$ be the world effective number of mature individuals of the pathogen, just before R is introduced (the use of $2N_e$ rather than N_e is a convenient device for harmonizing results for haploids with those for diploids; for a diploid species, the world effective population size would be written as N_e). If $4N_e\mu$ is less than one, the frequency of the virulent could well be zero, or very close to zero, even with only mild selection against the virulent. It should be noted that, for the very few species for which N_e has been estimated, the values obtained are surprisingly small, e.g. 3.3 million in *Drosophila melanogaster* (Kreitman, 1983). While one might question the assumptions on which such estimates are based, it would be wrong to dismiss them out of hand.

Conclusions

I have suggested two factors which could delay the initial advance of a virulent strain. (i) relatively low fitness of the virulent in the early stages of its spread, the observed high fitnesses being evolved at a later stage, and (ii) chance extinction. The latter will be particularly important if virulent individuals differ substantially in fecundity, and/or subpopulation extinction and replacement happens fairly often. A combination of (i) and (ii) could have a very marked effect. However, the relevance of these factors will be uncertain until the appropriate empirical data become available.

Acknowledgements

The author is indebted to many participants at the Conference, and especially to Dr C.E. Caten, for information and very helpful discussion.

References

Charlesworth B. (1980) *Evolution in Age-Structured Populations*. Cambridge University Press, Cambridge.

Ellingboe A.H. (1982) Genetical aspects of active defence. In: *Active Defence Mechanisms in Plants* (Ed. by R.K.S. Wood), pp. 179–92. Plenum Press, New York.

Ewens W.J. (1967) The probability of survival of a new mutant in a fluctuating environment. *Heredity* **22**, 438–43.

Fincham J.R.S., Day P.R. & Radford A. (1979) *Fungal Genetics* 4th ed. Blackwell Scientific Publications, Oxford.

Fisher R.A. (1930) *The Genetical Theory of Natural Selection*. Clarendon Press, Oxford.

Haldane J.B.S. (1927) A mathematical theory of natural and artificial selection, part V: selection and mutation. *Proceedings of the Cambridge Philosophical Society* **23**, 838–44.

Kimura M. (1957) Some problems of stochastic processes in genetics. *Annals of Mathematical Statistics* **28**, 882–901.

Kimura M. (1964) Diffusion models in population genetics. *Journal of Applied Probability* **1**, 177–232.

Kojima K. & Kelleher T.M. (1962) Survival of mutant genes. *American Naturalist* **96**, 329–46.

Kreitman M. (1983) Nucleotide polymorphism at the alcohol dehydrogenase locus of *Drosophila melanogaster*. *Nature* **304**, 412–17.

Maruyama T. (1977) *Stochastic Problems in Population Genetics*. Springer-Verlag, Berlin.

Maruyama T. & Kimura M. (1980) Genetic variability and effective population size when local extinction and recolonization of subpopulations are frequent. *Proceedings of the National Academy of Sciences, USA* **77**, 6710–14.

Moran P.A.P. (1961) The survival of a mutant under general conditions. *Proceedings of the Cambridge Philosophical Society* **57**, 304–14.

Nei M. & Murata M. (1966) Effective population size when fertility is inherited. *Genetical Research* **8**, 257–60.

Parlevliet J.E. (1981) Stabilizing selection in crop pathosystems: an empty concept or a reality? *Euphytica* **30**, 259–69.

Wright S. (1931) Evolution in Mendelian populations. *Genetics* **16**, 97–159.

Wright S. (1937) The distribution of gene frequencies in populations. *Proceedings of the National Academy of Sciences, USA* **23**, 307–20.

Wright S. (1938) Size of population and breeding structure in relation to evolution. *Science* **87**, 430–1.

6 Analysis of virulence diversity in populations of plant pathogens

J.V. GROTH and A.P. ROELFS
Department of Plant Pathology, University of Minnesota and United States Department of Agriculture Cereal Rust Laboratory, respectively, St Paul, MN 55108, USA

Introduction

The quantitative measurement of genetic diversity in populations of plant pathogenic fungi is essential in our goal of understanding the dynamics of virulence shifts as they influence the durability of resistance. Host-specific (intraspecific) virulence genes have been the most abundant, and hence the most commonly employed, markers in attempts to describe either overall virulence diversity (Roelfs & Groth, 1980; Lebeda, 1982) or some aspect of the distribution of virulence genes or races in populations (Browder & Eversmeyer, 1977; Simons *et al.*, 1979; Groth & Roelfs, 1982; Wolfe, this volume; Crute, this volume). Other kinds of markers, particularly isozymes, must be and are being employed (Burdon *et al.*, 1982) in obtaining a more complete and more objective picture of pathogen diversity. However, the direct importance and increasing abundance of virulence markers will ensure that they will continue to serve a key role in describing plant pathogen populations. Markers of all kinds will be needed to fully describe genetic structure of populations of plant pathogens and subsequently ecological and evolutionary events of importance to economic plant pathology.

The use of ecological diversity indexes at an intraspecific level has only recently been advocated and used as a means of combining into one easily understood variable two important aspects of genetic diversity: the number of races (race abundance) and the homogeneity/heterogeneity of the frequencies of different races (race frequency evenness) (Lebeda, 1982; Groth & Roelfs, 1982). A number of different indexes are available for use (Whittaker, 1972; Lebeda, 1982). The Shannon index is perhaps the most popular and simple to use, either in its pure form or, for finite populations, as the Brillouin's Index (Pielou, 1977). Since we are not describing sampled populations in most of the following development, we use the Shannon index:

$$D = \sum_{i} p_i \log_e (p_i) \quad [1]$$

Where D = the Shannon index, p_i = the frequency of the ith phenotype or race. For simplicity $\log_e$ = ln. The use of this index as a means of quantitatively linking various distributional properties of virulence genes will become clear in the description to follow.

Wolfe M.S. & Caten C.E. (1987) *Populations of Plant Pathogens: their Dynamics and Genetics.* Blackwell Scientific Publications, Oxford.

We have been interested in the means by which differential cultivars of a host species can be used to detect genetic diversity for virulence in plant pathogens. There are several aspects of the distribution of either resistance genes in the host differentials, or of virulence genes in the pathogen, which influence how much diversity a differential host set can reveal in a particular pathogen population. These influences must be quantitatively accounted for in order to present as complete a picture as possible of how differential host lines function. We will attempt to deal with the important aspects in this chapter.

Differential hosts divide the pathogen population into races (distinct virulence phenotypes). Each differential line that is added is capable of doubling the number of races. Because of its logarithmic nature, the Shannon index is linearly related to the number of differential lines when all detected virulences are polymorphic in the pathogen population, and at the same frequency. This linearity is not found for the other common index, the Simpson index of similarity, whose complement or inverse are used as diversity indexes. Because of this linearity and the potential value of the hierarchical use of the Shannon index, whereby amount of difference between phenotypes can be partly accounted for (Pielou, 1977, p. 311), we chose to use it exclusively in this analysis. Briefly, we use the index as a means of quantifying and directly comparing the effects of three different constraints on the capacity of a differential set to detect population diversity in the pathogen. These constraints are non-independence of resistance genes in differential lines, near fixation of virulence genes and associations of virulence genes.

The distribution of resistance genes

Differential host lines fall into two types, 'Single-gene' differentials, and those for which the number of resistance genes (and corresponding virulence genes) is not known. In a few instances the latter kind of differential may unknowingly be of the single-gene type. What is important is the number of *detectable* resistance genes, that is, the number that will be detected because the pathogen population possesses some phenotypes which are avirulent. On the other hand, if a resistance gene is unmatched by virulence, not only will the particular cultivar and gene be of no use in revealing genetic diversity of the pathogen, but also, because of the epistatic masking of their effects, any other genes in the cultivar will be rendered undetectable. Frequencies of virulences in the pathogen population corresponding to the masked resistance will have no effect on this loss of information. At times this loss may be reduced because the masking resistance gives a different, identifiable reaction from that of those which are masked. Usually, however, the masking of one gene by another will be complete. Qualitatively stated, the result is that for a given number of differentials used to measure diversity in a population, the best possible case is where each differential contains one gene only.

Any degree of clustering in distribution of the resistance genes will detract from the capacity of the differential host set to detect diversity. The worst case is an absurd one where one host line contains all of the resistance genes while all others contain none. In such an instance, if all virulences are independent and at 0.5 frequency, only two races would be detected, and only 0.5^n of the population would appear as a different race, where n is the number of genes for resistance and virulence.

Intermediate cases between the best and worst can be of different types. Taking a simple case, where $n = 4$, the number of resistance genes in each host line can be placed in four classes of decreasing degree of clustering (the number of *functioning* differential lines is in parentheses): (a) all four genes in one line and none in the other three (1 line); (b) three genes in one line and one in one of the other three, or two genes in each of two lines (2 lines); (c) two genes in one line and one in each of the other two lines (3 lines); and (d) one resistance gene in each line (4 lines). Figure 6.1 shows the effects of clustering on the Shannon index. There is an almost linear decrease in the measured diversity as the resistance genes are concentrated in fewer host lines, until only a tenth of the diversity is being detected. Maximum diversity is detected (or minimum information is lost) in the single-gene case (d). If we can generalize from this simple example, we must conclude that single-gene differentials are the most effective at detecting diversity. This is admittedly no great

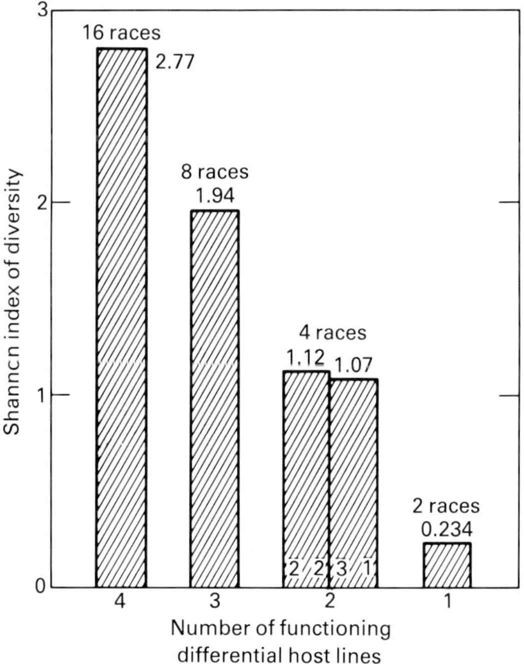

Figure 6.1. Potential diversity for virulence, as measured by the Shannon diversity index, in a pathogen population that is divided using differential host lines containing four resistance genes, where the genes are, from left to right, increasingly non-independent in distribution, being included in 4, 3, 2, and 1 of the lines, respectively. All corresponding virulences are assumed to be independent with a 0.5 frequency.

revelation as a qualitative fact. Nevertheless, it is worthwhile to begin to express how much is being lost as this assumption is relaxed more and more.

Virulence properties–polymorphism and independence

Using the Shannon index of diversity, we will look separately at the two properties of virulence that are responsible for loss of information or loss of efficiency of a set of differential lines in detecting diversity: (1) degree of polymorphism of virulence, and (2) association of virulences. For simplicity in this initial analysis, only the case involving single-gene differentials will be explored.

With respect to the two components of diversity, i.e. race abundance and race-frequency evenness, that are considered by the diversity indexes, maximum diversity (Figure 6.2) occurs in a hypothetical pathogen population that has (a) each virulence at 0.5 frequency and (b) each virulence independent of every other, i.e. no epistasis, allelism, or linkage disequilibrium (including association due to vegetative propagation of a small number of races). It follows then that departures from the above conditions provide the only two ways of decreasing diversity.

The effect of virulence frequency on diversity

The first property, virulence frequency, is the easiest to account for because, unlike virulence association, frequency is a property solely of a single virulence. In the analysis, we must first examine the ideal case for a set of differential lines, in which the effects of adding differential lines are examined cumulatively. Frequency of virulence is most efficiently expressed as departure from 0.5, which we call d. Retaining the independence assumption, the relationship between the Shannon index D and the number of differential lines for values of d decreasing in 0.1 increments is expressed as a set of lines (Figure 6.2). The assumption is that frequency of each virulence is the same in these linear cases. If this is not so, each added differential line increases the value of D differently, and the slope of each is described by:

$$\Delta D_i = -((0.5 + d_i) \ln (0.5 + d_i) + (0.5 - d_i) \ln (0.5 - d_i)) \quad [2]$$

where ΔD_i = the change in Shannon index by addition of the ith differential, d_i = the departure from frequency of 0.5 of virulence of the ith differential.

This will reduce to $-\ln (0.5)$ if $d_i = 0$.

Equation [2] allows a simple measurement of the influence on diversity of departure from 0.5 of each virulence, so that the information needed to account for this constraint is easily incorporated into the analysis. If the only source of loss in diversity is departure from 0.5 the total diversity is simply the sum of all ΔD_i's.

The effect of virulence associations on diversity

The other effect to consider is that of associations (in coupling or repulsion)

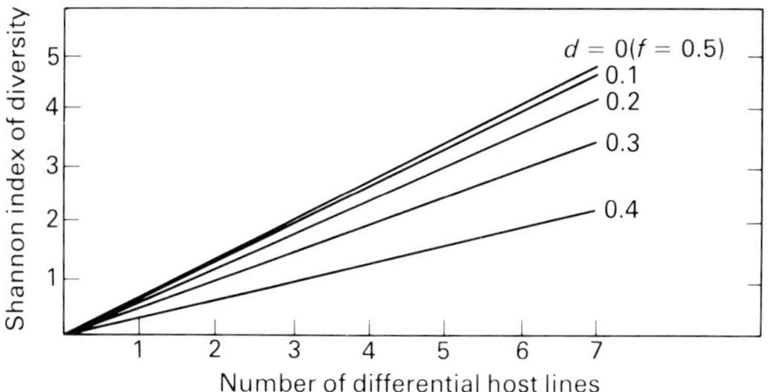

Figure 6.2. The relationship between the number of differential host lines and increase in Shannon diversity index for five virulence frequencies. The departure (d) of the virulence frequency from 0.5 is shown above each line and is the same for each virulence. All virulences are independent.

between virulences, whether they be due to lack of recombination, linkage, or something more obscure. This constraint on variation is not simply expressed algebraically, in fact we will not completely detail it in this treatment. The problem is that each virulence can be associated with some or all of the others in different degrees. At the most complex level, there exist extensive networks or matrices of associations. The magnitude of associations may or may not be mathematically related, depending on the biological basis of the association. For example, the linearity and approximate additivity of distances between genes in linkage groups might provide a mathematical relationship between associations in a few cases, provided that genetic linkage is the basis of association. Likewise, knowledge of the complete phenotypes of prevalent races with respect to pertinent virulences may allow associations to be related in pathogens that are predominantly asexual and possess a relatively small number of races. Confounded with real associations in such a case is the effect of selection on diversity. For some populations, association may also include various kinds of non-random combinations of virulence and background genotypes, all resulting in reduced diversity (Wolfe & Knott, 1982).

Two special cases can be explored with respect to associations. The first is the case (considering all virulences at 0.5 frequency) where each of the identified virulence genes is equally associated with every other. The effect of this case on the increment of diversity is defined by the relationship:

$$\Delta D = \frac{-\left((1 + a) \ln \left(\frac{1 + a}{2}\right) + (1 - a) \ln \left(\frac{1 - a}{2}\right)\right)}{2} \quad [3]$$

where ΔD = the change in Shannon index due to association when a differential line is added and a = the coefficient of association such that the products of frequencies of two virulences are multiplied by $1 + a$ if associated in coupling or $1 - a$ if associated in repulsion. This case can be described graphically as a series of lines in the same way as levels of departure from 0.5 frequency (Figure 6.3).

The second special case is where only one pair of the virulences that are being

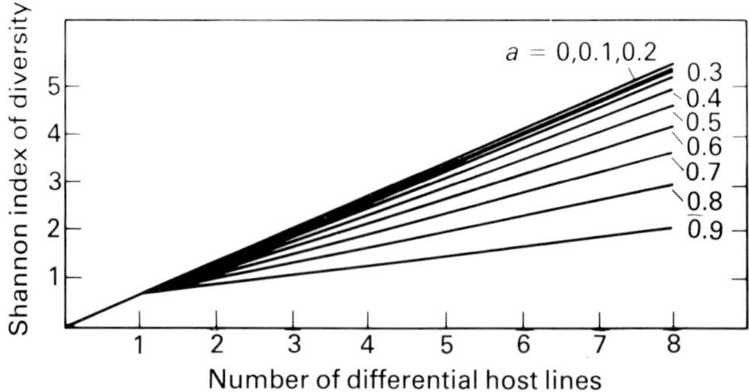

Figure 6.3. The relationship between the number of differential host lines and increase in Shannon diversity index for ten different degrees of association or disassociation of the corresponding virulences. The degrees of association are shown by the coefficients *a* above each line. All virulence frequencies are 0.5.

detected is associated. Mathematically, the increment of diversity is generally described as:

$$\Delta D = \frac{-2\left(\ln\left(\frac{1}{2^n}\right) + (1+a)\ln\left(\frac{1+a}{2^n}\right) + (1-a)\ln\left(\frac{1-a}{2^n}\right)\right)}{2^n} \qquad [4]$$

where ΔD = the change in Shannon index when a differential line is added, a = the coefficient of association, and n = the number of differential lines.

The case when all associations are different is not so clear, and no simply stated relationship for increments of Shannon index has been developed as yet. All of the myriad associations will have to be accounted for as each virulence is added, resulting in a cumbersome algebraic statement if all associations are explicitly included.

Since there are only two factors (virulence frequency and virulence associations) which influence diversity in our development, the lack of ability to generalize about many different associations is not a major problem. Direct calculation of the change in Shannon index when a new differential host line is added requires only that race numbers and frequencies be known both before and after adding the new line. The value obtained, which is the 'actual' line in Figure 6.4, will nearly always be less than the maximum possible, which is a slope of $-\ln(0.5)$ per differential line. Explicitly, the part of the difference between actual and maximum increment (a in Figure 6.4) that is due to virulence frequency of the newly detected virulence can be calculated using equation [2]. The value calculated is an increment or slope that will fall somewhere between, or rarely equal, either the actual or the maximum increments. The difference between $-\ln(0.5)$ and the actual calculated increment is the total effect of both virulence frequency (which is explicit) and associations (whose overall effects are known only implicitly). The difference (b in Figure 6.4) between the increment allowing for virulence frequency alone and the actual calculated slope

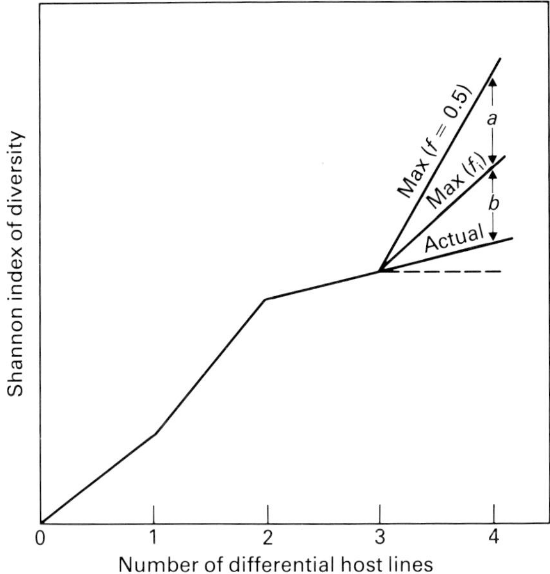

Figure 6.4. The increment of virulence diversity obtained when a hypothetical 4th host line is added to a hypothetical differential set, each member of whose increments is shown. Actual = the increment observed; Max ($f = 0.5$) = the estimated increment assuming the frequency of matching virulence is 0.5; Max (f_i) = the estimated increment assuming the actual frequency of matching virulence. Segment *a* represents the portion of diversity (information) lost due to the departure of the virulence frequency from 0.5. Segment *b* represents the portion of diversity lost because of non-random association of virulences.

must then be due to associations. So, while the associations cannot be separated, their overall influence can be determined. In so doing, we can make quantitative comparisons of the relative importance of degree of association and virulence frequency. The above analyses can be made as each differential host line is added. The linear relationship between the number of differential lines and the value of the Shannon index ensures that results are simply additive.

Application of the theory

To illustrate the uses to which these analyses can be put, we will use wheat stem rust data from 1975 collections made in the United States. The overall expectation that the amount of diversity detected is a function of the differential set used is best illustrated by an already published example of Young & Prescott (1977) for leaf rust of wheat, *Puccinia recondita* f. sp. *tritici*. Using two different sets of differential host lines, they obtained very different numbers and frequencies of races for each of two populations. The differences were illustrated and Shannon indexes calculated in a later analysis by Groth & Roelfs (1982).

Analyses were carried out on the 1975 sexual (Washington and Idaho) and asexual (east of the Rocky Mountains) populations, where 16 single-gene differential lines of wheat were used to characterize virulence diversity (Roelfs & Groth, 1980). Since virulence was not detected in either population to *Sr*13, this differential line was not included. Two other resistance genes, *Sr*5 and *Sr*16, added nothing to diversity in the asexual population because virulence to both was fixed. They were retained, however, because virulence to neither was fixed in the sexual population.

The sexual population was more diverse, with 39% of the maximum potential diversity (D_{max} = 10.4) being accounted for, as opposed to only 12% in the asexual population (Figure 6.5). As also might have been expected, a larger percentage of the maximum possible diversity (39%) was not detected due to virulence associations in the asexual population than in the sexual population (14%). The remaining diversity loss, due to virulence frequency was nearly equal, being 47% and 49% for the sexual and asexual populations, respectively (Figure 6.5).

In stem rust, we have examined selected sets of four differential host lines. For both asexually and sexually reproducing populations of the fungus two sets of four single-gene differential lines each were selected (Table 6.1). Set one includes four *Sr* genes whose corresponding virulences were as near to fixation as was possible without being actually fixed. Set two includes four *Sr* genes selected because the matching virulence in the rust population is polymorphic, and as close to 0.5 frequency as was available in the 16 single-gene differentials examined. An additional case, case 3, was included for the asexual population; the four genes were selected on the basis that they appeared to detect a larger number of races than other sets of four. Table 6.1 shows maximum possible and actual diversity indexes, as well as calculated loss in diversity due to degree of near-fixation of the four genes in each of the five cases. These analyses are presented graphically in Figures 6.6 and 6.7. The percentage of potential diversity represented by actual diversity ranges from 19% in the less polymorphic sexual case (Set 1) to 83.1% in the more polymorphic sexual case (Set 2) (Figure 6.7). These losses are further split into loss due to

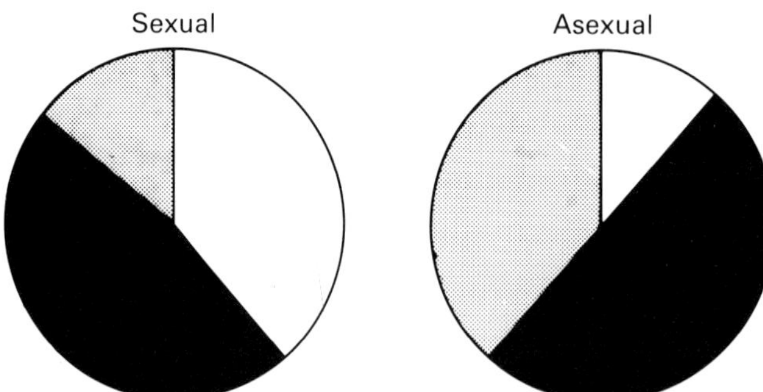

Figure 6.5. Components of virulence diversity of two 1975 populations of stem rust of wheat in the US. The sexual population occurs where the alternate host, barberry, is common while the asexual population effectively lacks barberry. White = detected diversity (on 15 single-gene differential host lines); black = loss in diversity due to departure of virulence from 0.5 frequency; stippled = loss in diversity due to associations of virulences.

Table 6.1. Values of components of virulence diversity for five sets of four single-gene differential wheat lines used to subdivide two 1975 populations of wheat stem rust in the US

Set	*Sr* gene series[d]	Shannon index of diversity[e]	
		Actual *D*	Loss due to departure from 0.5 frequency
Asexual			
Set 1[a]	*9d*(2363), *3c*(2148), *9b*(293), *6*(254)	.967	1.706
Set 2[b]	*Tmp*(1860), *9e*(1855), *17*(463), *7b*(414)	.682	0.768
Set 3[c]	*11*(1980), *6*(254), *8*(2077), *9a*(400)	1.171	1.150
Sexual			
Set 1[a]	*6*(4), *9b*(5), *10*(42), *15*(399)	.489	2.275
Set 2[b]	*5*(296), *9d*(296), *8*(92), *9a*(270)	2.234	0.364

[a] Four wheat lines whose corresponding virulences were less polymorphic.
[b] Four wheat lines whose corresponding virulences were more polymorphic.
[c] Four wheat lines that subdivided the pathogen population into a large number of races.
[d] Number of virulent isolates indicated in brackets.
[e] Maximum possible diversity (*D*) for each each set is 2.773.

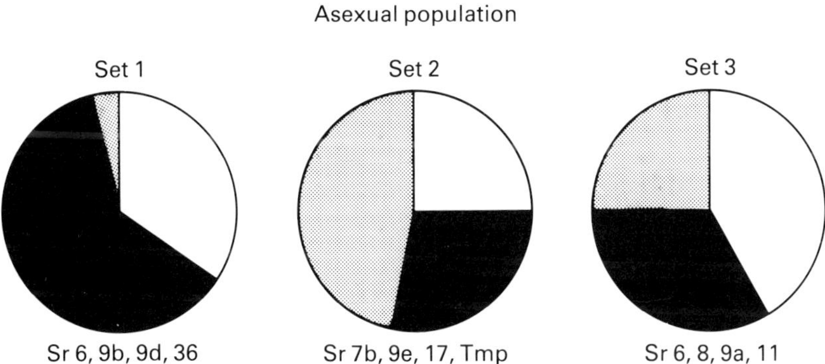

Figure 6.6. Components of virulence diversity of an asexually reproducing population of stem rust of wheat from the US sampled in 1975. Each circle is based on a different set of four single-gene differential host lines (see Table 6.1). White = detected diversity; black = loss in diversity due to departure of virulence from 0.5 frequency; stippled = loss in diversity due to associations of virulences.

virulence frequency, which ranged from 14% of the total possible diversity in set 2 sexual (Figure 6.7) to 82% of the total in set 1 sexual, and loss due to virulence associations ranging from 1% of total possible diversity in set 1 sexual to 48% in set 2 asexual. Both total amount of diversity detected and proportion of diversity loss due to frequency vs. association must be considered in order to obtain a complete picture of diversity.

Similar analyses can be done on single differentials. The amount of information

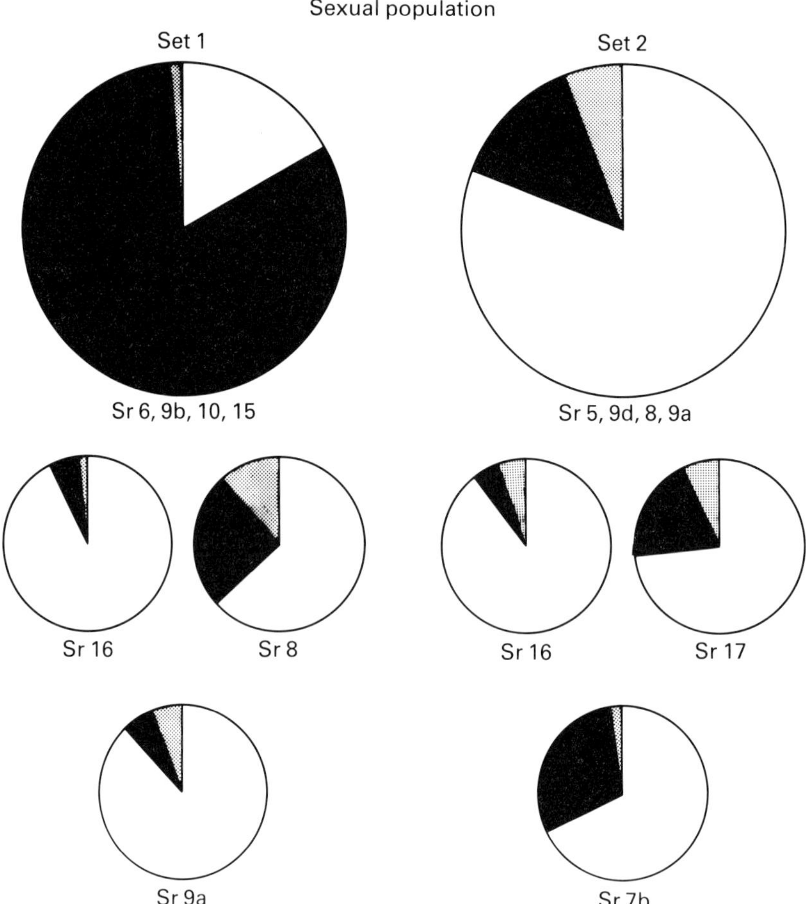

Figure 6.7. Components of virulence diversity of a sexually reproducing population of stem rust of wheat from the US sampled in 1975. Each large circle is based on a different set of four single-gene differential host-lines (see Table 6.1). The smaller circles each represent a fifth line which is being added to the set above. White = detected diversity; black = loss in diversity due to departure of virulence from 0.5 frequency; stippled = loss in diversity due to associations of virulences.

lost due to near-fixation is independent of the identities of the resistance genes already present in the differential host set to which a line is being added. This is not true for either the loss in information due to associations or the increment in diversity when the line is included; these will be influenced by the genes already present. To illustrate the effects of the addition of different genes to existing four-gene sets, separate analyses were made of the addition in diversity when new genes are added, singly, to each of the two sets of the sexual population (Table 6.2 and Figure 6.7). The proportion of loss in detected diversity due to departure from 0.5 virulence frequency of the newly-introduced resistance gene is directly proportional to the magnitude of this departure (Table 6.2), which is the single independent variable of equation [2]. The loss of diversity thus obtained is independent of

Table 6.2. Values of components of virulence diversity when single-gene differential lines are added, as the fifth differential host, to two sets of four lines that were used to divide a 1975 sexual population of stem rust of wheat in the US. Maximum possible diversity added by a single line is 0.693

Added gene	Virulence frequency	Shannon index of diversity: Actual increment	Shannon index of diversity: Loss due to departure from 0.5 frequency	Number of races added
Set 1[a]				
*Sr*16	.63	.647	.036	4
*Sr*8	.22	.440	.171	4
*Sr*9a	.63	.617	.036	3
Set 2[b]				
*Sr*16	.63	.630	.036	13
*Sr*17	.76	.513	.137	12
*Sr*7b	.19	.474	.207	13

[a] Four single-gene lines whose corresponding virulences were less polymorphic; set 1 identified 6 races.
[b] Four single-gene lines whose corresponding virulences were more polymorphic; set 2 identified 14 races.

the specific four genes to which the fifth is being added, as seen by the identical sizes of the black slices for the addition of *Sr*16 to the two different four-line sets (Figure 6.7). In addition, *Sr*9a, which shows the same virulence frequency as *Sr*16, shows an identical proportion of lost diversity as *Sr*16. As has been indicated more generally earlier, the loss in diversity due to association is dependent on the specific genes already present in a differential host set, when a new gene is added. This can most clearly be seen for *Sr*16 where the size of the stippled slice is larger when this gene is added to set 2 than when it is added to set 1 (Figure 6.7).

The remaining pie charts illustrate that various relative and absolute amounts of detected diversity are possible when new genes are added, and that the loss in diversity can be more the result of departure of virulence frequencies from 0.5 (*Sr*7b), association (*Sr*16 added to set 2), or both (*Sr*9a) (Figure 6.7). As previously pointed out, association was not as important a cause of loss in the sexual as in the asexual population (Figure 6.5, and cf. Figure 6.6 with Figure 6.7). A degree of non-random association of virulences in the sexual population has been shown to exist using other methods (Alexander *et al*, 1984).

Conclusions

The intent of this development and demonstration is to suggest a way in which binomial expectations can be used in conjuction with a diversity index to permit a more complete and objective analysis of how differential host lines operate. A number of changes or improvements in the analysis are possible. One problem is that the correspondence between the ideal cases outlined in the first part of the chapter, and the real cases for which such an analysis is intended, is not perfect. We have not addressed sampling problems here; small sample sizes and small numbers of races are not conducive to analysis using the above methods. Asexual populations, in particular, may show very large association effects. We have not tried explicitly to define virulence associations alone, much less the effects of selection when virulences are 'trapped' in specific genetic backgrounds. The contribution of association therefore is, and probably will contine to be, the most elusive component of diversity. A more analytical treatment of associations will provide further insight into phenomena important in the adaptation of plant pathogen populations to the use of host resistance. Integration of background genotype and race-specific virulence complement, as well as other interactions between virulence and natural selection, are examples of such phenomena.

Eventually such analyses should be applied to differential host lines that do not contain a single resistance gene each. It seems likely to us, however, that rather extensive knowledge of the genetics of resistance will be necessary if the model is to be valid.

Acknowledgements

This work was supported in part by USDA Competitive Grant 59-2271-1-1-687-0, and by the Minnesota Agricultural Experiment Station.

References

Alexander H.M., Roelfs A.P. & Groth J.V. (1984) Pathogenicity associations in *Puccinia graminis* f. sp. *tritici* in the United States. *Phytopathology* **74**, 1161–6.

Browder L.E. & Eversmeyer M.G. (1977) Pathogenicity associations in *Puccinia recondita tritici*. *Phytopathology* **67**, 766–71.

Burdon J.J., Marshall D.R., Luig H.H. & Gow D.J.S. (1982) Isozyme studies on the origin and evolution of *Puccinia graminis* f. sp. *tritici* in Australia. *Australian Journal of Biological Sciences* **35**, 231–8.

Groth J.V. & Roelfs A.P. (1982) Effect of sexual and asexual reproduction on race abundance in cereal rust fungus populations. *Phytopathology* **72**, 1503–7.

Lebeda A. (1982) Measurement of genotypic diversity of virulence in populations of phytopathogenic fungi. *Zeitschrift für Pflanzenkrankheiten und Pflanzenschutz* **89**, 88–95.

Pielou E.C. (1977) *Mathematical Ecology*. Wiley, New York.

Roelfs A.P. & Groth J.V. (1980) A comparison of virulence phenotypes in wheat stem rust populations reproducing sexually and asexually. *Phytopathology* **70**, 855–62.

Simons M.D., Rothman P.G. & Michel L.J. (1979) Pathogenicity of *Puccinia coronata* from buckthorn and from oats adjacent to and distant from buckthorn. *Phytopathology* **69**, 156–8.

Whittaker R.H. (1972) Evolution and measurement of species diversity. *Taxon* **21**, 213–51.

Wolfe M.S. & Knott D.R. (1982) Populations of plant pathogens: some constraints on analysis of variation in pathogenicity. *Plant Pathology* **31**, 79–90.

Young H.C. & Prescott J.M. (1977) A study of race populations of *Puccinia recondita* f. sp. *tritici*. *Phytopathology* **67**, 528–32.

7 The analysis of host and pathogen populations in natural ecosystems

A. DINOOR and N. ESHED
Faculty of Agriculture, Rehovot, The Hebrew University of Jerusalem, Israel

Introduction

Pathogen populations are analysed in relation to their hosts. Unlike agroecosystems where the hosts are uniform and usually known genetically, hosts in the wild are heterogeneous and genetically unknown. Analysis of pathogen populations in the wild is intimately related to the analysis of their hosts.

The interest in the genetics of host–parasite relationships in the wild was initiated by the search for novel plant germplasm in centres of crop origin (e.g. Harlan, 1976; Leppik, 1970). Resistance genes were identified and extracted from wild plants without analysis of the whole genetic make-up of those plants (Dinoor, 1970; Leppik, 1970; Segal *et al.*, 1980). Later on it was envisaged that defence strategies in the wild, if analysed and described, may be adopted for agricultural crops (Browning, 1974; Dinoor, 1970; Dinus, 1974; Segal *et al.*, 1980). Studies on pathogen virulence are being conducted through race surveys, overflowing from crops into wild plants. The common race surveys are based on unbalanced sampling of pathogen isolates. Balanced, random sampling was suggested and conducted on a very limited scale (Dinoor, 1973). The development of the mobile nursery system (Eyal *et al.*, 1973) was an important improvement in the survey of virulence. It was further developed into a mobile spore-trap aided nursery (Schwarzbach, 1979; Wolfe *et al.*, 1981).

While the technology has greatly improved, basic concepts are still in dispute. The differential varieties and/or the genetic markers used in virulence surveys are not directly related to the pathogen populations being studied. Should they carry universally identified genes, or host genotypes representative of the region or the location being studied? Attempts to study wild host and pathogen populations are faulty in one way or another. Host populations are characterized by pathogen cultures from other locations (including other countries) (e.g. Harper, 1977; Segal *et al.*, 1980), or at least away from their natural surroundings (Burdon, 1980 a, b and unpublished). The failure, so far, to study natural populations *in situ*, and in concert with all factors involved, was lately discussed (Dinoor & Eshed, 1984) and was the basis for some of our research programmes described herein.

Two host–parasite systems are currently being investigated; in each system we proceed via a different route but the ultimate goal is the same. The two systems are different. The wild barley–powdery mildew system (*Hordeum spontaneum*–

Wolfe M.S. & Caten C.E. (1987) *Populations of Plant Pathogens: their Dynamics and Genetics.* Blackwell Scientific Publications, Oxford.

Erysiphe graminis) involves an annual host with discrete seasons of activity, and a haploid parasite with an active phase depending on the existence of the host and a dormant sexual stage to ensure continuity in the next season. The sexual stage of the parasite contributes appreciably to pathogen variation. The wild limonium–rust system (*Limonium sinuatum–Uromyces savulescui*) involves a perennial host with continuous activity and a dikaryotic–haploid parasite with continuous activity, but with the involvement of different stages of the life cycle depending on seasonal conditions (Eshed & Dinoor, 1983).

The wild barley–powdery mildew system

Five locations across a transect from the sea-shore to the mountains were chosen for this study (Figure 7.1). These sites differ in their altitude and distance from the sea and represent different climatic conditions. Seasonal surveys have shown differences in plant development and disease development between the locations (Table 7.1). Germination was late near the coast and east of the divide. Mildew developed earlier on the mountains west of the divide and reached high levels in scattered sites in these two locations (Kesalon and Qiryat Anavim). Plant development slowed down higher in the mountains, and disease level was high till the end of the season (Qiryat Anavim). Mildew developed quite late in Jerusalem (east of the divide) and

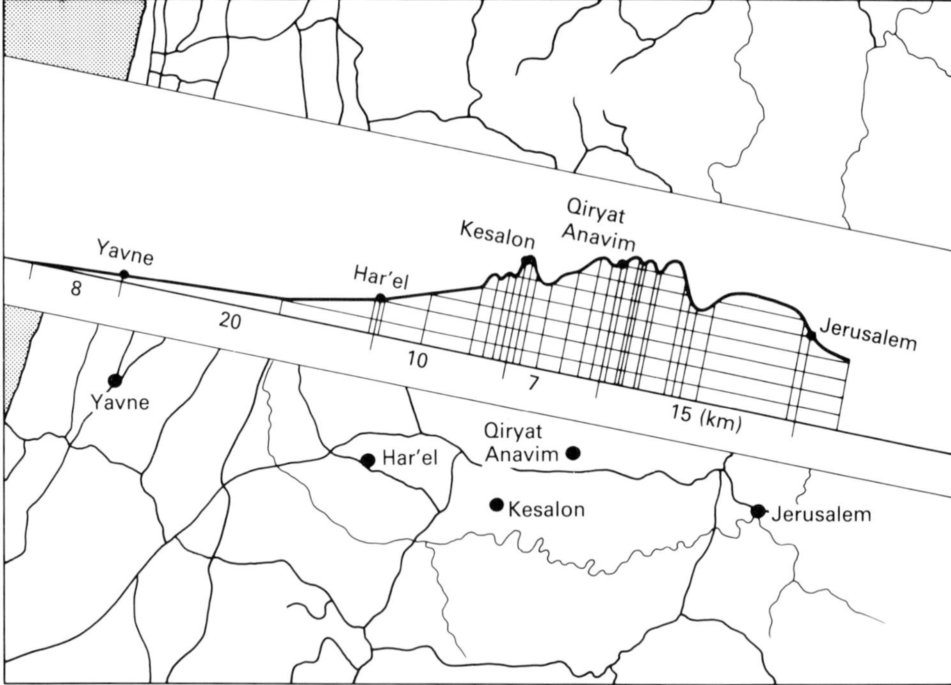

Figure 7.1. Topographic cross-section superimposed on the map showing the five locations from which wild barley and mildew were collected.

Table 7.1. Seasonal surveys of wild barley and mildew in 1979/80 at five locations. Upper row in each date indicates plant developmental stage according to Feeke's scale. Lower row indicates maximal disease severity level observed. The locations are:
Yavn – Yavne, Harl – Har'el, Ksal – Kesalon, Anav – Qiryat Anavim, Jrlm – Jerusalem (see Figure 7.1)

	Location				
Date	Yavn	Harl	Ksal	Anav	Jrlm
16.10	0	1	1	1	0
		0	0	0	
11.11	1	1	1	2	1
	0	0	2	2	0
11.12	1	1	3	2	1
	0	0	50	40	0
17.01	2	2	3	3	2
	0	0	30	20	0
19.02		7	7	3	3
		30	30	30	1
23.03		10.5	10.1	10.1	10.3
		40	30	30	30
20.04	10.5	11.4	11.3	11.4	11.1
	0	2	50	60	10

did not develop at all in Yavne (coastal plain). In each site a random sample of 50 spikes and a random sample of mildew cultures were collected. The plant material was propagated for experimentation. The mildew isolates were subcultured and maintained on detached barley leaves.

Three hundred and eighty-five isolates were characterized by inoculation on detached leaves of 36 barley cultivars. Virulence towards most of the resistance genes used in Europe is very high (Table 7.2) despite the absence here of these genes in cultivated barley. It is not known whether these resistance genes are present in wild barley. Only the virulence towards R6 is quite rare. The virulence towards *mlo* is also low or very low, but these cultures could not be further propagated on *mlo*. The data for the last nine cultivars are not uniform (Table 7.2). There are marked differences in frequencies of some virulences between locations. The fact that the frequency of virulence in Jerusalem towards some of the cultivars is higher than in the other locations, or at least higher than in one of the locations in the mountains, should be noted for the discussion on the geographical distribution of resistance.

Table 7.2. Frequency of virulence towards resistance genes used in Europe among isolates of *Erysiphe graminis* collected from five locations in Israel. See Table 7.1 for names of locations

Cultivar	Resistance gene	Location				
		Yavn	Harl	Ksal	Anav	Jrlm
Golden Promise	R0	0.95	0.90	0.93	0.95	0.88
Senta	R1	0.98	0.98	1.00	0.91	1.00
Zephyr	R2	0.35	0.38	0.32	0.51	0.41
Midas	R3	0.96	0.94	0.89	0.99	0.96
Lofa Abed	R4	1.00	0.95	0.96	0.97	1.00
Hassan	R5	1.00	0.97	0.91	0.99	0.98
Wing	R6	0.07	0.03	0.07	0.13	0.02
Tyra	R7	0.75	0.81	0.63	0.83	0.78
Algerian	*Mla1*	0.95	0.89	0.98	0.88	0.96
Russian	*Mla2*	1.00	0.97	0.99	0.93	0.96
Ricardo	*Mla3*	1.00	0.92	0.99	0.97	0.94
Australian	*Mla4*	1.00	0.97	0.98	0.97	1.00
Heils Hanna	*Mla8*	1.00	0.97	0.94	0.96	1.00
Durani	*Mla10*	0.89	0.57	0.41	0.57	0.53
A-222	*Mla11*	0.96	0.94	0.98	0.99	0.98
H D	*mlo*	0.02	0.05	0.02	0.15	0.12
HLN	*mlo*	0.33	0.35	0.25	0.46	0.41
Mazurka	R2 + 6	0.11	0.14	0.06	0.24	0.16
Triumph	R6 + x	0.12	0.27	0.04	0.20	0.23
Akka	R2 + 8	0.18	0.55	0.81	0.14	0.69
Jupiter	R3 + 4	0.91	0.84	0.12	0.93	0.90
Belfor	R4 + 6	0.09	0.08	0.06	0.13	0.86
Engledow India		0.72	0.32	0.28	0.29	0.37
Vagabond		0.70	0.38	0.30	0.39	0.43
Rupee		0.47	0.25	0.23	0.25	0.39

A computerized analysis (Dinoor & Peleg, 1972) of the host–parasite interactions was used to estimate the number of virulence genes in each culture. A frequency distribution of cultures from each location according to the number of virulence genes is presented in Table 7.3. The striking fact is that in each location, except Yavne, the cultures carrying genes for virulence at all the loci involved comprise the largest fraction. The data are unbalanced since the number of cultures

Table 7.3. Frequency distribution of cultures (%) according to the number of virulence genes that they possess (data from computerized analysis of host–parasite interactions of 36 cultivars and the cultures from the particular location). Note that the upper number in each column indicates the number of cultures possessing a virulence gene in all the loci involved in the interactions at that location. See Table 7.1 for names of locations

Number of virulence genes	Location				
	Yavn	Harl	Ksal	Anav	Jrlm
23			45		
22			34		
21			12		
20			4		
19			1		
17					49
16	16	63			23
15	39	24	1	53	16
14	30	9	2	35	6
13	9	2		9	6
12	4	2		3	
10	2				
Cultures tested	57	63	139	75	51
Virulence genes identified	16	16	23	15	17

tested is not equal between locations. Thus we do not know whether the larger number of genes apparently involved in the interactions in Kesalon is simply a reflection of the larger number of cultures tested. But we may conclude that the number of genes involved in the interactions in Jerusalem is larger than in the interactions in Yavne, Har'el or Qiryat Anavim, since the number of cultures tested from them was larger than those from Jerusalem. This fact should be borne in mind for the discussion on geographical distribution of resistance, and it fits well with the idea (E. Schwarzbach, personal communication) that wider virulence range may compensate for poorer climatic conditions.

Seed from the five locations was planted in 10 sites scattered over the farm of the Faculty of Agriculture. At each site two plots were sown, one of them shaded by black net transmitting 40% of the light. At each site spreader rows were inoculated with one mildew isolate, two isolates from each location being used over the 10 sites. Mildew was scored on three dates during the season and the results of the last score were used for the following analysis.

Frequency distributions of plants with different disease levels were drawn for each plant sample in each of the 20 plots. The distributions for plants from Har'el in each of the 20 plots are presented in Figure 7.2. The emphasis should be on the shape of the distributions; some of the distributions are close to normal, some are skewed towards resistance (to the left), some are skewed towards susceptibility (to the right) and some are almost flat. A comparison between reactions in the shade and in the open shows that in most cases the reactions in the shade are more towards susceptibility. A comparison between cultures within and between locations does not show a clear effect of the origin of the culture. The reaction to one culture from Yavne (upper left) is more similar to the reaction to one of the cultures from Har'el (upper, second from left) than to that of the second culture

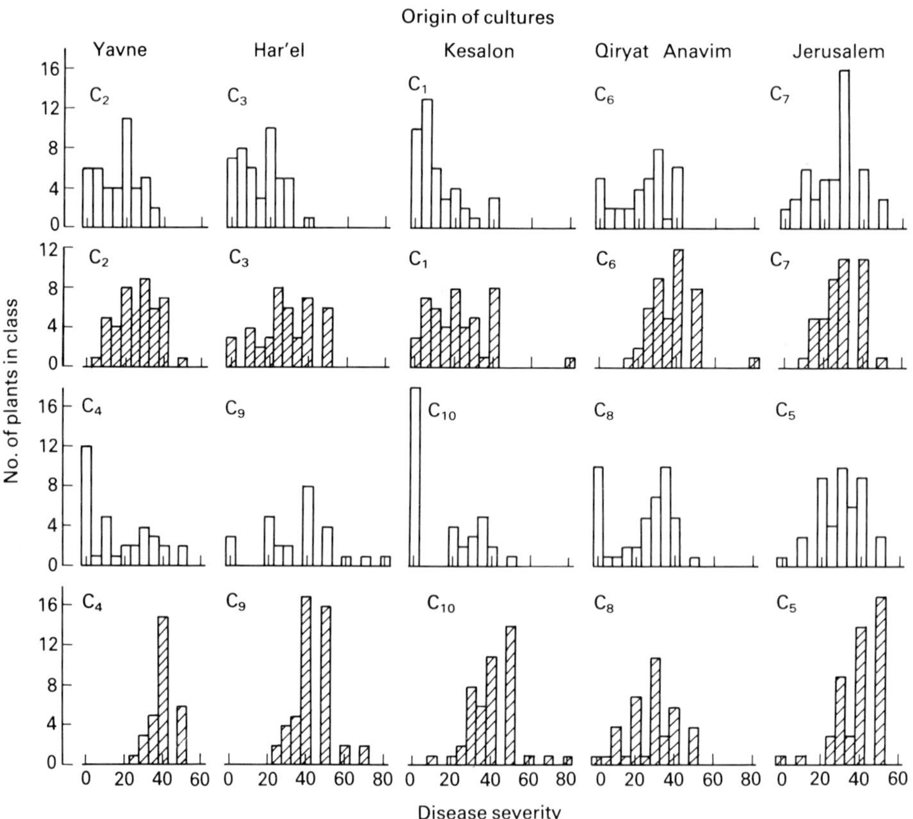

Figure 7.2. Frequency distribution of plants from the Har'el population, according to their reactions to 10 mildew cultures in the open and in the shade. The ten test cultures comprised two from each location in the following order, from left to right: Yavne, Har'el, Kesalon, Qiryat Anavim and Jerusalem. In each column the upper two graphs describe the reactions to one culture from that location and the lower two graphs those to the other culture. In each pair of graphs the upper describes the reactions in the open and the lower those in the shade.

from Yavne (lower left). There is a clear difference between the reactions to different cultures, sometimes even to two cultures from the same location.

Correlations were calculated for the reactions to the same culture in the shade and in the open. In some cases this correlation was significant (Figure 7.3.) while in others it was not, probably indicating interactions with the environment. Correlations were also examined for the reactions to two different cultures in the same environment. In some cases the correlation was significant, as for the reactions in the shade of plants from Har'el to cultures from Har'el and from Jerusalem (Figure 7.4). The correlation here means that the culture from Jerusalem is less aggressive on plants infected up to 24%, and more aggressive on plants infected more than 24%,

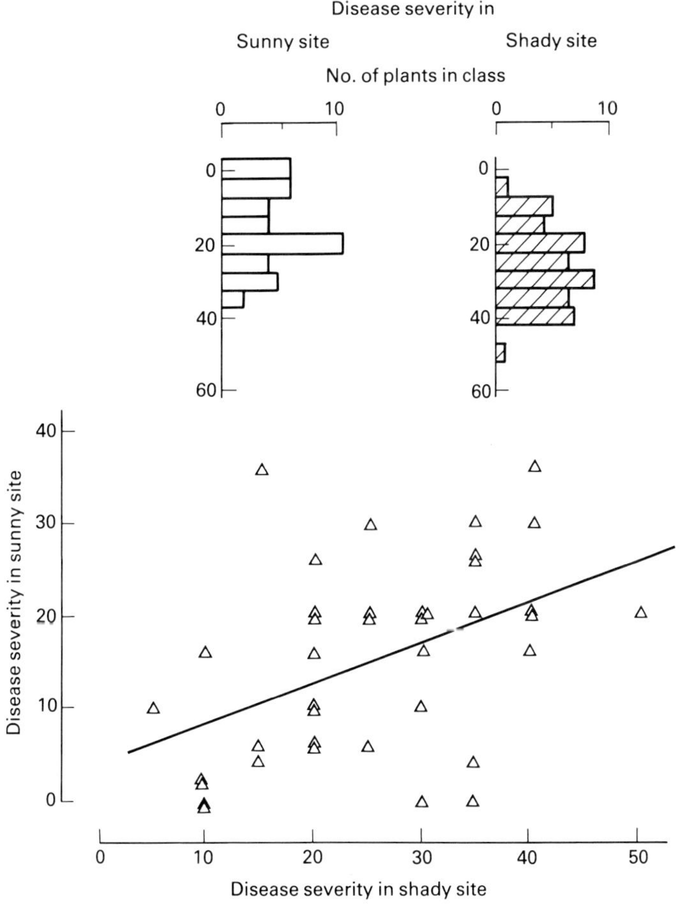

Figure 7.3. The correlation between reactions of barley plants from Har'el to a mildew culture from Yavne in the open (y in the plot, and the left distribution graph) and in the shade (x in the plot, and the right distribution graph). $y = 4.00 + 0.43x, P > 0.01$.

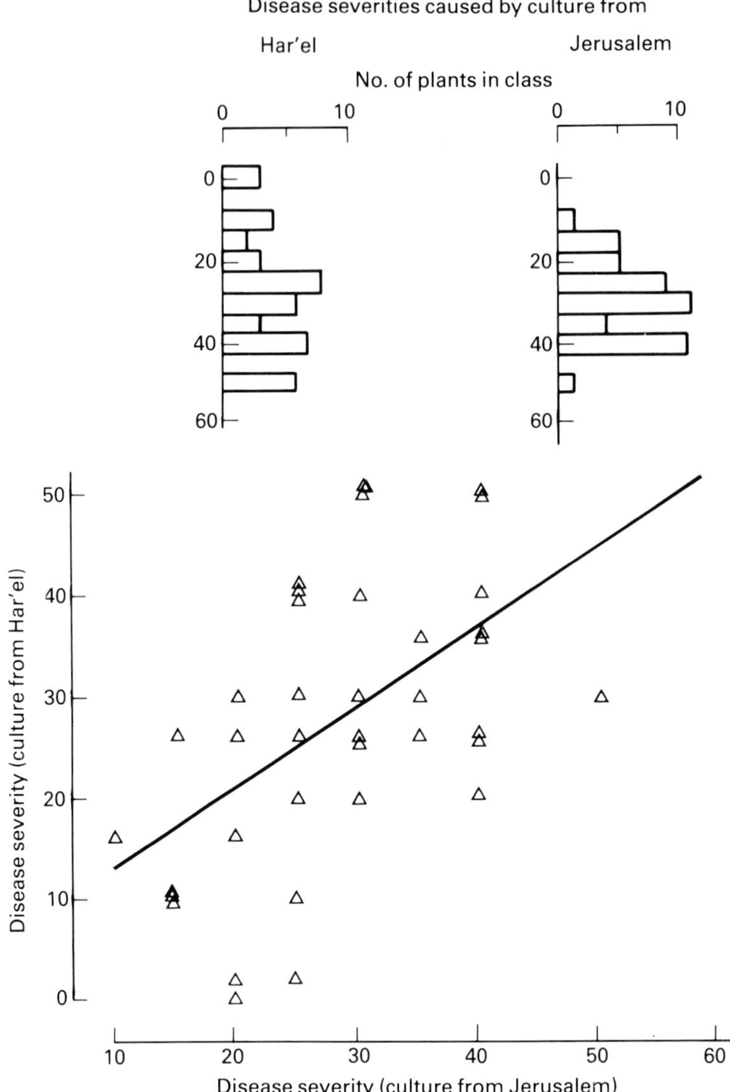

Figure 7.4. The correlation between reactions of barley plants from Har'el to mildew cultures from Har'el (y in the plot, and the left distribution graph), and from Jerusalem (x in the plot, and the right distribution graph). $y = 5.40 + 0.78x$, $P > 0.01$.

relative to the aggressiveness of the culture from Yavne. It also means that in these cases performance of the plants towards one culture may be predicted from the performance to another culture. But there are other cases where there is no correlation and it cannot a priori be reliable to predict from one selection test to another.

In a combined analysis of the results, the illumination factor had a significant effect on disease level. The culture factor also had a significant effect. There was a significant difference between the two cultures from Yavne and from Har'el, both in the shade (Table 7.4) and in the open. The difference between the two cultures from Kesalon was significant only in the shade. In the shade, plants from Yavne were significantly more susceptible than the plants from the other locations (Table 7.4). In the open, plants from Jerusalem were significantly more resistant than the plants from Kesalon or Yavne. On the average, plants from Yavne were significantly more susceptible than those from the other locations, and plants from Jerusalem were more resistant than those from Kesalon. There was no location effect on mildew aggressiveness or plant resistance apart from the fact that plants from Yavne were the most susceptible to the cultures from Yavne (at the same time they were more susceptible to all the cultures), and plants from Jerusalem were the most resistant to the cultures from Jerusalem (but were also the most resistant to the other cultures as well).

The Limonium – rust system

Three locations were chosen for the study of this system, two on calcareous hills near the coast, and one on the second range of calcareous hills about 15 km from the sea. Several sites were selected at each location; 25 plants at each site were marked and the level of disease on each plant recorded monthly. The results were summarized in the form of frequency distribution curves. The disease situation changed throughout the year and there were also marked differences between sites in the same location. At Tel Aviv, near the coast, four sites were studied. In one site,

Table 7.4. Mean disease level of powdery mildew on wild barley in the shade. The results are of 10 cultures on plant samples from 5 locations. Symbols: y_1 and y_2 = two cultures from Yavne, h_1 and h_2 = cultures from Har'el, k_1 and k_2 = cultures from Kesalon, q_1 and q_2 = cultures from Qiryat Anavim, j_1 and j_2 = cultures from Jerusalem; Y, H, K, Q, J = plants from Yavne, Har'el, Kesalon, Qiryat Anavim and Jerusalem, respectively. Frames indicate response of plants to cultures from their own location. Means with identical letters beside them are not significantly different (lower case for culture means and capital letters for lcoation means)

	Cultures										
Locations	y_1	y_2	h_1	h_2	k_1	k_2	q_1	q_2	j_1	j_2	Mean
Y	40.3	46.6	38.2	46.6	40.3	47.6	46.4	37.8	51.2	45.5	41.1A
H	26.3	39.7	28.6	42.9	21.3	40.9	36.6	28.2	39.0	29.1	33.3B
K	20.8	37.1	28.5	41.4	21.7	41.0	32.7	29.3	35.9	31.7	32.0B
Q	30.2	38.2	32.2	36.0	27.6	35.1	31.6	34.8	47.0	31.5	34.4B
J	16.1	37.9	28.8	31.2	36.2	36.2	30.8	18.9	31.4	24.6	29.2B
Mean	26.7b	39.9a	31.3b	39.6a	29.4b	40.2a	35.6ab	29.8b	40.9a	32.5ab	34.6

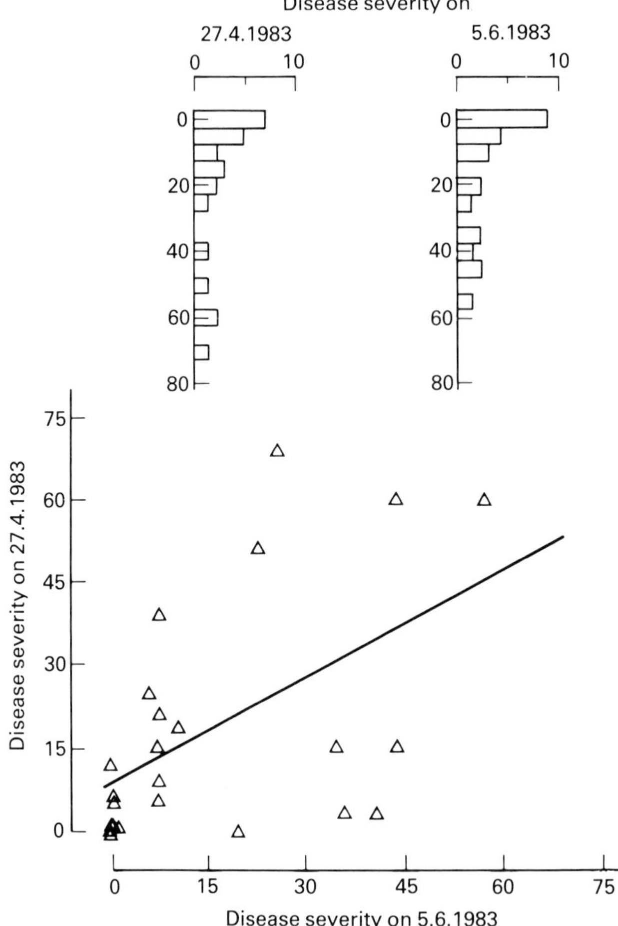

Figure 7.7. The correlation between the disease reactions of *Limonium* plants in Tel Aviv, Site D, on 27 April 1983 (y in the plot, and the left distribution graph), and on 5 June 1983 (x in the plot, and the right distribution graph). $y = 7.73 + 0.65x, P > 0.01$.

powdery mildew of wild barley, causing the barley population as a whole to be more susceptible. We do not know whether it is due to increased host susceptibility, or to improved conditions for mildew development, or both. Microclimatic factors and niche differentiation may have a detrimental effect on disease development. We have encountered it in wild barley with mildew, and in *Limonium* with rust. Both near the coast and away from it we found sites where almost no rust developed on *Limonium* plants throughout a whole year. Preliminary glasshouse studies have shown that some of the disease-free plants were susceptible to the rust cultures from that particular site or close by.

Beyond the niche differentiation and microclimatic effects on host–parasite relationships, there are the macroclimatic effects. It is implied from Wahl *et al.*

(1978) and from Segal *et al.* (1980) that in the arid zones, mildew development is retarded or impossible. They show that sources of resistance are very scarce in dry habitats and claim that this is due to reduced selection pressure by the pathogens. E. Schwarzbach (personal communication) found more virulence in some of the drier regions of Israel and was of the opinion that due to the adverse climatic conditions, mildew has to be more aggressive in order to survive. We found that the wild barley population from east of Jerusalem was more resistant than the other populations, and that cultures from that area possessed more genes for virulence. These findings are in line with the theory of compensation between factors demonstrated by Rotem *et al.* (1971).

New light was shed on natural host–parasite relationships by plant ecologists (Burdon, 1982; Burdon & Shattock, 1980; Harper 1977), who emphasized the important role of plant community on the restriction of disease development. They also pointed out that the inclination of plant pathologists to study single species and single pathogens rather than composed populations, tends to underestimate the natural mechanisms of protection and defence.

Burdon launched the idea that the frequency distribution of plants in a population according to their response to a disease has some important meaning in understanding the co-evolution of host and parasite. Frequency distributions used in ecological studies to describe population composition in relation to fixed traits were adopted to describe resistance to disease (Burdon 1980 a, b). He suggested that a frequency distribution skewed towards susceptibility indicates that this particular disease has a strong selection pressure on the host. The frequency distributions presented here show a great variety of forms. They also show that the same host population may respond in an entirely different way to different cultures of the same pathogen, or to the same culture under different ecological conditions. We have recently expressed our reservations regarding the meaning of frequency distribution graphs (Dinoor & Eshed, 1984) and what they represent. Maybe they are more informative when reflecting the behaviour of plants *in situ*, as with the *Limonium* populations, but even then the distributions change with the season and with the ecological niche.

Natural ecosystems are important for several reasons, for example as sources of novel germplasm and of ideas for balanced plant protection. Unfortunately, they face the danger of extinction under many circumstances through human activities. It is not only expected that we should contribute to the preservation of nature (Dinoor, 1981), but some of us think that this is our obligation (Frankel, 1974). Studies of natural ecosystems and knowledge about their management will greatly facilitate their salvation for future generations.

Acknowledgements

The authors acknowledge the support of two projects: (1) Studies on natural populations of barley and powdery mildew supported initially by the Israeli National

Academy of Sciences (the Fund for Basic Research), and then mainly by the Agricultural and Food Research Council of the UK and by Marks and Spencer; (2) Studies on natural populations of *Limonium* and rust, supported by the Fund for Encouragement and Initiation of Research at the Ministry of Science and Development in Israel. Much of the data presented here are from these projects.

References

Browning J.A. (1974) Relevance of knowledge about natural ecosystems to development of pest management programs for agro-ecosystems. *Proceedings of the American Phytopathological Society* **1**, 191–9.

Burdon J.J. (1980a) Intra-specific diversity in a natural population of *Trifolium repens*. *Journal of Ecology* **68**, 717–35.

Burdon J.J. (1980b) Variation in disease-resistance within a population of *Trifolium repens*. *Journal of Ecology* **68**, 737–44.

Burdon J.J. (1982) The effect of fungal pathogens on plant communities. In: *The Plant Community as a Working Mechanism* (Ed. by E.I. Newman), pp. 99–112. Blackwell Scientific Publications, Oxford.

Burdon J.J. & Shattock R.C. (1980) Disease in plant communities. *Applied Biology* **5**, 145–219.

Dinoor A. (1970) Sources of oat crown rust resistance in hexaploid and tetraploid wild oats in Israel. *Canadian Journal of Botany* **48**, 153–61.

Dinoor A. (1973) Comments on race surveys. *Cereal Rusts Bulletin*, Vol. **1**, part 2, 26.

Dinoor A. (1981) Dynamic preservation of germplasm in nature reserves. *Kidma* **24**, 22–5.

Dinoor A. & Eshed N. (1984) The role and importance of pathogens in the natural plant communities. *Annual Review of Phytopathology* **22**, 443–466.

Dinoor A. & Peleg N. (1972) The identification of genes for resistance or virulence without genetic analyses, by the aid of the 'Gene-for-gene' hypothesis. *Proceedings of the European and Mediterranean Cereal Rusts Conferences. Praha, Czechoslovakia*, Vol. II, 115–19.

Dinus R.J. (1974) Knowledge about natural ecosystems as a guide to disease control in managed forests. *Proceedings of the American Phytopathological Society* **1**, 184–90.

Eshed N. & Dinoor A. (1983) Effect of temperature on the life cycle of *Statice* rust caused by *Uromyces savulescui*. *Abstracts of the Fourth International Congress of Plant Pathology, Melbourne, Australia*, p 181.

Eyal Z., Yurman R., Moseman J.G. & Wahl I. (1973) Use of mobile nurseries in pathogenicity studies of *Erysiphe graminis hordei* on *Hordeum spontaneum*. *Phytopathology* **63**, 1330–4.

Frankel O.H. (1974) Genetic conservation: An evolutionary responsibility. *Genetics* **78**, 53–65.

Harlan J.R. (1976) Diseases as a factor in plant evolution. *Annual Review of Phytopathology* **14**, 31–51.

Harper J.L. (1977) *Population Biology of Plants*. Academic Press, London.

Leppik E.E. (1970) Gene centers of plants as sources of disease resistance. *Annual Review of Phytopathology* **8**, 323–44.

Rotem J., Cohen Y. and Putter J. (1971) Relativity of limiting and optimum inoculum loads, wetting durations and temperatures for infection by *Phytophthora infestans*. *Phytopathology* **61**, 275–8.

Schwarzbach E. (1979) A high throughput jet trap for collecting mildew spores on living leaves. *Phytopathologische Zeitschrift* **94**, 165–71.

Segal A., Manisterski J., Fischbeck G. & Wahl I. (1980) How plant populations defend themselves in natural ecosystems. In: *Plant Disease: an Advanced Treatise* (Ed. by J.G. Horsfall & E.B. Cowling), Vol. 5, 76–102. Academic Press, London.

Wahl I., Eshed N., Segal A. & Sobel Z. (1978) Significance of wild relatives of small grains and other wild grasses in cereal powdery mildews. In: *The Powdery Mildews* (Ed. by D.M. Spencer), pp. 83–100. Academic Press, London.

Wolfe M.S., Minchin P.N. & Slater S.E. (1981) Powdery mildew of barley. *Report of the Plant Breeding Institute Cambridge, for 1980*, pp. 88–92.

Wolfe M.S. & Schwarzbach E. (1978) Patterns of race changes in powdery mildews. *Annual Review of Phytopathology* **16**, 159–80.

Section 2
The dynamics of pathogen populations

8 Modelling the dynamics of pathogen populations

M.J. JEGER*
Department of Plant Pathology and Microbiology,
Texas A & M University, College Station, TX 77843, USA

Introduction

The modelling of plant pathogen dynamics, although generally less developed than other areas of population biology, has recently received increased attention and will be reviewed in this chapter. My title is unusual for a paper concerned primarily with the dynamics of plant disease epidemics, but is entirely justified within the context of this book. Let me first stress that my primary interests are in experimental epidemiology and not in mathematics. This, I believe, is no hindrance: there is more to modelling than the particular mathematical techniques brought to bear, and there are strong affinities of modelling with many experimental approaches. Let me also stress that I cannot hope to do justice to all aspects of modelling in this chapter. Accordingly, I have been selective, possibly biased, in my choice of topics. In particular I shall concentrate on theoretical approaches to modelling plant disease epidemics. The use of models in plant disease management is an important topic in its own right and is not discussed here.

Characteristics of plant disease epidemics

It seems self-evident that the dynamics of plant pathogens and of disease epidemics are closely related. First, consider some characteristics of plant disease epidemics. What is an epidemic? Despite the innocuousness of this question, it is surprisingly difficult to find a universally accepted and unambiguous definition (Gilligan, this volume). Most attempts fall into two categories: intuitive, observational definitions of disease increase in time and space; or more operational, dynamic definitions in terms of two interacting populations, the host and pathogen. If nothing else, the latter version has the advantage of stressing the level of integration at which epidemiologists work. A hierarchical perspective is important for modelling and, indeed, more generally. Are the 'principles' that describe processes at the whole-plant level sufficient to explain all processes at the population level? Are they even necessary? These important questions have been examined within the plant sciences

*Present Address: Tropical Development and Research Institute, 56/62 Gray's Inn Road, London WC1X 8LU, UK.

Wolfe M.S. & Caten C.E. (1987) *Populations of Plant Pathogens: their Dynamics and Genetics.* Blackwell Scientific Publications, Oxford.

(Passioura, 1979; Thornley, 1980; Waggoner, 1983a, b), but can barely be touched upon in this chapter.

I will be concerned mostly with foliar epidemics caused by airborne fungal pathogens, typified by the powdery mildews. Epidemics caused by soil-borne pathogens (Gilligan, this volume) and by plant viruses (Thresh, this volume; Hill, this volume) are considered elsewhere and will not be discussed in detail. Foliar epidemics occur because the pathogen is able to complete successive cycles of infection, growth, sporulation and dispersal. There is now a considerable literature on epidemiological aspects of these components, reviewed by Berger (1977), especially on infection processes and spore production. Spore dispersal has always been studied independently as a physical process (Waggoner, 1983a, b), and the relationship of transient air spora with disease increase in time and space remains unclear (McCartney & Fitt, this volume), and is the weakest link in our chain of knowledge (Waggoner, 1983b).

Approaches to modelling

Though quantification of the components of epidemics is certainly involved, modelling demands a conceptual framework, or level of abstraction, that need have nothing to do with the quantity of data. Many classifications of models have been proposed (e.g. Kranz & Royle, 1978) but, for the purposes of this chapter, I will describe three, and concentrate on two, approaches to population modelling. These are descriptive, systems and theoretical approaches: the former are usually regression models of events within the epidemic cycle, based on weather (Butt & Royle, 1974), and the latter two roughly correspond to trends in population ecology noted recently by McIntosh (1980). Descriptive models have proved very useful in succinctly summarizing data, in exploratory data analysis, and to some extent in decision making. They have also been considered purely empirical, despite the fact that the *modus operandi* of many systems models is an excess of empirical relationships. A single empirical relationship at least has the merit that when it fails there is only one candidate as to why this is so.

Complex systems models

Systems models are not synonymous with systems analyses (see Kranz & Hau, 1980); the latter represent one approach to investigating multiple cause–effect relationships, and can be purely qualitative in nature. Apart from their heuristic or instructional value, I can see only limited use for the type of model that attempts to reconstruct complexity by incorporating as many of the known or 'guessed' aspects of reality as possible. The relational diagram of EPIVEN, a model of epidemics caused by the conidial stage of the apple scab fungus (Kranz *et al.*, 1973), and similar diagrams (Shrum, 1978; Waggoner, 1978), typify the approach. Various justifications for complex models have been proposed: our basic lack of knowledge is exposed, our thoughts are clarified, our experiments are no longer necessary, our priorities can be set. Some of the fallacies of these justifications for complex models in the plant sciences were, despite a recent re-assessment (Loomis *et al.*, 1979), exposed by Passioura (1973). I have yet to be convinced of any example where complex models have expanded our basic understanding of epidemics, assisted in the development of an experimental programme, led to profound changes in the ways in which we view epidemics, or, on a more practical level, found a useful role in crop disease management.

Indeed, complex models may negate the essence and purpose of modelling as I understand it. For example, the modelling of plant physiological processes, growth, and development has proceeded along lines different from the modelling of plant disease epidemics; and by being largely concerned with individual plants in phytotrons, operates at a different level of integration. Some effort is now being given to incorporate disease 'sub-models' into complex host growth models in a range of crops (Loomis & Adams, 1983). This effort is understandable where the object of study is the physiology of yield of diseased plants and hence the derivation of mechanistic yield-loss models (Rouse, 1983), but such models should not become the means of attacking *all* crop-related problems. It has also been argued that disease and other crop 'sub-models' should be documented and exchanged (Bloomberg, 1980), so that a structured model can readily be assembled. Such developments, I would argue, are rarely appropriate when the dynamics of disease in populations is the primary concern. My view is that modelling should start with a limited and well-defined problem, and only the minimal degree of complexity necessary for the purpose at hand should be employed. This view is in direct opposition to the inductive view of models as synthesizers of all available knowledge. The situation is worsened by the terminology that has developed with systems models: they are somehow verified or validated (Teng, 1981), even proven true, thus disregarding the concept of falsifiability that is central and unique to experimental science. Indeed, according to Zadoks (1978), complex models are hardly falsifiable.

Theoretical population models

One view of a system is simply a set of state variables (X) and a set of rules, possibly involving variables representing the physical environment (E), that determine the rates of transition from one state to another. This need be little more than a set of linked differential equations as shown in Table 8.1. I would argue that it is possible to obtain insight into disease dynamics, sometimes by analytical and sometimes by numerical means, by very simple formulations of these equations. These formulations may often exclude the physical environment altogether, unless that is the particular concern of study. The systems approach then merges into what I am calling the theoretical approach.

The theoretical approach also has its dangers, and not just in the accessibility of mathematical symbolism. One unfortunate dichotomy that has crept into modelling is that between explanatory and predictive models (Fleming & Bruhn, 1983). Science

Table 8.1. System of linked differential equations describing the dynamics of n variables ($X_1 \ldots X_n$) as functions of the variables and m external variables ($E_1 \ldots E_m$) influencing dynamics. The X variables might be different categories of disease; the E variables, elements of the physical environment

$$dX_1/dt = f_1(X_1, X_2 \ldots X_n; E_1, E_2 \ldots E_m)$$
$$dX_2/dt = f_2(X_1, X_2 \ldots X_n; E_1, E_2 \ldots E_m)$$
$$\vdots$$
$$dX_n/dt = f_n(X_1, X_2 \ldots X_n; E_1, E_2 \ldots E_m)$$

is ultimately concerned with explanation rather than description, and hence theories in science should provide an explanation of some natural phenomenon. What is less clearly stated is that explanation necessarily leads to prediction. A model that purports to explain, by definition makes predictions that can and should be tested in the real world. Some theoreticians have argued for a wider interpretation of theory in which theory and experiment freely cross-fertilize (Levin, 1981a), and that the models that arise in this process are not necessarily suited for prediction, or for testing. This view is valid up to a point, but a theoretical model should not then be termed or viewed as explanatory. Development of this kind of theoretical model is then an exploratory activity; in much the same way that few experiments are key tests of hypotheses or theories, but rather explore a range of possibilities. At some stage, however, the model as explanation and the experiment as test should coincide, lest both activities result in sterile self-justification. Indeed, the interaction of modelling with experimentation, the provision of a rigorous analytical tool and of a theoretical framework for experimental studies, still remains, in my view, the basic justification for modelling.

Simple population models

Having clarified my position on modelling, the remainder of this chapter concerns two topics. These are the development of population dynamic models in plant pathology, especially during the last five years or so (and which has been accompanied, incidentally, by a decline in the rate of emergence of systems models) and areas in which further theoretical developments are an urgent requirement. Two further restrictions will be made: models of genetic change in populations (Barrett, this volume; Leonard, this volume) and spore transport (McCartney & Fitt, this volume) will not be considered in any detail.

Population dynamics of what?

The first question to be posed, and often begged, is: 'what are the individuals that comprise the population?' (see also Gale, this volume). With plant diseases we are concerned with the interaction of two populations, of a host and a pathogen, although this essential feature has rarely been accounted for in conventional disease assessment practices. Disease has traditionally been measured on a proportion or percentage scale. Measurement of absolute values, such as numbers or areas of disease lesions, is now technically feasible, if expensive, due to recent technical advances (Nilsson, 1980; Lindow & Webb, 1983). This is a healthy development for epidemiology and for modelling in particular. Lesion numbers, or their equivalent in areas, can be the unit of the population and, because they reflect both the host and pathogen populations, are more useful than either conventional disease assessments or counts of trapped spores.

The fundamental equation of population dynamics states that the increase in population numbers (N) in a unit time interval is given by births + immigrants−deaths−emigrants. In fact, where disease lesions are the unit of the population, the only concern is with births and 'emigration' (e.g. by premature fall of diseased leaves); deaths, or sterile lesions, are normally considered part of the disease population, and 'immigration' of lesions does not normally occur. Two versions of the basic equation can be defined: discrete-time models, usually in the form of difference equations, in which generations, as with temperate-zone insects for example, do not overlap,

$$N_{t+1} = f(N_t), \qquad [1a]$$

and continuous-time models, defined by differential equations, with overlapping generations,

$$dN/dt = f(N). \qquad [1b]$$

The logistic equation

Most of the epidemics I am concerned with have been modelled by differential equations, although difference equations have been used (Leonard, 1969; Shiyomi, 1970; Barrett, 1978; Kable & Jeffery, 1980; Jeger *et al.*, 1981a). Usually the function f is a density-dependent, or non-linear function of the population. The simplest form of the function arises in the logistic equation

$$dN/dt = rN\,(1 - N/K) \qquad [2]$$

which states that population increase is proportional to the level of the population but constrained linearly by the term $1 - N/K$ as N increases from initially low values. The parameter K has been termed the carrying capacity or maximum population the environment can support; mathematically, it is the value approached asymptotically for arbitrary large time values. The parameter r is the fractional rate of population increase at small values of N.

The logistic equation, when rescaled such that $x = N/K$, gives the familiar equation proposed by Vanderplank (1963), but used earlier to describe plant disease epidemics (Large, 1945),

$$dx/dt = rx\,(1 - x). \qquad [3]$$

Equation [3] states that disease increase is related to the proportion of the host population diseased (x), and to the proportion healthy ($1 - x$). If the rate parameter r is a constant, then the equation can be solved, linearized by logits, and r is given by the slope of the line:

$$\ln\,[x/(1 - x)] = a + rt \qquad [4]$$

where a is a constant introduced by solving the equation. Vanderplank (1963, 1982) decried the use of the logistic equation as a model of disease progress, mainly because of the requirement for r to be constant. In fact much of his analysis depends upon r being a constant for at least some part of the epidemic.

Objections to the logistic equation

There are many implicit assumptions in the logistic equation (Barrett, 1983). As a consequence, two kinds of objection have been raised. First, there are empirical objections in that solutions, symmetric sigmoid curves, do not always fit epidemic data. Secondly, there are more fundamental objections in that an instantaneous latent period is assumed, r is difficult to interpret biologically, and disease eventually goes to completion, i.e. approaches asymptotically a value of 1. The empirical objection is exemplified by the different kinds of disease progress curves that have been observed (Figure 8.1). The logistic equation, on a proportion scale can only correspond to curve A. The other curves represent epidemics in which an asymptote less than 1 is approached (B), disease increase is not smoothly increasing (C), and decrease as well as increase occurs (D). A full discussion of these and other types of epidemic curves is given by Kranz (1978). In fact this restriction is one of the *strengths* of the logistic equation and of the assumption of a constant r, rather than a weakness. The equation serves as an unambiguous standard of comparison for data, and for other models. One criterion that can then be made for the power and flexibility of a model is the ability to mimic these general patterns of epidemic increase.

Vanderplank's differential–difference equation

The fundamental objections to the equation were partly answered by Vanderplank (1963) in his differential–difference equation;

$$dy_t/dt = R\,[y_{t-p} - y_{t-i-p}]\,(1 - y_t) \qquad [5]$$

in which a latent period (p), an infectious period (i), and a rate parameter (R) with some biological meaning, are included. Further it can be shown that disease, under

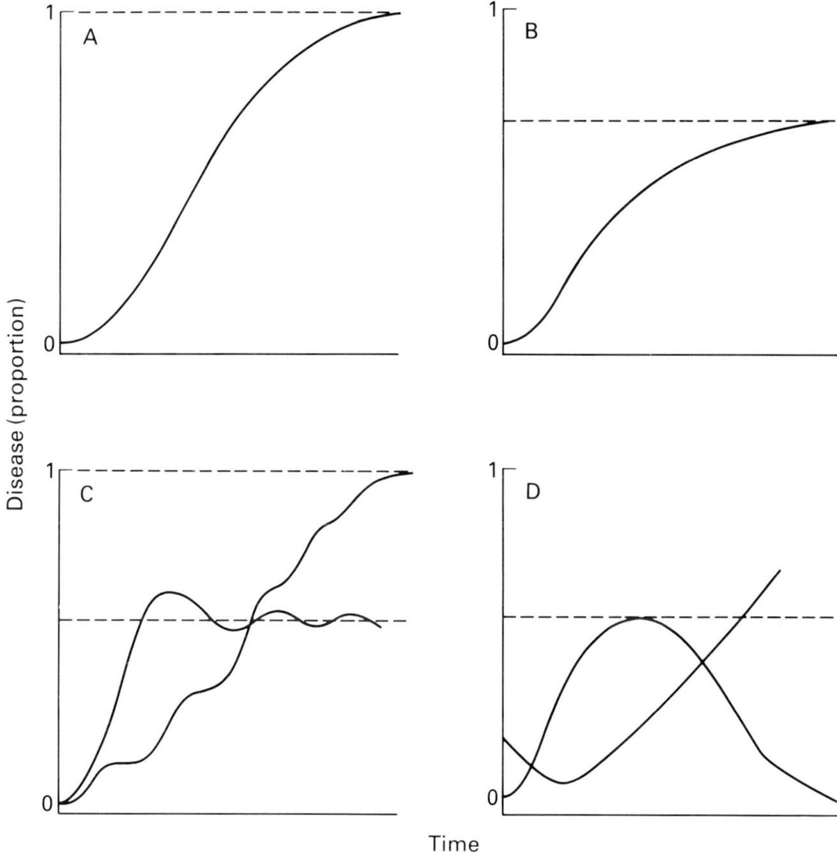

Figure 8.1. Idealized representation of observed patterns of epidemic progress, see discussion in text for Key A–D.

this model, approaches an asymptote strictly less than 1. Equation [5] is intractable analytically, although a qualitative treatment of a similar equation, in a different context, was given by Wangersky & Cunningham (1956). This more realistic model, however, places us in a dilemma. If the differential–difference equation is an accurate description of epidemic progress, what criteria must be met if, as does occur, increase is to appear logistic? It turns out that r can be approximately constant for a finite period of time only (during which time R must decrease); and further, where the length of the latent period is less than the doubling time of the epidemic (i.e. $p < \ln 2/r$), there are severe constraints on the length of the infectious period (Jeger, 1984b).

Other modifications to the logistic equation

The logistic equation can be modified in several ways to meet the fundamental objections raised. The constant rate parameter r can be replaced by a time-dependent function; for example the simple linear decrease as proposed by Kiyosawa (1972) to represent adult plant resistance, or as more complex sinusoidal forms to represent periodicities and other cyclical trends caused by a varying environment (Waggoner, 1978; Kranz, 1978). Provided the function can be integrated, then the basic structure of the equation remains the same, i.e. disease increases in logits on the integrated time scale (cf. equation [4]),

$$\ln [y/(1-y)] = a + \int f(t)\, dt. \quad [6]$$

Similar comments apply when the time scale is in physiological or heat sum units. Another modification is to account for the non-random nature of the distributions of disease lesions on foliage by including, for example, a term derived from the negative binomial function (Waggoner, 1983b).

Growth curves

The most expedient means of improving fits to data is to write the asymptote explicitly and include an additional power term m to obtain a range of asymmetric curves. This gives a generalized logistic, or Richards equation (Richards, 1959), developed originally in plant growth studies;

$$dN/dt = rN\,[1 - (N/K)^m]. \quad [7]$$

The logistic function is then a special case of the Richards function with $m = 1$. The differences in the curves can best be shown by plotting the fractional rate of disease increase $(1/N)\, dN/dt$, against disease, (N), for a range of values for the parameter m. Following this procedure with published data, a full range of responses was found (Jeger, 1980) including some not corresponding to the form of the Richards equation. Despite the power and flexibility of the additional parameter, the Richards equation does not account for all epidemic data.

An example of how to fit such functions to reasonably well-behaved epidemic data was given by Jeger (1982a). Interpretation of the fitted curve was difficult; the most valuable use of the procedure was in reliable estimation of the disease asymptote, strictly less than 100%. Other workers have developed similar generalizations of the logistic equation. For example, Turner *et al.* (1969) introduced a time-dependent asymptote. Gilpin *et al.* (1976) attempted to interpret the curvative parameter in terms of territoriality, but a convincing analogy for plant pathogens can hardly be made.

New theoretical approaches

We have looked at some of the ways in which the logistic equation has been modified to meet the objections raised against its use. In each case a single equation resulted. Although it is possible to derive single equations of disease progress from other assumptions (Oort, 1968; Fleming & Holling, 1982), another way of incorporating realistic features is by considering more than one equation. This will be illustrated by considering some features missing from the logistic equation: age structure in the disease population; the influence of host growth on disease dynamics; and conversely, the influence of disease on host growth.

Categories of disease

There is no provision for age structure in the logistic equation. Vanderplank's differential–difference equation does take into account the different categories of disease, i.e. disease that is successively latent, infectious and post-infectious, but it is intractable analytically, rather rigid assumptions are made, and it is difficult to draw conclusions as to the age structure of the disease population. A more flexible approach is to specify linked differential equations for each category of disease (Jeger, 1982b). The first point to be noted is that total disease (x) is simply the sum of latent (l), infectious (s) and post-infectious (r) disease. Hence, by the rule for differentiating a sum (Table 8.2), we simply have to specify equations for each category of disease to completely describe disease dynamics.

In the early stages of an epidemic it is possible to obtain explicit solutions for each category of disease, and for the proportion of disease in each category, i.e. the age structure (Jeger, 1982b). When equations are specified over the whole range of an epidemic, then explicit solutions cannot in general be found. However, many of the qualitative aspects of dynamics, certainly asymptotic behaviour, can be deduced. In particular, the proportion of total disease that is infectious and the proportion that is post-infectious can be characterized as disease increases. The important

Table 8.2. Fundamental rules of calculus used in developing models of disease progress. See text for discussion of examples in parentheses

(a) Rule for differentiating a sum (categories of disease):
if $x = l + s + r$
then $dx/dt = dl/dt + ds/dt + dr/dt$
where x = total disease, l = latent disease,
s = infectious disease, and r = post-infectious disease.

(b) Rule for differentiating a quotient (host growth):
if $x = X/P$
then $dx/dt = (P\,dX/dt - X\,dP/dt)/P^2$
where X and P are measures of the disease and host population in comparable units (e.g. absolute numbers or areas; x is disease on a proportion scale.

(c) Rule for differentiating a product (colony expansion):
if $x = NQ$
then $dx/dt = Q\,dN/dt + N\,dQ/dt$
where N = number of colonies, and Q = mean colony area; x is disease as an absolute area.

feature is that for some value of disease strictly less than 1, all disease is post-infectious. But at this level of disease, by definition, there can be no further increases in the amount of disease since the epidemic has indeed run out of spores. Even in a fully susceptible host, with optimal environmental conditions, and infinite time available, disease does not go to completion. The disease asymptote (y^*) is then given by the transcendental equation

$$y^* \sim 1 - \exp(-k^* y^*), \qquad [8]$$

where k^* is a dimensionless parameter that is independent of the length of the latent period. This provides independent corroboration of a result first noted by Vanderplank (1963), but who did not provide a formal derivation. That the same conclusion should arise from two quite different models is a convincing demonstration of the robustness of the result.

Host growth

Can effects on disease due to host growth, and reciprocal effects of disease on host growth, be incorporated into simple theoretical epidemic models (Rouse, 1983)? The linked equations describing the categories of disease (Jeger, 1982b) can indeed be extended by including a category of healthy plant tissue, and a compact self-contained description of disease dynamics is then obtained. Another approach is to use linked differential equations for host and pathogen dynamics in the form of Lotka–Volterra equations (Jowett *et al.*, 1974). However, for simplicity, let us look at host growth afresh. First, does host growth affect disease progress? Unquestionably yes; for example, Jeger & Butt (1983) plotted progress of powdery mildew on different apple cultivars and noted a marked divergence in disease with time; yet the cultivars were virtually indistinguishable with respect to components of resistance measured in the greenhouse or outside. The divergence in time was attributed largely to patterns of shoot growth and hence foliage production on the different cultivars. Variations in host growth are a primary reason for disease asymptotes less than unity (Kranz, 1978), and such asymptotes are the rule rather than the exception (Kranz, 1977). Vanderplank (1963) suggested a means of correcting for host growth in calculating 'infection' rates, and this was expanded on by Kushalappa & Ludwig (1982), but such attempts merely scratch at the surface; new analytical tools are necessary.

The effects of host growth on disease, measured as a proportion, can be handled by reference to another elementary rule from calculus – the procedure for differentiating a quotient (Table 8.2). If equations describing host growth in absolute units (P) and disease increase in comparable units (X) can be specified, then an equation for disease on a proportionate scale necessarily follows. Various possibilities for disease and host dynamics, and their consequences, can then be explored.

For example, a basic equation for disease dynamics is

$$dX/dt = kX\,(1 - X/P). \qquad [9]$$

where k is a rate parameter and P, host growth, a time-variable. Suppose now that host growth is exponential and unaffected by disease;

$$dP/dt = \alpha P \qquad [10]$$

where α is the relative growth rate of the host. By using the quotient rule, an equation is derived that leads either to disease decreasing, on a proportionate scale, or increasing to a stable equilibrium value < 1.0 given by

$$x^* = 1 - \alpha/k \qquad [11]$$

where $\alpha < k$. In a more realistic example, suppose that host growth is constrained logistically by a maximum size P_{max}, and that new growth is sustained only by non-diseased tissue, i.e. the difference between healthy and diseased units,

$$dP/dt = \alpha\,(P - X)(1 - P/P_{max}). \qquad [12]$$

In this case, an equation is derived which predicts that disease on a proportionate scale will increase to unity, or decrease to a lower limit when the host population has reached a certain size, given by

$$P = P_{max}\,(1 - \alpha/k) \qquad [13]$$

where $\alpha < k$. This lower limit is not an equilibrium value however, but a turning point, and disease subsequently increases. Numerical solutions of a range of similar examples, due to Okuno, were reported by Kato (1974). The whole question of reciprocal effects of host and disease dynamics, especially on the equilibria of disease, should be explored further (for a more detailed discussion see Jeger, 1986) and procedures developed for estimating and testing these effects.

In a similar vein, other theoretical questions can be approached using elementary rules from calculus. Rouse (1985) used the product rule (Table 8.2) to analyse the dynamics of disease where colony growth was explicitly incorporated, and found a more rapid increase in the early stages of the epidemic than the logistic model would suggest. In more sophisticated models, competition, and other interactions between (Hau & Kranz, 1978; Fleming, 1980; Putter, 1982; Brittain, 1983) and within (Skylakakis, 1982; Barrett, 1983) pathogen species can be modelled by use of the Lotka–Volterra (Wangersky, 1978) or other equations.

Applications of simple models

I will conclude with discussion of two areas in which there is an urgent need for theoretical advances. These areas are chosen mainly because of my own familiarity and involvement with the work. Other equally valid areas could be chosen.

Heterogeneity

There has been considerable modelling of the effects of crop heterogeneity on disease dynamics (e.g. Barrett, 1978; Østergaard, 1983; McCartney & Fitt, this volume). This work, quite naturally in view of the context in which it arose, has emphasized highly specialized pathogens such as the rusts and powdery mildews in multilines or mixtures of cereal cultivars (Wolfe, this volume). I think it now emerges that this aspect is but one part of the much wider question of crop heterogeneity, and that race-specialized pathogens interacting in a gene-for-gene relationship with host populations are but one special case. For example, in the case of the relatively unspecialized *Septoria nodorum*, there was no *a priori* expectation, on the basis of models proposed for specialized pathogens, of any benefit from mixtures of wheat cultivars differing in their level of resistance. Simple models proved an invaluable tool in defining, investigating and developing an experimental programme (Jeger *et al.*, 1981a, b). It was shown that, whatever the outcome in the early stages of an epidemic, the long-term outcome was that disease in mixtures was equal to the mean of the pure stands (Jeger *et al.*, 1982). There is usually some benefit with regard to unspecialized pathogens in mixtures during the early stages of an epidemic but, given time (which in practice, of course, does not occur), disease increases to the same level as the mean of the pure stands, e.g. see Sitch & Whittington (1983). This qualitative response was also found with barley powdery mildew (Chin, 1979; J.A. Barrett, personal communication), where the cultivars in the mixture did not discriminate among pathogen races. Lancashire (1983) has given a preliminary report on detailed studies of the long-term behaviour of *S. nodorum* in mixtures.

The spatial component of disease increase in the mixtures was also examined (Jeger *et al.*, 1983) despite a lack of analytical tools to quantify the effect. There had been some modelling of the spatial dynamics of disease in mixtures and other host populations, from both theoretical (Zadoks & Kampmeijer, 1977; Fleming *et al.*, 1982; McCartney & Fitt, this volume) and practical (Allen, 1978; Chin, 1979; Chin & Wolfe, 1984) standpoints, but none was appropriate in this instance. Accordingly, models of disease increase in time and space were developed pragmatically by considering the rate of isopath movement (equal amounts of disease) from a focus (Jeger, 1983);

$$\partial x/\partial t = -[f_t/f_x]\ (x,\ t), \qquad [14]$$

where f_t and f_x are the partial derivatives of a function which gives values of disease in both time (t) and space (x). The parameters of spatial spread and temporal progress were estimated simultaneously and good fits to data were obtained. A more extensive treatment of spatial spread has since been proposed (Minogue & Fry, 1983a, b). In my opinion, the spatial dynamics of disease as affected by heterogeneity, interpreted in the widest sense (i.e. crop, cropping system, physical and biotic environment, and management practice), is the outstanding problem in the theory of plant disease epidemics at the present time and will require hitherto unused mathematical tools (Levin, 1981b). Insights, which can at least be formalized using mathematics, are urgently needed to give direction and impetus to this area of research (McCartney & Fitt, this volume).

Epidemiology of fungicide action

The other area in which advances are needed is in the epidemiology of fungicide action. In some ways the situation is analogous to spore dispersal. There are good physical models of fungicide persistence (Bruhn & Fry, 1982), but very primitive insights into how fungicides modify epidemic progress. For example, the relationships between disease increase and the numbers of spores cumulatively trapped within canopies can be modelled by noting that disease increase (y) is proportional to the increase in spores (z) in a given interval but constrained by the amount of healthy tissue (Jeger, 1984a). That is

$$dy/dt = k_s \, dz/dt \, (1 - y/y_{\max}). \quad [15]$$

The dimensionless parameter k_s is a measure of inoculum efficiency (disease units/spore trapped). The equation can be solved to give disease as a function of the number of spores cumulatively trapped – a curve which asymptotically approaches $y_{\max}$ (Figure 8.2). The model was fitted to data on apple powdery mildew epidemics in sprayed and unsprayed orchard plots (Jeger, 1984a), and estimates of the parameters obtained. By far the biggest influence of spraying was on the disease asymptote, presumably because the protectant activity of the fungicide 'removed' otherwise susceptible foliage from the epidemic. There was little effect noted on the k_s parameter, suggesting that the fungicides were not affecting the viability of spores that did land on unprotected tissue. Virtually all work on fungicide effects on epidemics has been concerned with initial inoculum or with rates (Fry, 1977; Skylakakis, 1983); equal attention should now be given to asymptotes. This will be particularly important, I believe, in consideration of the development of fungicide insensitivity. It has been ignored thus far.

Many other examples of the use of models with respect to fungicides can be given. How can one quantify the effects of fungicides with different epidemiological modes of action (Jeger & Butt, 1983)? How is it best to model the integration of fungicide with cultivar partial-resistance (Wolfe, 1981), and perhaps explain examples of non-additive effects (Jeger & Butt, 1983)? How can one explain higher rates of disease increase in sprayed than in unsprayed plots for equivalent levels of disease (Berger, 1975; Griffiths, 1978)? Most workers with fungicides are aware of such aberrant phenomena, but they are rarely reported, being considered defects of experimental methodology or application rather than real effects.

The epidemiological consequences of fungicide usage need further attention, both with respect to insensitivity (Skylakakis, 1982; Georgopoulos, this volume; Skylakakis, this volume), and as an area of modelling of direct relevance in the practice of disease management.

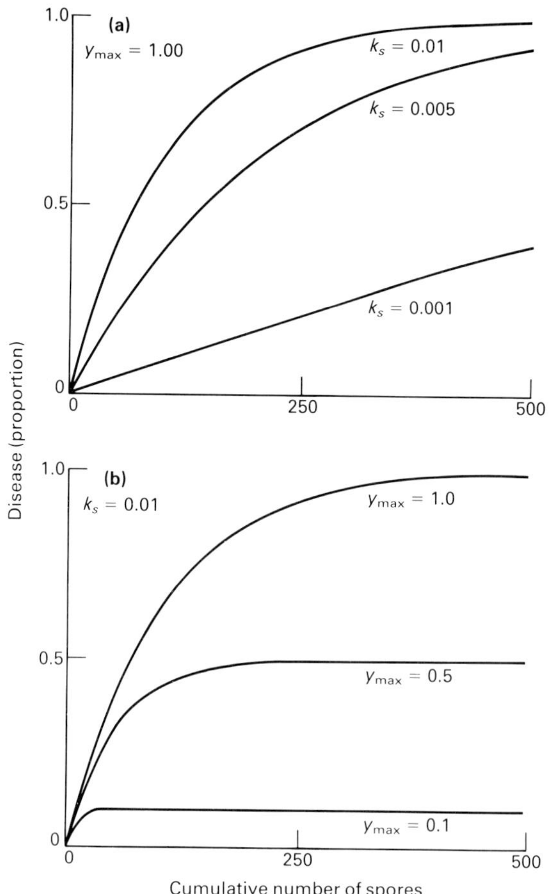

Figure 8.2. Schematic representation of solutions to equation [15] in text: progress of foliar disease in relation to number of spores cumulatively trapped within a canopy for (a) three values of k_s (constant y_{max}), and (b) three values of y_{max} (constant k_s).

Acknowledgement

R.D. Berger pointed out the work of Okuno reported by Kato (1974).

Note added in proof: The asymptotic and other qualitative behaviour of equations (4) were discussed in detail in Jegar M.J. (1986) *Plant Pathology* **35**, 355–61.

References

Allen R.N. (1978) Spread of bunchy top disease in established banana plantations. *Australian Journal of Agricultural Research* **29**, 1223–33.

Barrett J.A. (1978) A model of epidemic development in variety mixtures. In: *Plant Disease Epidemiology* (Ed. by P.R. Scott & A. Bainbridge), pp. 129–37, Blackwell Scientific Publications, Oxford.

Barrett J.A. (1983) Estimating relative fitness in plant parasites: some general problems. *Phytopathology* **73**, 510–12.

Berger R.D. (1975) Rapid disease progress in early epidemic stages. (Abstr) *Proceedings of the American Phytopathological Society* **2**, 35.

Berger R.D. (1977) Application of epidemiological principles to achieve plant disease control. *Annual Review of Phytopathology* **15**, 165–83.

Bloomberg W.J. (1980) A case for structured, modular simulation models in plant pathology research. *Protection Ecology* **2**, 209–13.

Brittain E.G. (1983) A model system of four species with a unique equilibrium point. *Ecological Modeling* **19**, 199–211.

Bruhn J.A. & Fry W.E. (1982) A mathematical model of the spatial and temporal dynamics of *chlorothalonil* residues on potato foliage. *Phytopathology* **72**, 1306–12.

Butt D.J. & Royle D.J. (1974) Multiple regression analysis in the epidemiology of plant diseases. In: *Epidemics of Plant Diseases: Mathematical Analysis and Modelling* (Ed. by J. Kranz), pp. 78–114. Springer-Verlag, Berlin.

Chin K.M. (1979) Aspects of the epidemiology and genetics of the foliar pathogen, *Erysiphe graminis* f. sp. *hordei* in relation to infection of homogeneous and heterogeneous populations of the barley host (*Hordeum vulgare*). *Ph.D. thesis*, University of Cambridge.

Chin, K.M. & Wolfe, M.S. (1984) The spread of *Erysiphe graminis* f. sp. *hordei* in mixtures of barley varieties. *Plant Pathology* **33**, 89–100.

Fleming R.A. (1980) The potential for control of cereal rust by natural enemies. *Theoretical Population Biology* **18**, 374–95.

Fleming R.A. & Bruhn J. (1983) The role of mathematical models in plant health management. In: *Challenging Problems in Plant Health* (Ed. by T. Kommendahl & P.H. Williams), pp. 368–78. American Phytopathological Society, St Paul, Minnesota.

Fleming R.A. & Holling C.S. (1982) A comparison of disease progress equations for cereal rust. *Canadian Journal of Botany* **60**, 2154–63.

Fleming R.A., Marsh L.M. & Tuckwell H.C. (1982) Effect of field geometry on the spread of crop disease. *Protection Ecology* **4**, 81–108.

Fry W.E. (1977) Management with chemicals. In: *Plant Disease: an Advanced Treatise, Vol. I. How Disease is Managed* (Ed by J.G. Horsfall & E.B. Cowling), pp. 213–38. Academic Press, New York.

Gilpin M.E., Case T.J. & Ayala F.J. (1976) Theta selection. *Mathematical Biosciences* **32**, 131–40.

Griffiths E. (1978) Plant disease epidemiology – retrospect and prospect. In: *Plant Disease Epidemiology* (Ed. by P.R. Scott & A. Bainbridge), pp. 3–9. Blackwell Scientific Publications, Oxford.

Hau B. & Kranz J. (1978) Modellrechnungen zür Wirkung des Hyperparasiten *Eudarluca caricis* auf Rostepidemien. *Zeitschrift für Pflanzenkrankheiten und Pflanzenschutz* **85**, 131–41.

Jeger M.J. (1980) Choice of disease progress model by means of relative rates. *Protection Ecology* **2**, 183–8.

Jeger M.J. (1982a) Using growth curve relative rates to model disease progress of apple powdery mildew. *Protection Ecology* **4**, 49–58.

Jeger M.J. (1982b) The relation between total, infectious, and post-infectious diseased plant tissue. *Phytopathology* **72**, 1185–9.

Jeger M.J. (1983) Analyzing epidemics in time and space. *Plant Pathology* **32**, 5–11.

Jeger M.J. (1984a) Relating disease progress to cumulative numbers of trapped spores: apple powdery mildew and scab epidemics in sprayed and unsprayed orchard plots. *Plant Pathology* **33**, 517–23.

Jeger M.J. (1984b) Relation between rate parameters and latent and infectious periods during a plant disease epidemic. *Phytopathology* **74**, 1148–52.

Jeger M.J. (1986) The potential of analytical compared with simulation approaches to modeling in plant disease epidemiology. In: *Plant Disease Epidemiology: Population Dynamics and Management* (Ed. by K.J. Leonard & W.E. Fry), pp. 255–81. MacMillan, New York.

Jeger M.J. & Butt D.J. (1983) Using partial resistance in the integrated control of apple powdery mildew. *IOBC Bulletin* VI (4), 111–22.

Jeger M.J., Griffiths E. & Jones D.G. (1981a) Disease progress of non-specialized fungal pathogens in intraspecific mixed stands of cereal cultivars. I. Models. *Annals of Applied Biology* **98**, 187–98.

Jeger M.J., Griffiths E. & Jones D.G. (1982) Asymptotes of disease caused by non-specialized fungal pathogens in intraspecific mixed stands of cereal cultivars. *Annals of Applied Biology* **101**, 459–64.

Jeger M.J., Jones D.G. & Griffiths E. (1981b) Disease progress of non-specialized fungal pathogens in intraspecific mixed stands of cereal cultivars. II. Field experiments. *Annals of Applied Biology* **98**, 198–210.

Jeger M.J., Jones D.G. & Griffiths E. (1983) Disease spread of non-specialized fungal pathogens from inoculated point sources in intraspecific mixed stands of cereal cultivars. *Annals of Applied Biology* **102**, 237–44.

Jowett D., Browning J.A. & Cournoyer Haning B. (1974) Non-linear disease progress curves. In: *Epidemics of Plant Diseases: Mathematical Analysis and Modelling* (Ed. by J. Kranz), pp. 115–36. Springer-Verlag, Berlin.

Kable P.F. & Jeffery H. (1980) Selection for tolerance in organisms exposed to sprays of biocide mixtures: A theoretical model. *Phytopathology* **70**, 8–12.

Kato H. (1974) Epidemiology of rice blast disease. *Review of Plant Protection Research* **7**, 1–20.

Kiyosawa S. (1972) Effect of addition of field resistance to variety with true resistance. *Japanese Journal of Breeding* **22**, 140–6.

Kranz J. (1977) A study in maximum severity in plant diseases. *Travaux dedies a G. Viennot-Bourgin*, 1977, pp. 169–73.

Kranz J. (1978) Comparative anatomy of epidemics. In: *Plant Disease: an Advanced Treatise, Vol. II. How Disease Develops in Populations* (Ed. by J.G. Horsfall & E.B. Cowling), pp. 33–62. Academic Press, New York.

Kranz J. & Hau B. (1980) Systems analysis in epidemiology. *Annual Review of Phytopathology* **18**, 67–83.

Kranz J., Mogk M. & Stumpf A. (1973) EPIVEN – ein Simulator für Apfelschorf. *Zeitschrift für Pflanzenkrankheiten und Pflanzenschutz* **80**, 181–7.

Kranz J. & Royle D.J. (1978) Perspectives in mathematical modelling of plant disease epidemics. In: *Plant Disease Epidemiology* (Ed. by P.R. Scott & A. Bainbridge), pp. 111–20. Blackwell Scientific Publications, Oxford.

Kushalappa A.C. & Ludwig A. (1982) Calculation of apparent infection rate in plant disease: development of a method to correct for host growth. *Phytopathology* **72**, 1373–7.

Lancashire P. (1983) When are mixtures beneficial? *Abstracts of Papers and Posters: Populations of Plant Pathogens: Their Dynamics and Genetics, BSPP Meeting*, University of Leeds, 19–22 December 1983. p. 10.

Large E.C. (1945) Field trials of copper fungicides for the control of potato blight. I. Foliage protection and yield. *Annals of Applied Biology* **32**, 319–29.

Leonard K.J. (1969) Factors affecting rates of stem rust increase in mixed plantings of susceptible and resistant oat varieties. *Phytopathology* **59**, 1845–50.

Levin S.A. (1981a) The role of theoretical ecology in the description and understanding of populations in heterogeneous environments. *American Zoologist* **21**, 865–75.

Levin S.A. (1981b) Models of population dispersal. In: *Differential Equations and Applications in Ecology, Epidemics and Population Problems* (Ed. by S.N. Busenberg & K.L. Cooke), pp. 1–18. Academic Press, New York.

Lindow S.E. & Webb R.R. (1983) Quantification of foliar plant disease symptoms by microcomputer-digitized video image analysis. *Phytopathology* **73**, 520–4.

Loomis R.S. & Adams S.S. (1983) Integrative analysis of host–pathogen relations. *Annual Review of Phytopathology* **21**, 341–62.

Loomis R.S., Rabbinge R. & Ng E. (1979) Explanatory models in crop physiology. *Annual Review of Plant Physiology* **30**, 339–67.

McIntosh R.P. (1980) The background and some current problems of theoretical ecology. *Synthese* **43**, 195–225.

Minogue K.P. & Fry W.E. (1983a) Models for the spread of disease: model description. *Phytopathology* **73**, 1168–73.

Minogue, K.P. & Fry W.E. (1983b) Models for the spread of plant disease: some experimental results. *Phytopathology* **73**, 1173–76.

Nilsson, H.E. (1980) Remote sensing and image processing for disease assessment. *Protection Ecology* **2**, 271–4.

Oort A.J.P. (1968) A model of the early stage of epidemics. *Netherlands Journal of Plant Pathology* **74**, 177–80.

Østergaard H. (1983) Predicting development of epidemics on cultivar mixtures. *Phytopathology* **73**, 166–72.

Passioura J.B. (1973) Sense and nonsense in crop simulation. *Journal of the Australian Institute of Agricultural Science* **39**, 181–3.

Passioura J.B. (1979) Accountability, philosophy and plant physiology. *Search* **10**, 347–50.

Putter C.A.J. (1982) An epidemiological analysis of the *Phytophthora* and *Alternaria* blight pathosystem in the Natal midlands. *Ph.D. thesis*, University of Natal, 192 pp.

Richards F.J. (1959) A flexible growth function for empirical use. *Journal of Experimental Botany* **10**, 290–300.

Rouse D.I. (1983) Plant growth models and plant disease epidemiology. In: *Challenging Problems in Plant Health* (Ed. by T. Kommendahl & P.H. Williams), pp. 387–98. American Phytopathological Society, St Paul, Minnesota.

Rouse, D.I. (1985) Construction of temporal models: I. Disease progress of air-borne pathogens. In: *Advances in Plant Pathology Vol. 3: Mathematical Modelling of Crop Disease* (Ed. by C.A. Gilligan), pp. 11–29. Academic Press, London.

Shiyomi M. (1970) Mathematical models representing the increase in the number of disease lesions. *Bulletin of the National Institute of Agricultural Science, Japan A* **17**, 103–16.

Shrum R.D. (1978) Forecasting of epidemics. *In: Plant Disease: an Advanced Treatise, Vol. II. How Disease Develops in Populations* (Ed. by J.G. Horsfall & E.B. Cowling), pp. 223–38. Academic Press, New York.

Sitch L. & Whittington W.J. (1983) The effect of variety mixtures on the development of swede powdery mildew. *Plant Pathology* **32**, 41–6.

Skylakakis G. (1982) The development and use of models describing outbreaks of resistance to fungicides. *Crop Protection* **1**, 249–62.

Skylakakis G. (1983) Theory and strategy of chemical control. *Annual Review of Phytopathology* **21**, 117–35.

Teng P.S. (1981) Validation of computer models of plant disease epidemics: a review of philosophy and methodology. *Zeitschrift für Pflanzenkrankheiten und Pflanzenschutz* **88**, 49–63.

Thornley J.H.M. (1980) Research strategy in the plant sciences. *Plant Cell and Environment* **3**, 233–6.

Turner M.E., Blumenstein B.A. & Sebaugh J.L. (1969) A generalization of the logistic law of growth. *Biometrics* **25**, 577–80.

Vanderplank J.E. (1963) *Plant Diseases: Epidemics and Control*. Academic Press, London.

Vanderplank J.E. (1982) *Host–pathogen Interactions in Plant Disease*. Academic Press, London.

Waggoner P.E. (1978) Computer simulation of epidemics. In: *Plant Disease: an Advanced Treatise, Vol. II. How Disease Develops in Populations* (Ed. by J.G. Horsfall & E.B. Cowling), pp. 203–22. Academic Press, New York.

Waggoner P.E. (1983a) Quantifying the effect of the physical environment. In: *Challenging Problems in Plant Health* (Ed. by T. Kommendahl & P.H. Williams), pp. 215–25. American Phytopathological Society, St Paul, Minnesota.

Waggoner P.E. (1983b) The aerial dispersal of the pathogens of plant disease. *Philosophical Transactions of the Royal Society London B* **302**, 451–62.

Wangersky P.J. (1978) Lotka-Volterra population models. *Annual Review of Ecology and Systematics* **9**, 189–218.

Wangersky P.J. & Cunningham W.J. (1956) On time lags in equations of growth. *Proceedings of the National Academy of Science, USA* **42**, 699–702.

Wolfe M.S. (1981) Integrated use of fungicides and host resistance for stable disease control. *Philosophical Transactions of the Royal Society London B* **295**, 175–84.

Zadoks J.C. (1978) Methodology of epidemiological research. In: *Plant Disease: an Advanced Treatise, Vol II. How Disease Develops in Populations* (Ed. by J.G. Horsfall & E.B. Cowling) pp. 64–96. Academic Press, New York.

Zadoks J.C. & Kampmeijer P. (1977) The role of crop populations and their deployment, illustrated by means of a simulator EPIMUL 76. *Annals of the New York Academy of Sciences* **287**, 164–90.

9 Spore dispersal gradients and disease development

H.A. McCARTNEY and B.D.L. FITT
Rothamsted Experimental Station, Harpenden, Herts AL5 2JQ, UK

Introduction

Disease in crops frequently occurs in patches. Such patches may be due to variation in the environmental factors affecting disease development or to variation in the distribution of pathogen inoculum (Gregory, 1973). Patches due to the variation in inoculum distribution are termed disease foci for epidemic pathogens spread by air-borne or splash-borne spores. The decrease with distance from the focus (or source) in the number of pathogen spores deposited is referred to as the spore dispersal gradient. When viable infective spores are dispersed in a population of susceptible hosts under environmental conditions favourable for infection, a spore dispersal gradient will cause a disease gradient. Thus the spatial distribution of pathogen spores can determine the rate of disease spread and therefore affect the development of an epidemic.

This chapter considers methods for measuring spore dispersal gradients and the implications of such gradients for disease development, with particular reference to the dispersal of air-borne spores of *Erysiphe graminis* DC. f. sp. *hordei* and splash-borne spores of *Pseudocercosporella herpotrichoides* (Fron) Deighton.

Measurement of spore dispersal gradients

To measure spore dispersal gradients, it is necessary to sample numbers of spores at various distances from a source, which may be an infected plant (point source), a hedge (line source), or a field (area source) (Gregory, 1968). Gradients may be measured in a particular direction from the source (e.g. downwind) or averaged over all directions. The ideal sample of spores at each distance would be the number of infective spores deposited per unit area of susceptible host. However, such samples are difficult to collect; it is not always possible to use bait plants and it is difficult to remove spores from plant surfaces, especially in a way that does not kill the spores. A combination of bait plants, subsequently incubated and scored for disease symptoms, and artificial samplers to estimate the spore concentration, can provide a useful compromise in practice.

A wide range of devices have been evaluated as samplers for air-borne spores (Gregory, 1973). A sampler which collects spores by inertial impaction on to some surface is usually preferable to a sampler which collects spores by sedimentation.

Wolfe M.S. & Caten C.E. (1987) *Populations of Plant Pathogens: their Dynamics and Genetics.* Blackwell Scientific Publications, Oxford.

For example, in a comparison of samplers fewer spores were collected by sedimentation on to horizontal slides than by impaction on to vertical cylinders or Hirst samplers (Table 9.1). A volumetric sampler is preferable for small spores at low concentrations and low wind speeds; for example, Sutton & Jones (1976) collected more ascospores (14 × 7 μm) of *Venturia inaequalis* with rotorod and Burkard samplers than with vertical cylinders. The simpler, sticky cylinders become more useful with larger spores at higher concentrations; e.g. conidia (30 × 15 μm) of *E. graminis* (Jenkyn, 1974).

Few devices have been evaluated as samplers for splash-borne spores (Fitt & Bainbridge, 1983). The size of the spore-carrying droplets, rather than the spores themselves, determines the distance that they travel from the source. For sampling large, ballistic spore-carrying droplets which follow trajectories that are little affected by turbulence, funnels or horizontal microscope slides under rainshields may be appropriate (Fitt & Bainbridge, 1983). For smaller spore-carrying droplets a suction sampler, such as a pre-impinger, is necessary (Faulkner & Colhoun, 1977); this may be connected to a rain-activated switch to limit sampling to periods of rainfall (Fitt *et al.*, 1982).

Models of spore dispersal gradients

The relationships between numbers of spores deposited (y) and distance (x) from a source are usually concave curves, which are difficult to compare directly between experiments (Gregory, 1968). Two empirical models commonly used to describe spore dispersal gradients are the power law model (Gregory, 1968) and the exponential model (Kiyosawa & Shiyomi, 1972), represented respectively by the equations:

$$y = ax^{-b} \quad [1]$$

and

$$y = c\mathrm{e}^{-dx} \quad [2]$$

where e (= 2.718) is the base of natural logarithms and the constants a and c are related to the source strength. Regression analyses of the appropriate log-

Table 9.1. Collection of dry air-borne spores by samplers at 2 m height, June–Sept 1951 (spores/cm^2)

Organism	Spore diameter	Horizontal slide	Vertical cylinder	Hirst sampler
Cladosporium	12 × 5 μm	59	376	8930
Erysiphe	30 × 15 μm	2	69	100
Pollen	20 – 80 μm	13	490	181

Gregory *et al.*, quoted in Hirst (1959).

transformed equations can be used to estimate values of b or d, which are measures of the dispersal gradient. With the exponential model (equation 2), the gradient can also be expressed as a half-distance, α, which is the distance from the source in which deposition is halved. The half-distance is related to the gradient:

$$\alpha = \frac{0.693}{d}. \qquad [3]$$

These empirical models have been compared by McCartney & Bainbridge (1984) and McCartney & Fitt (1985); generally they fit dispersal gradient data equally well, although the power law model tends to overestimate and the exponential model to underestimate spore deposition near the source. Both are descriptive, rather than interpretative (Gregory, 1968) and give no information about the processes underlying spore dispersal, such as spore removal, transport and deposition.

As an alternative, physical models of spore dispersal have been constructed using theories of turbulent diffusion. For dispersal above crops, models have been based on Sutton's statistical diffusion theory (Gregory, 1973). Within crops, eddy-diffusion theory has been used to model deposition of *E. graminis* spores on to barley by sedimentation and impaction (Legg & Powell, 1979). A more complex but more accurate model, based on Markov chain simulation of individual spore trajectories in turbulent flow, is discussed by Legg (1983). However, the development of such models is still at an early stage.

Interpretation of spore dispersal gradients

The principles for interpreting dispersal gradients have been well summarized by Gregory (1968, 1973, 1982), but nevertheless they are frequently misinterpreted. Gradients from line or area sources are generally shallower than gradients from point sources. With a strong prevailing wind, gradients downwind of a point source are generally shallower than those in other directions. Small values of b or d suggest proximity to a large area source, or spore dispersal from further generations of pathogen lesions, or a large background contamination (Gregory, 1968). Gradients of splash-dispersed spores are generally steeper than those of dry-air-dispersed spores, both over open ground (Stedman, 1979) and within crops (Stedman, 1980). Spore dispersal gradients are also affected by the method of expressing the number of spores collected and by crop structure and atmospheric turbulence. Thus it is difficult to interpret small differences in the value of b or d between pathogens or between different experiments with one pathogen.

Gradients of air-borne and splash-borne spores

McCartney & Bainbridge (1984) simulated dispersal of air-borne *E. graminis* spores within a barley crop by releasing 20-μm diameter droplets containing a chemical tracer (thiabendazole) from a point source (a May spinning disc) placed at mid-crop height. Deposition patterns were determined by measuring the amount of tracer deposited on 4 cm^2 plastic strips placed around the source. The total amounts of tracer deposited in narrow arcs at different distances from the source at ground level and at mid-crop height were determined by integrating the measured deposition rates over all directions. Power law and exponential models fitted the data equally well; in most cases the coefficient of determination, r^2, was greater than 0.8 (Table 9.2).

The gradients were affected by atmospheric turbulence. For example, gradients over open ground were quite different on two days with differing wind speed and turbulence (Table 9.2(a), experiments 1 and 2) and within the crop the steepest gradient was measured when there was greatest turbulence (Table 9.2(a), experiment

Table 9.2. Deposition gradients estimated by fitting exponential and power law models to observed deposits of 20 μm diameter droplets released at mid-crop height in a barley crop

Expt. number*	Crop height (cm)	Power law model†		Exponential model†	
		b	r^2	$d(cm^{-1})$	r^2
(a) Ground level					
Open ground					
1	–	0.50	0.93	0.007	0.79
2	–	2.28	0.85	0.030	0.72
Short open crop					
3	10	0.99	0.98	0.018	0.99
4	12–20	–	–	–	–
5	20–30	0.88	0.87	0.012	0.80
Short closed crop					
6	25–30	0.94	0.82	0.014	0.92
7	30–35	0.43	0.83	0.005	0.74
Tall closed crop					
8	50–58	1.47	0.85	0.019	0.95
9	85–90	0.64	0.82	0.007	0.91
10	90	0.21	0.48	0.003	0.59
(b) Mid-crop height					
Short open crop					
3	10	1.11	0.93	0.017	0.80
4	12–20	0.54	0.47	0.014	0.68
5	20–30	1.13	0.88	0.018	0.79
Short closed crop					
6	25–30	1.47	0.85	0.025	0.90
7	30–35	1.27	0.97	0.027	0.93
Tall closed crop					
8	50–58	1.00	0.83	0.018	0.98
9	85–90	0.95	0.84	0.017	0.94
10	90	0.91	0.83	0.016	0.93

* Full experimental details are given in McCartney & Bainbridge (1984).
† b and d are gradients estimated by regression analysis; r^2 is the coefficient of determination.

8). Gradients at ground level within crops tended to become shallower as the crop grew taller. At mid-crop height, gradients were often steeper than at ground level (Table 9.2), indicating that the centre line of the plume was angled downwards. In addition, the gradients were steepest when the crop was short and the canopy closed, demonstrating that crop structure is important in spore dispersal.

Gradients of the splash-dispersed spores of *P. herpotrichoides* were measured in experiments in a rain-tower/wind-tunnel complex. Experiments were done either in still air with single drops falling on to spore suspensions or in moving air with simulated rain falling on to naturally infected wheat straw. Gradients generally fitted the data considerably better when estimated by the exponential model rather than the power law model (Table 9.3). The gradients were much steeper than in the experiments with air-borne spores.

Effects of spore dispersal gradients on disease development

Observations

Observations of spore dispersal gradients have helped in the interpretation of observed patterns of disease spread. Fried *et al.* (1979) found a good correlation between gradients of spores of *E. graminis* deposited on sticky cylinders within wheat crops, and subsequent disease gradients measured by number of lesions per tiller, in crops of cultivar Chancellor and in multilines. Johnson and Powelson (1983)

Table 9.3. Dispersal gradients for spores of *P. herpotrichoides* estimated by fitting exponential and power law models to observed deposits of spores in rain-tower/wind-tunnel experiments

Experimental details*	Power law model†		Exponential model†		Reference
	b	r^2	$d(\text{cm}^{-1})$	r^2	
Still air, single drops, spore susp.					
5 mm drop, spores target	3.2	0.82	0.10	0.95	Fatemi & Fitt, 1983
4 mm drop, spores target	2.8	0.89	0.11	0.95	Fitt & Lysandrou, 1984
4 mm drop, spores drop	2.0	0.54	0.06	0.66	Fitt & Lysandrou, 1984
5 mm drop, spores target	3.8	0.78	0.12	0.93	Fitt & Lysandrou, 1984
5 mm drop, spores drop	3.5	0.87	0.09	0.94	Fitt & Lysandrou, 1984
Moving air, simulated rain, straw					
3 mm drops, upwind	7.3	0.99	0.10	0.99	Fitt & Nijman, 1983
3 mm drops, downwind	4.9	0.77	0.04	0.89	Fitt & Nijman, 1983
3 mm drops, downwind	5.1	0.81	0.05	0.93	Fitt & Bainbridge, 1984

* In still air single simulated raindrops of the diameter indicated fell on to target water films, with spores incorporated into either incident drops or target films; in moving air simulated rain fell on to naturally infected straw.

† b and d are gradients estimated by regression analysis; r^2 is the coefficient of determination.

found that gradients of bean (*Phaseolus vulgaris* L.) pod rot at harvest were as steep as were those of spores of *Botrytis cinerea* Pers. ex Fr. at full bloom, although by harvest the spore dispersal gradients had become shallower because of secondary disease spread. They concluded that the spores which infected the senescing blossoms were the primary cause of the pod rot.

Simulation of spore dispersal and disease development

A simulation model was constructed to investigate the effects of spore dispersal gradients on the spread of foliar diseases such as barley powdery mildew. The model assumed values for the leaf area available to the pathogen, lesion growth rate, age of lesion when sporulation starts (latent period), and proportion of spores which infect successfully. Diseased leaf area was calculated across a one-dimensional array of plants, equivalent to disease spread from a long infected strip of crop. This is not a realistic crop model but it does permit examination of the possible effect of deposition on disease spread. It was assumed that the spores released from infected plants were deposited equally on either side of the infected strip of plants and that the number deposited decreased exponentially with distance. Lesion growth was assumed to begin after a fixed incubation period (5 days) and to continue for a fixed period (10 days); during lesion growth, sporulation was assumed to be proportional to lesion size. The amount of disease on each plant was calculated on a daily basis from: (a) the number of new lesions initiated, (b) the number of lesions which had stopped producing spores, (c) the number of lesions which had become active, and (d) the daily growth of active lesions. The number of spores released by each plant was based on the total amount of active disease at the beginning of each day. The model allowed for multiple infections from subsequent generations and for self-infection of host plants.

Initially all plants were considered susceptible to the pathogen and at the beginning of each simulation, the middle row of plants in a 20 m wide strip was assumed to be infected. In addition to the amount of disease on each plant, the total disease in the crop was calculated. Figure 9.1 shows the distribution of disease with distance from the centre of the strip for two deposition gradients, with half-distances $\alpha = 5$ cm (a) and $\alpha = 50$ cm (b), after 3, 4, 5 and 6 generations (equivalent to 15, 20, 25 and 30 days). This figure suggests that, with steep gradients, disease will intensify rapidly but spread slowly, whereas with shallow gradients, disease will intensify more slowly but spread rapidly.

The model also indicated that the same total amount of leaf area becomes diseased irrespective of the spore dispersal gradient, until about 30% of the leaf area at the focus is infected. Subsequently, there is an increasing probability of multiple infections; i.e. some spores land on tissues that are already infected and thus do not contribute to disease increase. Since disease intensifies more quickly when the gradient is steep (α small), multiple infection effects occur sooner and the total amount of disease in the crop will increase less rapidly than when the gradient is

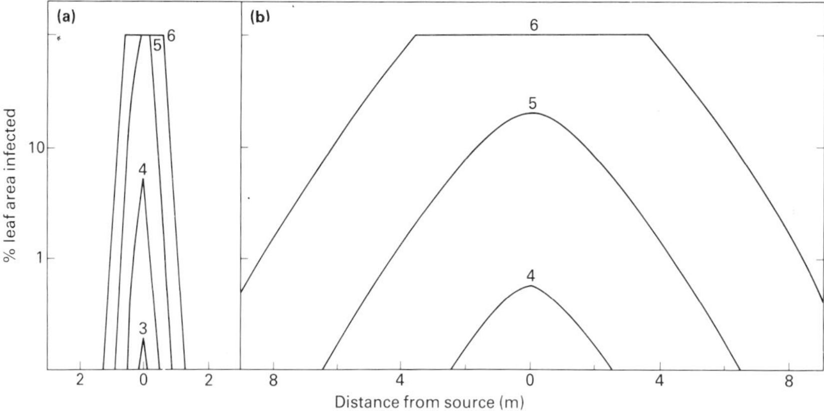

Figure 9.1. Predicted amount of disease per plant after three to six pathogen generations in a 20-m wide strip of crop with spore deposition gradient half-distances (α) of (a) 5 cm and (b) 50 cm. The simulation model assumes values for lesion growth rate etc. similar to those of *Erysiphe graminis* f. sp. *hordei*.

shallow. This is illustrated in Figure 9.2, which shows the total amount of disease in the 20 m strip after 4, 5 and 6 generations, plotted against deposition half-distance. For large α, total disease decreases as a result of loss of spores from the edge of the crop. However, because the model assumes that no spores are lost from the crop except at its edges, it may overestimate disease when gradients are shallow, since some spores will be lost to the atmosphere.

Effects of cultivar mixtures

The model was used to study the interaction between the effects of cultivar mixtures and the effects of spore dispersal gradients on disease development. The same assumptions were made except that plants were considered either wholly susceptible or wholly resistant to the disease. Susceptible plants were distributed randomly on either side of the middle row of infected plants and the ratio of susceptible to resistant plants in the crop was set to 1:1, 1:2 or 1:3. Twenty-five simulations with different random distributions of susceptible plants were carried out for each ratio and the total disease in the crop after each day was averaged over all runs. Figure 9.3 shows the average amount of disease in the mixtures, relative to that in the pure stand, plotted against α. The results suggested that mixtures would be most effective in decreasing disease caused by pathogens which have spores with shallow dispersal gradients. In these circumstances the rate of disease intensification would be considerably decreased by mixtures through loss of spores deposited on resistant plants. When deposition gradients are steep much of the disease in a focus appears on the initial infected plant (self-infection), thus limiting the effectiveness of mixtures in reducing disease spread.

A field experiment to investigate the development of disease foci in stands of mixtures of spring barley cultivars was carried out in 1983 (McCartney & Bain-

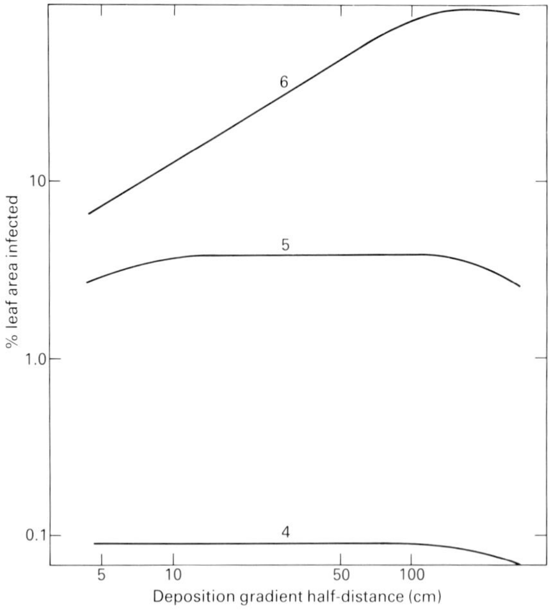

Figure 9.2. Predicted effect of spore deposition gradient on total amount of disease in a 20-m wide strip of crop after four to six pathogen generations for different deposition gradients (α). The simulation model assumes values for lesion growth rate etc. similar to those of *Erysiphe graminis* f. sp. *hordei*.

bridge, in preparation). Nine plots, 9 × 9 m (three replicates of three treatments) were sown with either cv. Koru, which was susceptible to powdery mildew (*E. graminis*), or with a mixture of cv. Koru and cv. Atem, a resistant cultivar, in the ratio 1:2 or 1:4. When the crop was about 10 cm tall each plot was inoculated at the centre with greenhouse-grown infected plants. The pattern of disease which developed in each plot was assessed at 2- or 3-weekly intervals. Table 9.4 shows average observed amounts of disease, assessed on leaf 8, for the two mixtures as a fraction of that in the pure stand and amounts of disease calculated by the model with $\alpha = 25$ and 50 cm which are typical values for barley mildew. The amounts of disease found in the plots were similar to those calculated by the model after five or six pathogen generations, and were considerably less in the mixtures than in the pure stand. Although the model does not attempt to simulate the exact conditions of the field experiment, the results shown in Table 9.4 suggest that its predictions are realistic.

Conclusions

Despite the crucial role which spore dispersal plays in epidemics of plant diseases caused by pathogens with air-borne or splash-borne spores, dispersal has been studied little by comparison with other stages in pathogen life-cycles. There is a need for more accurate models of spore dispersal gradients, especially of gradients

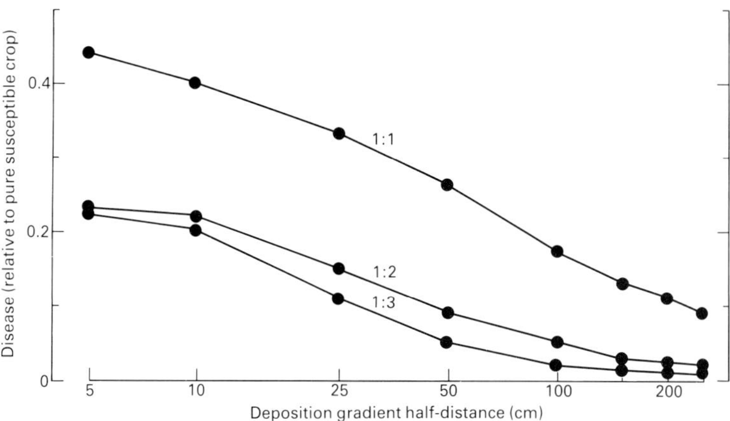

Figure 9.3. Predicted effect of spore deposition gradient (α) on amount of disease in mixtures of susceptible and resistant cultivars (1:1, 1:2, 1:3) after six pathogen generations. The simulation model assumes values for lesion growth rate etc. similar to those of *Erysiphe graminis* f. sp. *hordei.*

Table 9.4. Amounts of powdery mildew in mixtures of susceptible and resistant barley cultivars, relative to pure stands of a susceptible cultivar, as predicted by a simulation model and as observed in infected field plots

	Predicted				Observed on date*		
Mixture	α = 25 cm†		α = 50 cm†				
Sus.: Res.	generation‡		generation‡				
	5	6	5	6	22/6	6/7	18/7
1:2	0.07	0.15	0.06	0.09	0.05	0.08	0.10
1:4	0.04	0.08	0.02	0.03	0.01	0.02	0.03

* Disease assessed on leaf 8, mean of three replicates.
† α is the deposition gradient half-distance.
‡ The amounts of disease measured in the susceptible plots were similar to those calculated as occurring after five or six pathogen generations.

within crops, and for more experimentation to examine the influence of spore dispersal gradients on disease development (Jeger, this volume).

References

Fatemi F. & Fitt B.D.L. (1983) Dispersal of *Pseudocercosporella herpotrichoides* and *Pyrenopeziza brassicae* spores in splash droplets. *Plant Pathology* **32**, 401–4.

Faulkner M.J. & Colhoun J. (1977) An automatic spore trap for collecting pycnidiospores of *Leptosphaeria nodorum* and other fungi from the air during rain and maintaining them in viable condition. *Phytopathologische Zeitschrift* **89** 50–9.

Fitt B.D.L. & Bainbridge A. (1983) Dispersal of *Pseudocercosporella herpotrichoides* spores from infected wheat straw. *Phytopathologische Zeitschrift* **106**, 214–25.

Fitt B.D.L. & Bainbridge A. (1984) Effect of cellulose xanthate on splash dispersal of *Pseudocercosporella herpotrichoides* spores. *Transactions of the British Mycological Society* **82**, 570–1.

Fitt B.D.L. & Lysandrou M. (1984) Studies on mechanisms of splash dispersal of spores, using *Pseudocercosporella herpotrichoides* spores. *Phytopathologische Zeitschrift* **111**, 323–31.

Fitt B.D.L. & Nijman D.J. (1983) Quantitative studies on dispersal of *Pseudocercosporella herpotrichoides* spores from infected wheat straw by simulated rain. *Netherlands Journal of Plant Pathology* **89**, 198–202.

Fitt B.D.L., Rawlinson C.J. & Smith C.B. (1982) A comparison of two rain-activated switches used with samplers for spores dispersed by rain. *Phytopathologische Zeitschrift* **105**, 39–44.

Fried P.M., MacKenzie D.R. & Nelson R.R. (1979) Dispersal gradients from a point source of *Erysiphe graminis* f. sp. *tritici*, on Chancellor winter wheat and four multilines. *Phytopathologische Zeitschrift* **95**, 140–50.

Gregory P.H. (1968) Interpreting plant disease dispersal gradients. *Annual Review of Phytopathology* **6**, 189–212.

Gregory P.H. (1973) *The Microbiology of the Atmosphere*. Leonard Hill, London.

Gregory P.H. (1982) Disease gradients of windborne plant pathogens: interpretation and misinterpretation. In: *Advancing Frontiers of Mycology and Plant Pathology* (Ed. by K.S. Bilgrami, R.S. Misra & P.C. Misra), pp. 107–17. Today and Tomorrow's Publishers, New Delhi.

Hirst J.M. (1959) Spore liberation and dispersal. In: *Plant Pathology: Problems and Progress 1908–1958* (Ed. by C.S. Holton, G.W. Fischer, R.W. Fulton, H. Hart & S.E.A. McCallan), pp. 529–538. University of Wisconsin Press, Madison.

Jenkyn J.F. (1974) A comparison of seasonal changes in deposition of spores of *Erysiphe graminis* on different trapping surfaces. *Annals of Applied Biology* **76**, 257–67.

Johnson K.B. & Powelson M.L. (1983) Analysis of spore dispersal gradients of *Botrytis cinerea* and gray mold disease gradients in snap beans. *Phytopathology* **73**, 741–5.

Kiyosawa S. & Shiyomi M. (1972) A theoretical evaluation of the effect of mixing resistant variety with susceptible variety for controlling plant diseases. *Annals of the Phytopathological Society of Japan* **38**, 41–51.

Legg B.J. (1983) Movement of plant pathogens in the crop canopy. *Philosophical Transactions of the Royal Society, London B* **302**, 559–74.

Legg B.J. & Powell F.A. (1979) Spore dispersal in a barley crop: a mathematical model. *Agricultural Meteorology* **20**, 47–67.

McCartney H.A. & Bainbridge A. (1984) Deposition gradients near to a point source in a barley crop. *Phytopathologische Zeitschrift* **109**, 219–36.

McCartney H.A. & Fitt B.D.L. (1985) Construction of dispersal models. *Advances in Plant Pathology* **3**, 107–43.

Stedman O.J. (1979) Patterns of unobstructed splash dispersal. *Annals of Applied Biology* **91**, 271–85.

Stedman O.J. (1980) Splash droplet and spore dispersal studies in field beans (*Vicia faba* L.). *Agricultural Meteorology* **21**, 111–27.

Sutton T.B. & Jones A.L. (1976) Evaluation of four spore traps for monitoring discharge of ascospores of *Venturia inaequalis*. *Phytopathology* **66**, 453–6.

10 Epidemiology of soil-borne plant pathogens

CHRISTOPHER A. GILLIGAN
Department of Applied Biology, University of Cambridge, Pembroke Street, Cambridge, CB2 3DX, UK

Introduction

While research in botanical epidemiology is concentrated principally upon leaf-infecting pathogens (Jeger, this volume), soil-borne pathogens are winning increasing attention (Bloomberg, 1979; Campbell *et al.*, 1980; Gilligan, 1983b; Tomimatsu & Griffin, 1982). Accompanying this change is the realization that most soil-borne pathogens can properly be regarded as causing epidemics. This realization reflects in turn a change in understanding of the nature of epidemics (Jeger, this volume). Garrett (1970), whose work lays the foundation of much of the epidemiological investigation of soil-borne fungal pathogens (Garrett, 1956, 1960, 1970, 1979), defines an epidemic as an outbreak of disease characterized by a high ratio of secondary to primary infections. This definition is too restrictive: it implies that epidemics should be catastrophic, a feature that reflects the common rather than the scientific use of the term. Kranz (1974) offers a broader definition: an epidemic may be defined as any change (increase or decrease) in a plant disease in a host population in time and space. I prefer this definition because it emphasizes both the temporal and spatial aspects of epidemics and because a decrease in the level of disease is acknowledged as part of an epidemic. Pathologists have long monitored survival of infected host material, thereby acknowledging the importance of decline in the epidemiology of soil-borne pathogens.

To understand the population dynamics and genetics of soil-borne pathogens, we must first identify the components of epidemics. The objective of this chapter is to disentangle the complexities of epidemics in order to reveal these components. Discussion is restricted to fungi and is directed to biological rather than to mathematical aspects. Mathematical modelling of soil-borne pathogens is discussed elsewhere (Gilligan, 1983b).

I begin by discussing the concept of the infection chain as it relates to subterranean fungal pathogens. Components of the infection chain are each examined in detail. Emphasis is given to the parasitic phases of the infection chain but the temporal and spatial dynamics of inoculum are also discussed. Two disputed assumptions are considered briefly: that infection and disease occur at random; that most soil-borne pathogens incite simple-interest disease, *sensu* Van der Plank (1963).

Wolfe M.S. & Caten C.E. (1987) *Populations of Plant Pathogens: their Dynamics and Genetics.* Blackwell Scientific Publications, Oxford.

Infection chain

The infection chain, first proposed by Gaümann, (1951), consists of a number of infection cycles. Each infection cycle typically comprises: germination of a fungal propagule, infection, growth in the host, and production and release of more propagules. This classical description however, is shaped almost exclusively by consideration of aerial pathogens. Thus, much effort in above-ground epidemiology is directed at measurement of the latent and infectious periods and of the rate of spore production (see e.g. Jeger, 1982 and this volume; Mehta & Zadoks, 1970, Parlevliet, 1979).

Few quantitative investigations of the infection chains of soil-borne fungi have apparently been undertaken, and formal quantification of the components is difficult to find in the literature. Sewell (1959), however, effectively estimated the latent period of *Verticillium albo-atrum* on tomato by using root observation boxes. No sporulation was observed until degradation of the host root system was apparent to the naked eye. The latent period and rate of spore production of *Phytophthora* spp. are discussed briefly by Weste (1983). Byrt & Holland (1978) have demonstrated a latent period of only 24 h for *P. cinnamomi* on axenically grown eucalypt seedlings. The dearth of knowledge on factors such as latent, incubation and infectious periods of soil-borne fungi undoubtedly reflects the difficulties of such investigations. But it also reflects the questionable value of some of the concepts of above-ground epidemiology to below-ground epidemiology. While, for example, the period of latency before a lesion becomes infectious can be applied to a damping-off pathogen or even to a wilt-inducing fungus, it is of questionable use for pathogens that spread by ectotrophic mycelium. Similarly, difficulties can be envisaged in the estimation of infectious periods and of rate of production of propagules by many soil-borne fungi that are released slowly upon the decay of host tissue, or that spread by mycelial growth. While there is scope for further investigation and quantification, the danger exists that the quest might be motivated by considerations appropriate to an aerial rather than to a subterranean pathogen. For this reason I propose that the emphasis within the concept of the infection cycle can usefully be changed for soil-borne fungi.

Inoculum density, increase in the size of lesions or in the extent of infection, limitations on dispersal and the balance between parasitic and saprophytic activities assume great importance in the epidemiology of many soil-borne fungi. The infection chain of many soil-borne fungi, therefore, can better be considered as heterogeneous, consisting of several dissimilar phases or cycles, rather than as a polycyclic chain of uniform cycles of latent and infectious periods with instantaneous dispersal. First in the sequence is a non-parasitic phase that may involve an increase or a decrease in population density. This phase may involve cycles of saprophytic activity (Figure 10.1). The non-parasitic phase is followed by a parasitic phase in which primary infection is initiated from surviving inoculum. This may be followed by cycles of secondary infection, each arising from previous infections. Components of the infection chain are discussed in the rest of the chapter.

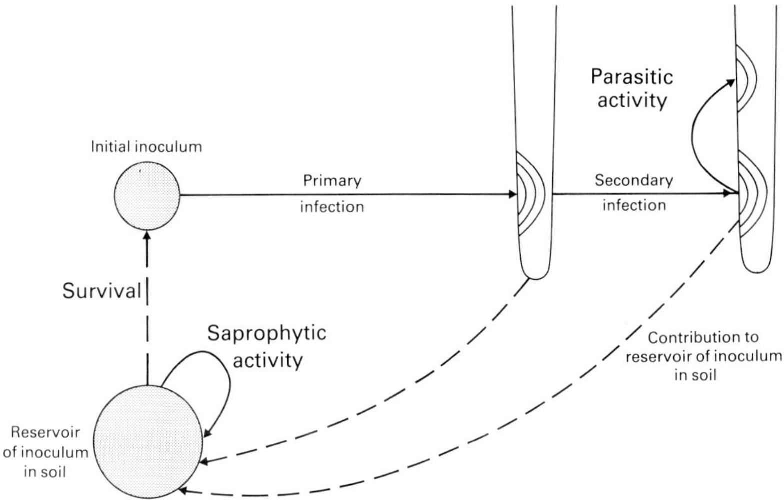

Figure 10.1. Simplified outline of infection chain of soil-borne, fungal pathogens.

Inoculum

Terminology

Inoculum is the viable material that can infect a host (Dimond & Horsfall, 1965). It comprises only a portion of the total biomass of a pathogen species in a given volume of soil. The total biomass consists of both living and dead material. The living material constitutes the soil population. This in turn is divided between dispersal and infection units *sensu* Zadoks & Schein (1979). A dispersal unit is any device for spread or survival that can be recognized and counted. It is synonymous with the term propagule that is normally used to describe the host-free unit of population of soil-borne pathogens. An infection unit is the mycelial structure that originates from a dispersal unit after infection of the host. It may be a lesion or a length of infected root. Of the dispersal units that are viable, not all are necessarily capable of infecting a host, even when favourably placed to do so. Garrett (1960) used the term 'effective inoculum' to distinguish the viable and infective fraction from the viable but non-infective fraction. The term is rendered tautologous by the later definition of Dimond & Horsfall (1965). It serves, however, as a reminder of the selectivity of the term inoculum. Inoculum density is frequently used to describe the entire soil population of dispersal units in inoculum density experiments. More correctly is meant relative inoculum density. Statistical strategies for the estimation of inoculum density are discussed elsewhere (Gilligan, 1983b; Pfender *et al.*, 1981).

Form and variability of inoculum

For the epidemiologist, there is a quest for some identifiable unit of inoculum, the propagule, that can be counted. The nature of such propagules varies enormously. Some pathogens produce spores that are uniform in size and, by inference, in nutrient status, others have sclerotia which sometimes incorporate fragments of host tissue and hence vary in nutrient status. Extreme variability is to be seen in pathogens such as *Gaeumannomyces graminis*, in which inoculum comprises fragments of colonized host tissue (Hornby, 1981).

The apparent uniformity in nutrient status of many spores may be deceptive. Phillips (1965) harvested macroconidia of *Fusarium roseum* f. sp. *cerealis* that had been grown on either a high- or a low-nutrient medium. Conidia grown on high-nutrient media were shorter, thicker and generally more uniform than conidia produced on low-nutrient. Percentage germination was similar in the two types, but the high-nutrient conidia caused more disease of carnations than did the others. A similar correlation between nutrition and infectivity of hypocotyls of cotton by *Pythium ultimum* has been shown by Johnson *et al.* (1981). If considerable variability in nutritional status of host substrates occurs in soil, as seems likely, then variability in the phenotype of propagules of inoculum is to be expected.

Variability in genotype amongst propagules is also to be expected. Parmeter (1970) writes '. . . genetic diversity is characteristic of most, if not all, root-infecting fungi'. According to Parmeter (1970), recurrent mutation alone is insufficient to account for observed variation in pathogenic and ecological behaviour. Recombination may also be important. Forms of asexual 'recombination' such as heterokaryosis and parasexuality have been demonstrated in a number of soil-borne fungi in culture (Parmeter, 1970), notably, *Verticillium* spp., *Fusarium oxysporum* and *F. solani*. Variability in genotype and nutrient status may be particularly important in the population dynamics of fungi in which relatively low densities of inoculum are encountered, as with *Rhizoctonia solani* (Benson & Baker, 1974), *Sclerotium cepivorum* (Crowe & Hall, 1980) and *G. graminis* (Hornby, 1981). Chance plays a comparatively greater part relative to natural selection in such small and isolated populations (Gale, this volume).

Population dynamics of inoculum

The density of inoculum changes due to birth, death and migration. Migration may be effected by growth, e.g. rhizomorphs of *Armillariella mellea* (Garrett, 1956), by transport in the soil water (Wallace, 1978) or perhaps by movement of soil fauna. Most inoculum, however, is static and birth and death assume most importance in the population dynamics of inoculum. The reservoir of inoculum in soil is increased or maintained by parasitic activity of the fungus. This is augmented in the case of facultative saprophytes, such as the damping-off and root-rotting fungi, by saprophytic activity (see e.g. Schlub *et al.*, 1981). Death of inoculum results from antagonism and autolysis.

Antagonism is a broad term that describes mechanisms that involve direct damage to the inoculum by biological agents other than the host (Baker, 1968) and includes predation, parasitism, competition and antibiosis. Baker (1965) was the first to discuss the effects of these factors on inoculum density. Predation of inoculum by microfauna has been widely reported (Cook & Baker, 1983) but to date few quantitative studies of its effects have been reported. Curl, 1979 (cited in Cook & Baker, 1983), however, reported a reduction in inoculum density of *R. solani* due to predation by two species of collembola. Parasitism, often referred to as hyperparasitism or, where appropriate, as mycoparasitism, has enjoyed a renaissance of interest amongst researchers as a potentially successful method of biological control (Baker, 1983; Cook & Baker, 1983). By implication parasitism is a factor of potential importance in the dynamics of inoculum density. Further epidemiological study and quantification of the components of the infection chains of hyperparasites is needed. Suitable factors for quantification are estimates of the minimum infective dose of hyperparasite necessary to cause death of a propagule, the latent and infectious periods, and the rate and amount of spore production of the hyperparasite.

Competition was defined by Clark (1965) as active demand in excess of immediate supply of material or condition on the part of two or more organisms. By condition was meant a site for saprophytic activity, though it might also be taken to refer to an infection site. Since an active demand is specified, competition does not cause reduction in inoculum that is quiescent (Baker, 1968, 1981). Competition for nutrients (Scher & Baker, 1983) following germination of spores may, however, reduce inoculum density by exhausting inoculum before new propagules are produced. The role of competition in the population dynamics of *Fusarium* spp. is discussed by Baker (1981).

Antibiosis, the inhibition or destruction of one organism by a metabolic product of another has been observed in the laboratory (Baker & Cook, 1974). The importance of antibiosis as a method of reducing inoculum density, however, has repeatedly been questioned by Baker (1968, 1983). Autolysis, due to a shortage of either endogenous or exogenous nutrients is the most important mechanism responsible for the loss of inoculum. Garrett (1965) puts it simply: '. . . the commonest cause of death in a micro-organism in whole or part is straight starvation.'

Spatial pattern of inoculum

Just as the mean density of inoculum is dynamic, so the variance of inoculum density within a given population is also dynamic. This variance describes the spatial pattern of inoculum. Randomness is the standard against which spatial pattern is usually measured (Gilligan, 1982, 1983a). By randomness is meant that the occurrence of one propagule at a particular site in soil is independent of the location of other propagules. Can this really be true of such relatively immobile organisms as soil-borne fungi? If the assumption of randomness is false, the consequences of aggregation of propagules, or even regularity of propagules imposed, for example; by regular spacing of the host, must be considered. The consequences of deviation from randomness vary with scale. Inoculum may, for example, be aggregated into relatively large foci, several metres in diameter, but within the focus the pattern of inoculum density per 10^4 cm^3 may be random. The hierarchical nature of spatial pattern should always be considered (Gilligan, 1982).

Although there have been numerous studies on variation in mean density of inoculum with depth (see e.g. Crowe *et al.*, 1980; Hornby, 1975) studies on the spatial pattern of inoculum are scarce. The consensus of this work, however, is that aggregation rather than randomness is the rule. Microsclerotia of *Cylindrocladium crotalariae* were shown to be aggregated in several studies of both cultivated soil (Taylor *et al.* 1981, Hau *et al.* 1982) and undisturbed soil (Roth & Griffin, 1981). Stanghellini *et al.*, (1982) demonstrated an aggregated pattern of oospores of *Pythium aphanidermatum* in sugar beet fields. They recorded very high numbers of oospores following infections of sugar beet in rhizospheres of the plants. Inoculum density ranged from 422 to 2500 spores/g within 1 mm of the root surface to < 5 spores/g 5 mm away. Stanghellini *et al.* (1982) speculate that the localized population increase may account for the aggregated pattern of infection encountered in the field. The statistical basis of testing for deviation from randomnness is discussed elsewhere (Gilligan, 1983b).

Primary infection

Probability of infection

Primary infection can be regarded as the result of a sequence of chance events (Gilligan, 1985). The sequence is initiated by the occurrence of a propagule close enough to a susceptible host to have some chance of infecting it. This is followed by growth or, if the propagule is motile, movement to the surface of the host, colonization, and infection. The sequence is one of conditional probabilities, say θ_i where $i = 1, \ldots 4$; θ_1 is the probability of occurrence of a propagule close to a host; θ_2 is the probability of germination of the propagule given that it occurs close to the host; θ_3 is the probability that the propagule reaches the host given that it germinates; θ_4 is the probability that infection is initiated given contact. The probability that a host is infected is the product of these probabilities, $\Pi\theta_i$. It is often easier, because of the problem of identifying multiple infection, to work with the probability that a host remains uninfected (Gilligan, 1983b). Thus, the term $(1 - \Pi\theta_i)^I$ is the probability that a host-occupying unit volume is uninfected when exposed to an inoculum density, I, (Gilligan, 1985).

The probability θ_i depends upon the inoculum density, the rate of growth of the host and the extent of 'rhizosphere-influence'. I have proposed elsewhere that the term, pathozone, be used instead of rhizosphere (Gilligan, 1985). The pathozone is the region of soil surrounding a host unit within which the centre of a propagule must lie for infection of the host unit to be possible. Reasons for preferring this term rather than previously used terms such as rhizosphere or spermosphere are threefold (Gilligan, 1985): it is a more general term that can be applied to different organs such as roots, seeds, hypocotyls; it does not carry the connotation of exudation of nutrients that may be associated with the term rhizosphere; and it allows for the possibility of spontaneous germination of propagules in the absence of direct stimuli from the host.

Increasing the inoculum density decreases the probability, $(1 - \theta_1)^I$, that a pathozone does not contain inoculum. Methods for calculating the risk of infection for random and aggregated patterns of inoculum are discussed elsewhere (Gilligan, 1985). The rate of growth of the host determines the rate of exploration of the soil. This rate is important, relative to the rate of decay of the reservoir of inoculum, in its effect on the probability of infection.

Of the remaining probabilities, θ_2 is simply estimated as the proportion of viable propagules within the pathozone. Once a propagule germinates, the probability, θ_3, that it reaches the host surface is influenced by the continuing availability of endogenous and/or exogenous nutrient. Autolysis, competition and parasitism can all reduce the magnitude of θ_3.

The probability, θ_4, that infection occurs, given contact between the pathogen and the host, depends primarily upon the susceptibility of the site of potential infection, and the nutritional status and genotype of the parasite. Susceptibility varies with site on the host: thus hypocotyls and seeds of pea differ in susceptibility to *Fusarium solani* f. sp. *pisi* (Cook & Snyder, 1965). Susceptibility of a single infection site also varies with time. Sometimes the change in susceptibility is rapid and so the period between germination of a propagule and arrival at the surface of the host may be critical in determining whether or not infection occurs (Benson & Baker, 1974; Huisman, 1982). The occurrence of previous infections may also affect suceptibility of a distant site of infection (Gilligan, 1980b).

The conditional probabilities θ_2, θ_3 and θ_4, influence the width of the pathozone. Estimates of this dimension have been published for numerous host–pathogen systems: see for example, Henis & Ben-Yephet (1970) for *R. solani* and cotton; Brown & Hornby (1971) and Gilligan (1980c) for *G. graminis* and wheat; Stanghellini & Hancock (1971), and Ferriss (1982) for *Pythium ultimum* on sugar beet and soy bean seeds, respectively. There is also limited evidence on the effects of antagonism on the width of the pathozone (Rouse & Baker, 1978; Gilligan, 1979). Estimation of the width of the pathozone has motivated the construction of mathematical models. For a discussion of the mathematical and biological assumptions of these models see Baker & Drury (1981), Gilligan (1979, 1983b, 1985), Ferriss (1981) and Grogan *et al.* (1980).

Efficiency of infection

So far we have considered a sequence of events leading to infection, and summarized it as a sequence of conditional probabilities. We can extend the sequence to the occurrence of a lesion and to the production of further inoculum, in order to consider the efficiency of infection. By efficiency I mean the ratio of output to input. Quantitative estimates of the efficiency of infection can be obtained from the ratio of successive components in the sequence (Table 10.1). This approach is analogous to the method of components analysis used by Zadoks (1972) for the uredial infection cycle of *Puccinia recondita* on wheat. Clearly the sequence may still be regarded as a series of conditional probabilities, although the relationships now become more complex. Reference to the idealized sequence of Table 10.1 reveals that numerous variables may be used to describe what is loosely referred to as infection efficiency. Terminology is confusing. The term 'infection efficiency' is usually taken to represent the ratio between two states separated by one latent period (Zadoks & Schein, 1979) e.g. the ratio of spores produced by a rust to spores applied to a leaf. Another term, 'infection frequency' is used to describe the ratio of sporulating lesions to infections (Parlevliet, 1979). To avoid confusion, I commend the simple procedure adopted by Tomimatsu and Griffin (1982). These authors used the term 'efficiency of inoculum for infection' to describe the ratio between the number of propagules of inoculum in the pathozone and the number of infections on a root. Other terms were 'efficiency of inoculum for necrosis' and 'efficiency of infection for necrosis' (Tomimatsu & Griffin, 1982).

There have been a number of studies on the efficiency of pathozone inoculum for germination. Reported values have been high, often close to 1.0 (see e.g. Griffin, 1969; Short & Lacy, 1974; Tomimatsu & Griffin, 1982). Differences in this component of efficiency have been shown for different infection courts, as for example by *Fusarium oxysporum* f. sp. *pisi* in seed and epicotyls of peas (Whalley & Taylor, 1976). Differences have also been shown amongst cultivars differing in resistance to the same pathogen (Short & Lacy, 1974). Few studies, however, have been reported on efficiency subsequent to germination. Recent work by Tomimatsu & Griffin

Table 10.1. Possible ratios for the estimation of infection efficiency of soil-borne plant pathogens

Ratios based on immediately sequential processes:
Pathozone propagules/Total propagules
Germinated propagules/Pathozone propagules
Propagules contacting host/Germinated propagules
Infections/Pathozone propagules
Lesions/Infections
Propagules/Lesions
Some derived ratios:
Infections/Pathozone propagules
Lesions/Total propagules
Propagules/Germinated propagules

(1982), Griffin & Tomimatsu (1983), and Stanghellini *et al.* (1983) elegantly redress the balance.

Tomimatsu & Griffin (1982) found the efficiency of pathozone inoculum of *Cylindrocladium crotalariae* for germination to be high (close to 1.0). The efficiency of inoculum for root lesions on peanuts (i.e. the number of lesions per propagule of inoculum that was close enough to infect a root), however, was very low (approximately 0.003). Similarly large differences between efficiency of inoculum for germination and for infection were reported by Stanghellini *et al.* (1983) for *Pythium aphanidermatum* on sugar beet. Efficiency of inoculum for infection may be expected to increase with the size of the unit of inoculum. Values close to 1.0 have been reported for *Sclerotium cepivorum* by Crowe *et al.* (1980), for *Sclerotium rolfsii* and petioles of sugar beet by Punja & Grogan (1981), and for *Rhizoctonia solani* on bean seeds by Henis & Ben-Yephet (1970).

Plant pathologists have wrestled for some time with the relationships between inoculum density, size and vigour of inoculum units, and the probability of infection. The terms inoculum potential and disease potential have been central to the discussion (Baker, 1965, 1978). Inoculum potential was originally identified closely with inoculum density but the term was redefined by Garrett (1956) as the energy of growth available for infection of the host at the surface of the host. The emphasis is shifted in this definition from the effect of the entire population of inoculum in a given volume of soil to the mean vigour of individual propagules. Disease potential is the ability of the host to contract disease (Grainger, 1956). The concept of inoculum potential has not attracted universal approval (Van der Plank, 1975; Wood, 1967). Wood (1967, p. 114) dismisses the term as meaning different things to different plant pathologists. Van der Plank (1975, p. 87) asserts that inoculum potential cannot be quantified, although he concedes (p. 85) that the widespread use of the term suggests the existence of 'some concepts that writers are trying to express.' The ability of units of inoculum to cause greater or lesser infection according to the nutritional status of the inoculum, the susceptibility of the host tissue, and the capacity of the environment to favour disease is, I believe, an important epidemiological concept. Moreover, the concept can be quantified by placing units of inoculum against an infection court, and measuring the amount of infection and/or disease that results (Henis & Ben-Yephet, 1970; Gilligan, 1980a). This is different from the relationship between disease on a population of plants and inoculum density which can properly be summarized by disease response curves (Baker, 1971; Gilligan, 1983b).

Inoculum potential, as discussed above, is closely related to efficiency of inoculum and infection. Inoculum potential and disease potential are also closely related to pathogenicity and resistance *sensu* Scott *et al.* (1980). Pathogenicity is the ability of a parasite to injure a host; resistance is the ability of a host to hinder a parasite (Scott *et al.*, 1980). Resistance and pathogenicity are matching terms. They are heritable variables that may vary qualitatively or quantitatively (Christ *et al.*, this volume). Examples of qualitative resistance and pathogenicity are relatively uncommon amongst soil-borne pathogens. However, quantitative variation in the expression of disease is frequent and can be resolved into components, *viz.* genotypic variance, environmental variance and genotype/environmental variance (Blanch *et al.*, 1981). It is probable that the environment and interaction components for inoculum potential are large for many soil-borne pathogens. Baker (1978) and Baker & Wijetunga (1981) have also emphasized the genotypic and environmental components of inoculum potential. I suggest that there is value in reconsidering the concepts of inoculum potential and disease potential as aspects of pathogenicity and resistance (*sensu* Scott *et al.*, 1980), respectively. Such reconsideration would offer a nominal, genetic complement to the consideration of efficiency of infection. A beginning is to be found, for example, in the work of Flentje (1970) on *Thanatephorus cucumeris*. Flentje (1970) showed that pathogenicity (for which he used the term virulence), was controlled by a large number of genes, most of which affected penetration growth, i.e. the initiation of infection, and only a small number affected growth within host tissue.

Secondary infection

Distinction between primary and secondary infection is heuristically simple: primary infection is that which is caused by initial inoculum in soil, and secondary infection is that which arises from other infections. In practice, however, the distinction is less clear. There is, nevertheless, an intuitive appeal in distinguishing between primary and secondary infection in relation to disease progress. Pathogens with no form of secondary infection within a single season are described as causing simple-interest disease, i.e. they are monocyclic; those with secondary infection are described as causing compound-interest disease, i.e. they are polycyclic. Van der Plank (1963) quite legitimately used Fusarium wilt to illustrate simple-interest disease, but from this illustration grew a mistaken impression that most soil-borne disease is simple interest in type (see the discussions of Pfender, 1982, and Huisman, 1982). Pfender (1982) also criticized the improper use of goodness-of-fit tests of disease progress data to distinguish between simple- and compound-interest disease. The model for disease progress should be selected on prior biological knowledge rather than upon goodness-of-fit to one of several mathematical models. Thus Pfender (1982) shows that the simple- rather than compound-interest model gives a better fit to data for *S. cepivorum* on onion, even though it is known that plant-to-plant spread occurs within a season (Crowe & Hall, 1980). The statistical treatment of disease progress curves of soil-borne pathogens is discussed elsewhere (Gilligan, 1983b).

So, caution is necessary in the selection of models for disease progress. More information is needed on the dynamic behaviour of pathogens on individual infection courts (Gilligan, 1980a), and between adjacent plants (Crowe & Hall, 1980). We need to look more carefully at what is meant by secondary infection and how it is measured. Lesions rather than infections are ordinarily observed in epidemiological investigations. The occurrence of secondary infection can be inferred from increase in lesion numbers over time in those diseases in which lesions frequently remain discrete. This evidence, however, is not unequivocal. New primary infections arise synchronously with secondary infections as roots grow and are exposed to more inoculum in soil. Further confusion arises with some relatively large units of inoculum which give rise to more than one lesion (Gilligan, 1980a).

Distinction between primary and secondary infection is confounded still further by growth of infections. Many soil-borne pathogens produce infections that, once initiated, may progess until all the contiguous host tissue is infected. These are characterized by the ectotrophic root-rots. This infection continues over time, is nutritionally independent of the initial inoculum, involves multiplication of the pathogen biomass, and ought properly to be considered as secondary infection. Hence, if we regard primary infection as originating from initial inoculum in soil, two types of secondary infection can be distinguished: (1) increase in number of infections; (2) increase in size of infected tissue. These would normally be estimated indirectly by increase in lesion number and increase in lesion size.

Further distinction within secondary infection can be made between reinfection of the same host and infection of another host (see also Wolfe, this volume). The consequences for spread of disease and effect on the host are likely to be quite different in each case (Gilligan, 1983b). The original concept of simple-interest disease treated the plant as a unit, with the consequence that almost all wilts could be regarded as causing diseases of this type. The not inconsiderable multiplication of fungal biomass within a single plant was overlooked. Attention is now shifting from the plant to the root as a unit in epidemiological studies. Thus, Tomimatsu & Griffin (1982) propose an infection rate for *C. crotalariae* that expresses the number of infections per unit length of root per unit time per unit of inoculum per g soil. This infection rate has the advantage of allowing for root growth over time which the classical apparent infection rates derived from the compound- and simple-interest models do not (but see Jeger, this volume). Much attention is being given to host growth (Buwalda *et al.* 1982) and to increase in size as well as number of infections (Smith & Walker, 1981; Walker & Smith, 1984).

Concluding remarks

I have attempted to identify the components of epidemics of soil-borne fungal pathogens with the object of assisting our understanding of the population dynamics of these organisms. The infection chain, appropriately revised for soil-borne fungi, is a useful concept upon which to order the components. Considerable experimental work has already been done on components of the infection chain. There is continuing need, however, to synthesize the results of work on survival, natural and artifically enhanced antagonism and infection into unifying theories of the epidemiology of soil-borne pathogens. Hypotheses such as, for example, the effect of competition on the dimensions of the pathozone (Rouse & Baker, 1978) must be tested. Gaps in our knowledge must be filled; e.g. to date much more is known about the dynamics of inoculum and primary infection than of secondary infection.

Study of the population genetics of soil-borne fungi is almost entirely untouched. Undoubtedly the complexity of the systems, including sexual and asexual genetic recombination, and the accompanying difficulties of experimentation have deterred investigation. A further deterrent is the paucity of examples in which monogenically inherited characters have been shown to have overriding influences on the fitness of these organisms. In particular, pathogenic race-specificity is comparatively uncommon amongst species of soil-borne pathogens. I hope that the increasing interest in epidemiology and the renaissance of interest in biological control will stimulate work on the population genetics as well as the dynamics of soil-borne pathogens.

References

Baker K.F. & Cook R.J. (1974) *Biological Control of Plant Pathogens*. Freeman, San Francisco.

Baker R. (1965) The dynamics of inoculum. In: *Ecology of Soil-Borne Plant Pathogens* (Ed. by K.F. Baker & W.C. Snyder), pp. 395–403 & 415–19. University of California Press, Berkeley.

Baker R. (1968) Mechanisms of biological control of soil-borne pathogens. *Annual Review of Phytopathology* **6**, 263–94.

Baker R. (1971) Analyses involving inoculum density of soil-borne plant pathogens in epidemiology. *Phytopathology* **61**, 1280–92.

Baker R. (1978) Inoculum potential. In: *Plant Disease: an Advanced Treatise*, Vol. II. (Ed. by J.G. Horsfall & E.B. Cowling), pp. 137–57. Academic Press, London.

Baker R. (1981) Ecology of the fungus *Fusarium*: competition. In: *Fusarium: Disease, Biology and Taxonomy* (Ed. by P.E. Nelson, T.A. Toussoun & R.J. Cook), pp. 245–9. Pennsylvania State University Press, University Park.

Baker R. (1983) State of the art: plant diseases. In: *Proceedings of the National Interdisciplinary Biological Control Conference* (Ed. by S.L. Battenfield), pp. 14–21. CSRS/USDA, Washington DC.

Baker R. & Drury R. (1981) Inoculum potential and soilborne pathogens: the essence of every model is within the frame. *Phytopathology* **71**, 363–72.

Baker R. & Wijetunga C. (1981) Population growth and regulation. In: *The Fungal Community* (Ed. by D.T. Wicklow & G.C. Carroll), pp. 249–62. Marcel Dekker, New York.

Benson D.M. & Baker R. (1974) Epidemiology of *Rhizoctonia solani* pre-emergence damping-off of radish: inoculum potential and disease potential interaction. *Phytopathology* **64**, 957–62.

Blanch P.A, Asher M.J.C. & Burnett J.H. (1981) Inheritance of pathogenicity and cultural characters in *Gaeumannomyces graminis* var *tritici*. *Transactions of the British Mycological Society* **77**, 391–9.

Bloomberg W.J. (1979) A model of damping-off and root rot of Douglas fir seedlings caused by *Fusarium oxysporum*. *Phytopathology* **69**, 74–81.

Brown M.E. and Hornby D. (1971) Behaviour of *Ophiobolus graminis* on slides buried in soil in the presence of wheat seedlings. *Transactions of the British Mycological Society* **56**, 95–103.

Buwalda J.G., Ross G.J.S., Stribley D.P. & Tinker P.B. (1982) The development of endomycorrhizal root systems. III. The mathematical representation of the spread of vesicular–arbuscular mycorrhizal infection in root systems. *New Phytologist* **91**, 669–82.

Byrt P.N. & Holland A.A. (1978) Infection of axenic *Eucalyptus* seedling with *Phytophthora cinnamomi* zoospores. *Australian Journal of Botany* **26**, 169–76.

Campbell C.L., Pennypacker S.P. & Madden L.V. (1980) Progression dynamics of hypocotyl rot of snapbean. *Phytopathology* **70**, 487–94.

Clark F.E. (1965) The concept of competition in microbial ecology. In: *Ecology of Soil-Borne Plant Pathogens* (Ed. by K.F. Baker & W.C. Snyder), pp. 339–45. University of California Press, Berkeley.

Cook R.J. & Baker K.F. (1983) *The Nature and Practice of Biological Control of Plant Pathogens*. American Phytopathological Society, St Paul.

Cook R.J. & Snyder W.C. (1965) Influence of host exudates on growth and survival of germlings of *Fusarium solani* f. *phaseoli* in soil. *Phytopathology* **55**, 1021–5.

Crowe F.J. & Hall D.H. (1980) Vertical distribution of sclerotia of *Sclerotium cepivorum* and host root systems relative to white rot of onion and garlic. *Phytopathology* **70**, 70–3.

Crowe F.J., Hall D.H., Greathead A.S., & Bagholt K.J. (1980) Inoculum density of *Sclerotium cepivorum* and the incidence of white rot of onions and garlic. *Phytopathology* **70**, 64–9.

Curl E.A. (1979) Suppression of *Rhizoctonia solani* and damping-off in cotton by mycophagous insects of the order Collembola. *Phytopathology* **69**, 526.

Dimond A.E. & Horsfall J.G. (1965) The theory of inoculum. In: *Ecology of Soil-Borne Plant Pathogens* (Ed. by K.F. Baker & W.C. Snyder), pp. 404–415. University of California Press, Berkeley.

Ferriss R.S. (1981) Calculating rhizosphere size. *Phytopathology* **71**, 1229–31.

Ferriss R.S. (1982) Relationship of infection and damping-off of soybean to inoculum density of *Pythium ultimum*. *Phytopathology* **72**, 1397–1403.

Flentje N.T. (1970) Genetical aspects of pathogenic and saprophytic behavior of soil-borne fungi: Basidiomycetes with special reference to *Thanatephorus cucumeris*. In: *Root Diseases and Soil-Borne Pathogens* (Ed. by T.A. Toussoun, R.V. Bega & P.E. Nelson), pp. 45–9. University of California Press, Berkeley.

Garrett S.D. (1956) *Biology of Root-Infecting Fungi*. Cambridge University Press, London.

Garrett S.D. (1960) Inoculum potential. In: *Plant Pathology: an Advanced Treatise*, Vol. III (Ed. by J.G. Horsfall & A.E. Dimond), pp. 23–56. Academic Press, New York.

Garrett S.D. (1965) Toward biological control of soil-borne plant pathogens. In: *Ecology of Soil-Borne Plant Pathogens* (Ed. by K.F. Baker & W.C. Snyder), pp. 4–17. University of California Press, Berkeley.

Garrett S.D. (1970) *Pathogenic Root-Infecting Fungi*. Cambridge University Press, London.

Garrett S.D. (1979) The soil-root interface in relation to disease. In: *The Soil-Root Interface* (Ed. by J.L. Harley & R.S. Russell), pp. 301–13. Academic Press, New York.

Gaümann E. (1951) *Pflanzliche Infektionslehre*. 2nd ed. Birkhauser, Basel.

Gilligan C.A. (1979) Modeling rhizosphere infection. *Phytopathology* **69**, 782–4.

Gilligan C.A. (1980a) Dynamics of root colonisation by the take-all fungus, *Gaeumannomyces graminis*. *Soil Biology & Biochemistry* **12**, 507–12.

Gilligan C.A. (1980b) Inoculum potential of *Gaeumannomyces graminis* var. *tritici* and disease potential of wheat roots. *Transactions of the British Mycological Society* **75**, 419–24.

Gilligan C.A. (1980c) Zone of potential infection between host roots and inoculum units of *Gaeumannomyces graminis*. *Soil Biology & Biochemistry* **12**, 513–14.

Gilligan C.A. (1982) Statistical analysis of the spatial pattern of *Botrytis fabae* on *Vicia faba*: a methodological study. *Transactions of the British Mycological Society* **73**, 193–200.

Gilligan C.A. (1983a) A test for randomness of soilborne infection. *Phytopathology* **73**, 300–3.

Gilligan C.A. (1983b) Modeling of soilborne pathogens. *Annual Review of Phytopathology* **21**, 45–64.

Gilligan C.A. (1985) Probability models for host infection by soilborne fungi. *Phytopathology* **75**, 61–7.

Grainger J. (1956) Host nutrition and attack by fungal parasites. *Phytopathology* **46**, 445–56.

Griffin G.J. (1969) *Fusarium oxysporum* and *Aspergillus flavus* spore germination in the rhizosphere of peanut. *Phytopathology* **59**, 1214–18.

Griffin G.J. & Tomimatsu G.S. (1983) Root infection pattern, infection efficiency, and infection density-disease incidence relationships of *Cylindrocladium crotalariae* on peanut in field soil. *Canadian Journal of Plant Pathology* **5**, 81–8.

Grogan R.G., Salt M.A. & Punja Z.K. (1980) Concepts for modeling root infection by soil-borne fungi. *Phytopathology* **70**, 361–3.

Hau F.C., Campbell C.L. & Beute M.K. (1982) Inoculum distribution and sampling methods for *Cylindrocladium crotalariae* in a peanut field. *Plant Disease* **66**, 568–71.

Henis Y. & Ben-Yephet Y. (1970) Effect of propagule size of *Rhizoctonia solani* on saprophytic growth, infectivity and virulence on bean seedlings. *Phytopathology* **60**, 1351–6.

Hornby D. (1975) Inoculum of the take-all fungus: nature, measurement, distribution and survival. *EPPO Bulletin* **5**, 319–33.

Hornby D. (1981) Inoculum. In: *Biology and Control of Take-All* (Ed. by M.J.C. Asher & P.J. Shipton), pp. 133–56. Academic Press, New York.

Huisman O.J. (1982) Interrelations of root growth dynamics to epidemiology of root-invading fungi. *Annual Review of Phytopathology* **20**, 303–27.

Jeger M.J. (1982) The relation between total, infectious, and postinfectious diseased plant tissue. *Phytopathology* **72**, 1185–9.

Johnson L.F., Hsieh C.-C. & Sutherland E.D. (1981) Effects of exogenous nutrients and inoculum quantity on the virulence of *Pythium ultimum* to cotton hypocotyls. *Phytopathology* **71**, 629–32.

Kranz J. (1974) Introduction. In: *Epidemics of Plant Disease: Mathematical Analysis and Modeling* (Ed. by J. Kranz), pp. 1–6. Springer-Verlag. Berlin.

Mehta Y.R. & Zadoks J.C. (1970) Uredospore production and sporulation period of *Puccinia recondita* f. sp. *triticina* on primary leaves of wheat. *Netherlands Journal of Plant Pathology* **76**, 261–76.

Parlevliet J.E. (1979) Components of resistance that reduce the rate of epidemic development. *Annual Review of Phytopathology* **17**, 203–22.

Parmeter J.R. (1970) Mechanisms of variation in culture and in soil. In: *Root Diseases and Soil-Borne Pathogens* (Ed. by T.A. Toussoun, R.V. Bega & P.E. Nelson), pp. 63–8. University of California Press, Berkeley.

Pfender W.F. (1982) Monocyclic and polycyclic root diseases: distinguishing between the nature of the disease cycle and the shape of the disease progress curve. *Phytopathology* **72**, 31–2.

Pfender W.F., Rouse D.I. & Hagedorn D.J. (1981) A 'most probable number' method for estimating inoculum density of *Aphanomyces euteiches* in naturally infested soil. *Phytopathology* **71**, 1169–72.

Phillips D.J. (1965) Ecology of plant pathogens in soil. IV. Pathogenicity of macroconidia of *Fusarium roseum* f. sp. *cerealis* produced on media of high and low nutrient content. *Phytopathology* **55**, 328–9.

Punja Z.K. & Grogan R.G. (1981) Mycelial growth and infection without a food base by eruptively germinating sclerotia of *Sclerotium rolfsii*. *Phytopathology* **71**, 1099–103.

Roth D.A. & Griffin G.J. (1981) Cylindrocladium root rot of black walnut seedlings and inoculum pattern in nursery soils. *Canadian Journal of Plant Pathology* **3**, 1–5.

Rouse D.J. & Baker R. (1978) Modeling and quantitative analyses of biological control mechanisms. *Phytopathology* **68**, 1297–302.

Scher F.M. & Baker R. (1983) Effect of *Pseudomonas putida* and a synthetic iron chelator on induction of soil suppressiveness to Fusarium wilt pathogens. *Phytopathology* **73**, 1567–73.

Schlub R.L., Lockwood J.L. & Komada H. (1981) Colonisation of soybean seeds and plant tissue by *Fusarium* species in soil. *Phytopathology* **71**, 693–6.

Scott P.R., Johnson R., Wolfe M.S., Lowe H.J.B., & Bennett F.G.A. (1980) Host-specificity in cereal parasites in relation to their control. In: *Advances in Applied Biology*, Vol. V (Ed. by T.H. Coaker), pp. 349–93. Academic Press, New York.

Sewell G.W.F. (1959) Direct observation of *Verticillium albo-atrum* in soil. *Transactions of the British Mycological Society* **42**, 312–21.

Short G.E. & Lacy M.L. (1974). Germination of *Fusarium solani* f. sp. *pisi* chlamydospores in the spermosphere of pea. *Phytopathology* **64**, 558–62.

Smith S.E. & Walker N.A. (1981) A quantitative study of mycorrhizal infection in *Trifolium*: separate determination of rates of infection and of mycelial growth. *New Phytologist* **89**, 225–40.

Stanghellini M.E. & Hancock J.G. (1971) Radial extent of bean spermosphere and its relation to the behaviour of *Pythium ultimum*. *Phytopathology* **61**, 165–68.

Stanghellini M.E., Stowell C.E., Kronland W.C. & von Bretzel P. (1983) Distribution of *Pythium aphanidermatum* in rhizosphere soil and factors affecting expression of the absolute inoculum potential. *Phytopathology* **73**, 1463–6.

Stanghellini M.E., von Bretzel P., Kronland W.C. & Jenkins A.D. (1982) Inoculum densities of *Pythium aphanidermatum* in soils of irrigated sugar beet fields in Arizona. *Phytopathology* **72**, 1481–5.

Taylor J.D., Griffin G.J. & Garren K.H., (1981) Inoculum pattern, inoculum density–disease incidence relationships, and populations of *Cylindrocladium crotalariae* microsclerotia in peanut field soil. *Phytopathology* **71**, 1297–302.

Tomimatsu G.S. & Griffin G.J. (1982) Inoculum potential of *Cylindrocladium crotalariae*: infection rates and microsclerotial density – root infection relationships on peanut. *Phytopathology* **72**, 511–17.

Van der Plank J.E. (1963) *Plant Diseases: Epidemics and Control*. Academic Press, New York.

Van der Plank J.E. (1975) *Principles of Plant Infection*. Academic Press, New York.

Walker N.A. & Smith S.E. (1984) The quantitative study of mycorrhizal infection. II. The relation of rate of infection and speed of fungal growth to propagule density, the mean length of the infection unit and the limiting value of the fraction of the root infected. *New Phytologist* **96**, 55–69.

Wallace H.R. (1978) Dispersal in time and space. In: *Plant Disease: an Advanced Treatise*, Vol. II (Ed. by J.G. Horsfall & E.B. Cowling), pp. 181–202. Academic Press, New York.

Weste G. (1983) Population dynamics and survival of *Phytophthora*. In: *Phytophthora: its Biology, Taxonomy, Ecology and Pathology* (Ed. by D.C. Erwin, S. Bartnicki-Garcia & P.S. Tsao), pp. 237–57. American Phytopathological Society, St Paul.

Whalley W.M. & Taylor G.S. (1976) Germination of chlamydospores of physiologic races of *Fusarium oxysporum* f. *pisi* in soil adjacent to susceptible and resistant pea cultivars. *Transactions of the British Mycological Society* **66**, 7–13.

Wood R.K.S. (1967) *Physiological Plant Pathology*. Blackwell Scientific Publications, Oxford.

Zadoks J.C. (1972) Modern concepts of disease resistance in cereals. In: *The Way Ahead in Plant Breeding* (Ed. by F.G.H. Lupton, G. Jenkins & R. Johnson), pp. 89–98. Proceedings 6th Congress of Eucarpia, Cambridge.

Zadoks J.C. & Schein R.D. (1979) *Epidemiology and Plant Disease Management*. Oxford University Press, Oxford.

11 The population dynamics of plant virus diseases

J.M. THRESH
Overseas Development Administration
Maidstone, Kent ME19 6BJ, UK

Introduction

Demographic studies are important in seeking to explain the abundance and distribution of plants and animals. This explains why such studies have long featured so prominently in ecology, whether the approach has been practical or theoretical. The widely used logistic equation of Verhulst (1804–1849) and the concept of habitats having a finite carrying capacity, can be traced through Charles Darwin (1809–1882) and Thomas Malthus (1766–1834) to even earlier studies of human mortality statistics in 16th-century London (Hutchinson, 1978). There is no equivalent tradition in studying the population dynamics of plant pathogens, and the main emphasis has until recently been on taxonomy, aetiology, life cycles and control. The 1963 book of Vanderplank was of seminal importance in stimulating an entirely new approach to collecting, analysing and interpreting data on plant disease progress.

An objective of recent work has been to relate studies of the population dynamics of plant pathogens to those of free-living organisms and of pathogens and parasites of animals. Practical investigations have progressed alongside theoretical studies intended to provide an overall conceptual framework (Jeger, this volume). Moreover, it is now apparent that co-evolutionary relationships between hosts and their parasites or pathogens are analogous to those between plants and herbivores or predators and prey.

Much of Vanderplank's 1963 book concerned virus diseases, whereas his later volumes and the ensuing discussion have mainly related to fungus diseases. There is continuing debate on the reality and concepts of 'horizontal' and 'vertical' resistance to phytopathogenic fungi, on the effectiveness and durability of such resistance, and on overall breeding strategies (Johnson, this volume). This mycological bias reflects the continuing preoccupation of plant virologists with aetiology and taxonomy, the current interest in the molecular biology of viruses, and the limited efforts made until recently in breeding for virus resistance.

Plant virus epidemiology has been unduly neglected and a reassessment of current attitudes is required to correct the present imbalance (Thresh, 1983a). Initial steps have been to consider the topic in the wider contexts of plant pathology, comparative epidemiology and ecology, and to assess the extensive literature on

Wolfe M.S. & Caten C.E. (1987) *Populations of Plant Pathogens: their Dynamics and Genetics.* Blackwell Scientific Publications, Oxford.

progress curves of plant virus diseases (Thresh, 1974a, 1978, 1980, 1983b). This chapter adds to the series of reviews, with continuing emphasis on a comparative, ecological approach rather than on mathematics.

Viruses as plant pathogens

Charles Darwin wrote in *The Origin of Species* 'we see beautiful adaptations everywhere and in every part of the organic world'. His comment arose from observations on bird ectoparasites, but applies equally to plant pathogens which are now known to comprise fungi, bacteria, algae, protozoa, rickettsias, mycoplasmas, spiroplasmas, viruses and viroids.

Viruses and viroids are exclusively obligate, with the simplest structural and physicochemical features of all pathogens, yet they resemble true living organisms in displaying great diversity, versatility and adaptability in exploiting varied habitats and different modes of perennation and spread. Some notable features of plant viruses are listed below and many are also characteristic of viroids. The latter are a separate group of submicroscopic systemic pathogens distinguished from viruses by their mode of replication, smaller genome and lack of coat or other protein components.

1. Plant viruses have a limited ability to enter intact host cells and mainly depend on animal or fungus vectors to gain entry.

2. Viruses can invade many of their host plants systemically from initial entry sites. Infections that remain localized in the roots or elsewhere seldom have very deleterious effects.

3. Viruses usually persist throughout the life of systemically infected plants and so pass to vegetative propagules, and in some instances, to a proportion of the pollen and seed. Viruses are seldom eliminated naturally from systemically infected plants and there is no recovery phenomenon equivalent to the immune response of mammals.

4. Few viruses remain viable for long outside living tissues and their survival is largely dependent on a continuous sequence of hosts. Thus there are likely to be close co-evolutionary relationships between viruses and their host plants. This possibility has received little attention, although it is apparent that plant viruses are seldom lethal.

5. Some plant viruses can infect their insect vectors and so persist for long periods in vector populations, and in some instances through many generations, even in the absence of susceptible host plants.

6. Viruses as obligate intracellular parasites largely avoid competition with other micro-organisms. However, important interactions occur between dissimilar viruses or between different strains of the same virus. 'Dependent' viruses are not transmissible by vectors in the absence of a distinct 'assistor' virus and a few 'satellite' viruses only multiply in the presence of an appropriate unrelated 'helper' virus.

7. Viruses can be spread widely by vectors or other means and sometimes

influence plant populations far beyond the usual range of gene flow spanned by pollen or seed.

8. Differences in host range and in the susceptibility or response of plants to infection suggest that viruses act as selective, discriminating pathogens exerting important ecological effects on competition within and between species in plant communities.

9. The severity of the damage caused by viruses is seldom closely related to virus content. Some viruses reach uniformly high concentrations yet have barely detectable effects, whereas others that occur in much smaller amounts or mainly in phloem tissue are extremely damaging.

These distinctive features of plant viruses and viroids explain their great economic importance. They also influence the 'natural history' of viruses and the population dynamics of the diseases they cause. Before discussing these topics it is appropriate to consider some aspects of crop plants and the habitats they provide.

Host populations

Biologists considering the population dynamics of animal pests or weeds of crops can utilize the experience and concepts derived by ecologists studying natural habitats. Crop pathologists seldom have this advantage because there have been few quantitative studies of fungal pathogens in natural stands of the type discussed by Dinoor (this volume). There is even less information on bacteria and viruses, which is a serious limitation of the literature. It has led to a biased 'agrocentric' perspective in plant pathology that has until recently largely overlooked the profound differences between crops and wild vegetation (Browning, 1981). These differences are discussed by Burdon & Shattock (1980), who emphasize how many crops are planted in monoculture and more or less synchronously in dense, regular arrays of uniform genotype.

Virtually all published data on virus spread were obtained from such plantings, with the emphasis on a few particularly important and mainly arable crops of temperate regions. For this reason and because virus incidence is usually expressed in percentages, there is a tendency to ignore the great differences between crops in plant population and in the duration and type of habitat they provide.

Longevity

Crops grown from seed may be harvested within only a few weeks (e.g. radish, lettuce) or months (e.g. cereals). In these circumstances there is only a brief opportunity for viruses to become prevalent and the only ones likely to be damaging are those that can spread rapidly or persist very effectively between successive crops. The opportunity for spread is greater in biennial crops and greatest with long-lived woody perennials. These include tree crops such as cocoa and coconut, in which the time scale of epidemics is relatively unimportant. They can develop over many years and ultimately cause serious damage even though annual spread is slow and mainly localized (Thresh, 1983b).

Vegetative propagation

The importance of longevity has been greatly enhanced by the widespread adoption of vegetative propagation to provide planting material of many important crops. This long-established practice may involve inter-grafting rootstocks and scions of different species or variety and is likely to become increasingly important with an extension of *in vitro* culture and micro-propagation techniques.

Vegetative propagation greatly facilitates crop establishment and the exploitation of select genotypes, but the epidemiological implications are profound. This is because grafting is a very effective means of transmitting viruses and of creating new virus combinations. Moreover, viruses and their vectors can be widely disseminated with vegetative propagules into entirely new regions and over distances far greater than those traversed naturally. Virus-infected plants so introduced into plantings become particularly dangerous foci of infection as they occur from the outset and tend to be randomly scattered. Furthermore, viruses may build-up over successive cycles of propagation, leading to the progressive degeneration of stocks that has long been familiar to potato and temperate fruit growers, and also occurs with such tropical crops as avocado, banana, cassava, citrus, yams, sweet potato and taro.

Viruses can ultimately become prevalent in vegetatively-propagated species, even if individual plantings are of short duration and virus spread is slow. Inevitably there has been selection of virus-tolerant genotypes that grow and yield satisfactorily despite virus infection. Selection has occurred unwittingly with many crops and some fruit cultivars have been grown successfully for over 100 years, even though all available clones are totally infected. The practice of replacing degenerated stocks of potato with healthier ones from less severely affected areas also emerged empirically and led to many subsequent certification schemes.

Mobility

Those comparing the epidemiology of plant and animal pathogens emphasize the immobility of rooted plants compared with the mobility and complex social behaviour of higher animals. This attitude is fully justified, as the differences are indeed great and have a crucial impact on virus spread. Nevertheless, there is a tendency to underestimate the epidemiological significance of plant movement, whether this occurs naturally or due to man.

Many viruses can be disseminated efficiently and far, in or on the seed of weeds or wild plants and this is particularly important in the epidemiology of some viruses with slow-moving nematode vectors (Murant, 1981). Furthermore, there is an extensive and expanding traffic in crop seed, seedlings or vegetative propagules due to the activities of commercial growers, international companies or research organizations, and to the use of specially favoured regions to produce material for use elsewhere.

Another aspect of plant mobility is the use of seed beds or nurseries to raise plants for transplanting to wider spacings at cropping sites that are sometimes far away. Viruses can spread at either location and new or extended opportunities occur as the plants are handled or when they become mixed during collection, handling, grading, distribution and transplanting. This disrupts any groups of infected plants so that new contacts are established with healthy neighbours and further spread is facilitated. Transplants are used widely to make the most effective use of the land, irrigation water or growing seasons available. This may permit sequential cropping or involve raising plants in glasshouses or plastic structures before transfer outside. Such practices are becoming increasingly widespread and further exacerbate disease problems (Thresh, 1982).

Growing seasons

Crops may be grown in continuous overlapping sequence where the climate is suitable throughout the year. This enables viruses and their vectors to become endemic, in the sense that they are always present and encounter few problems of survival or perennation. Elsewhere, inadequate rainfall or low temperatures lead to seasonal growth with distinct and sometimes prolonged breaks between successive plantings. Virus spread is then greatly restricted and the severity of the dry season or of the winters at higher latitudes has a crucial influence on the prevalence of many vector-borne viruses (Thresh, 1986a).

Seasonal crops that are only briefly accessible may largely escape infection, or become infected so late that losses are insignificant. Any serious outbreaks that develop are likely to be due to viruses with exceptional epidemiological 'competence', using this term for the ability of pathogens to sustain the continuous sequence of infection necessary for their survival (*sensu* Crosse, 1967). This can be due to local perennation or the capacity to colonize and then exploit new habitats quickly and sometimes from afar. The degree of competence required becomes less when the natural growing season is prolonged or extended by irrigation or some form of protected cropping (Thresh, 1982).

Strategies of virus spread

Considering the great range of crop species and habitats, it is hardly surprising that there is equivalent diversity amongst viruses in host range, means of spread and perennation. Some viruses are 'specialists' in an ecological sense, as they have a very restricted host range amongst a few closely related species, or they are transmitted by few vector species. Other viruses are 'generalists' with a very wide host range, or they are transmitted by many different vectors. Specialists are dependent on a few species of host or vector, but have more opportunity than generalists to become closely adapted to particular hosts or vectors in ways that facilitate survival.

Similar considerations could account for the great differences between viruses in mode of spread by contact, through pollen and/or seed and by animal or fungal vectors. No one virus is known to be transmitted by all these routes and it is unlikely that the small viral genome could carry sufficient information for this to occur. Thus each virus exploits only one or a combination of some of the possible modes of spread to achieve an effective distribution of inoculum over short distances within plantings and also over greater distances. The actual methods employed greatly influence the pattern and sequence of spread, and hence the overall dynamics of disease progress. They are discussed elsewhere (Thresh, 1974b, 1978, 1983b, 1986b) and are not considered further here.

Disease progress curves

In all demographic studies there are inevitable problems of mensuration as it is necessary to determine what, where and how to measure, and it may be appropriate to distinguish subclasses within populations, or to assess biomass or energy flow. Nevertheless, it is usually possible to plot some absolute measure of population size on a suitable time scale and estimate the carrying capacity of the habitat.

Various procedures are used to follow disease progress, including direct counts of bacteria, spores, pustules or lesions and estimates of the amount or proportion of tissue affected. There is often a general relationship between disease severity and the amount of pathogen present, but such relationships are seldom evident in studying virus diseases. An additional difficulty is that virus content may be difficult or impossible to assess. The available techniques include serology, electron microscopy, infectivity assays and various biochemical procedures. However, all have limitations and are inappropriate for following changes in virus populations in the field. Thus it is seldom possible to do other than record the number or proportion of virus-affected plants as determined by visual inspection for obvious symptoms. This can be misleading because plants infected late are seldom damaged severely and may be invaded less completely or become less infectious than those infected early. Inevitably there is little information on the spread of virus diseases causing symptoms that are inconspicuous or difficult to diagnose. Moreover, the published data mainly relate to a few notable diseases, and some were obtained at atypical sites where spread was in some way exceptional.

In presenting curves of virus disease progress, the cumulative incidence of disease (X) is usually plotted as a percentage of the total stand against time on the horizontal axis expressed as days, weeks, months or years, according to the longevity of the crop and the cultural practices adopted. Little use has been made of non-arithmetic time scales or physiological units such as heat summations or day-degrees above critical thresholds, although temperature greatly influences virus spread, virus content, vector activity and symptom expression. Moreover, early spread is of disproportionately great importance compared with that occurring later, when plants tend to be less vulnerable and there is only limited opportunity for further spread.

The use of a percentage scale to express disease incidence is convenient, but conceals great differences between close and widely spaced crops in the total number of successful infections required to cause a serious epidemic. A percentage scale also sets an upper limit to spread, which in ecological terms represents the carrying capacity of the habitat, and many curves of disease progress rise to an asymptote at this level (see also Jeger, this volume). 'Multiple infection' becomes increasingly important as increasing amounts of inoculum reach plants already infected and lead to disease 'saturation'. However, spread can be retarded or halted earlier if conditions become unfavourable or plants acquire 'mature plant resistance'.

Many curves of plant virus disease progress are initially concave if observations are begun sufficiently early. This phase is often referred to as 'logarithmic', although evidence is seldom presented that spread is truly autocatalytic and equivalent to the increase of capital by compound interest. Indeed, it is doubtful if such spread ever occurs, because it is restricted from the outset by multiple infection and by the limited distances over which most inoculum is distributed. These constraints are obviously more important with viruses and other pathogens causing systemic diseases than with those causing discrete pustules or lesions due to widely dispersed wind-borne inoculum. Thus many viruses cause 'crowd' diseases that do not spread far in any considerable amount (Vanderplank, 1948). They tend to occur in discrete patches that are a prominent feature of many virus diseases and not only those with slow-moving, soil-inhabiting vectors (Thresh, 1983b).

These considerations largely explain the sigmoid shape of many virus disease progress curves, as the amount of spread occurring in unit time increases to a maximum and then declines. There are numerous published examples relating to diverse crops, regions, time-scales and viruses (Thresh, 1974a, 1980, 1983b). In a few instances spread is due entirely to an influx of infective vectors from outside sources with no subsequent spread within crops. Such 'monocyclic' spread can be analysed in terms of 'simple interest' (Vanderplank, 1963). However, infected plants usually become infectious and lead to further spread which is therefore 'polycyclic' and often considered in terms of the equivalent 'compound interest' equation.

A feature of many polycyclic diseases is the rapidity of spread in herbaceous annuals compared with woody perennials. This can be explained largely by the overall resistance of woody plants to infection and by the prolonged duration of each infection cycle. There is an interval of months or even years between a successful inoculation of a tree or shrub and the opportunity for further spread. The equivalent period for herbaceous annuals is days or weeks, and many cycles of infection are possible within a single growing season, even if this is of limited duration or the crop is only briefly vulnerable. For example, rice tungro viruses are detectable in young plants within a few days of inoculation and can then be acquired and transmitted efficiently by their leafhopper vectors within hours. The infection cycle is longer with rice dwarf and several other viruses that propagate in their vectors, because of the temperature-dependent incubation period of days or weeks before viruliferous insects become infective. However, the total duration of each infection cycle is far less than the year or more that elapses before mites can transmit reversion virus from blackcurrant bushes. There are many other such examples, and the mechanisms whereby woody perennials restrict the damaging effects of viruses are of obvious survival value and have been discussed in co-evolutionary terms (Vanderplank, 1949; Thresh, 1983b).

Many of the great differences encountered in the amount and sequence of polycyclic spread are explicable in terms of three basic parameters: (i) the number of cycles of infection possible during the life of the crop (N); (ii) the rate of spread or average number of plants infected by an infectious individual during each cycle (R); and (iii) the initial amount of inoculum from which subsequent spread occurs (X_0).

The significance of X_0 has been discussed by Vanderplank (1963), Zadoks & Schein (1979) and Thresh (1983b), and a term comparable to R was used by Anderson (1981) who stressed the requirement that in the long term it must equal or exceed 1 to maintain the continuity of infection. R is particularly appropriate for use with viruses of mammals because of the distinct cycles of infection and generally brief period of viraemia during which spread can occur from infected individuals before death or recovery. However, the general concept is also valid for use with viruses of higher plants, and there are obviously great differences between diseases in overall values of R, even though they are difficult to quantify. Another problem is that for any one disease, R and the duration of the infection cycle are influenced by such factors as temperature, host cultivar, and the stage of growth when infection occurs.

Much spread is inevitable if the combined effects of N, R and X_0 are adequate, but particularly large values of any one parameter can compensate for small values of the others. Thus, viruses can be prevalent mainly because X_0 is large due to the use of infected planting material, and some vegetatively propagated crops are totally infected from the outset. Similarly, much spread can occur even if both X_0 and R are small, provided that individual crops are long-lived or stocks are propagated vegetatively to give large values of N. There are also instances of rapid rates of spread from a few initial foci in situations where X_0 and N are small. Clearly, epidemics occur for different reasons and various approaches to control are possible.

There are obvious advantages in delaying the onset of disease by using healthy propagules, or roguing, or by selecting sites where sources of infection are absent or have been removed. Such sanitation measures decrease X_0, and when totally effective decrease N by delaying disease onset, so shortening the total period over which spread can occur. Sanitation can be very effective even if there is little impact on final disease incidence, as crop losses are often related to the area under the disease progress curve rather than to the total amount of spread occurring (Thresh, 1983b).

Another means of decreasing N is by early harvesting or by otherwise terminating growth prematurely, as widely used in producing seed potato stocks, despite the yield penalty incurred. An alternative is to breed varieties that support little multiplication and are slow to become infectious, even if they have little resistance to primary infection. Short-duration varieties also decrease N, although their use can create further disease problems if X_ϕ is increased where additional crops are taken by planting in close or overlapping sequence, as in rice and fresh vegetable production.

R can be decreased by breeding plants with resistance to viruses or their vectors. Another possibility is to adopt planting dates that avoid exposing plants at a highly vulnerable stage of growth during the main period of vector activity (Hill, this volume). Other methods utilize barrier crops, intercrops, or mineral oils to impede spread, or pesticides or repellents to control vectors or decrease their activity.

Direct methods of controlling viruses with chemotherapeutants are not yet available. Thus, virologists have usually had little option but to integrate cultural practices with any other measures available in seeking to decrease crop loss. Many diseases are controlled satisfactorily in this way, whereas with many insect pests and fungus diseases far greater use has been made of either pesticides or resistant varieties. There is now general recognition of the need to consider the dynamics of plant pests and pathogens and to study their ecology and distribution in seeking integrated methods of control. Such analyses can provide new insights and improve the performance of resistant varieties or other possible control methods.

Virus strains and variation

Harper (1982) discussed contrasting approaches to population biology and emphasized how taxonomists seek stable, readily recognized features to delineate species, whereas ecologists are more concerned with the intraspecific variation that leads to important differences within and between populations. He also stressed that lists and maps of species distribution can be misleading in ignoring polymorphisms within populations and ecotypic differences between them.

The importance of intraspecific variation has been apparent in plant pathology since the introduction of varieties resistant to black stem rust of wheat and the recognition of resistance-breaking races of the pathogen (Caten, this volume). The use of differential hosts or other means of distinguishing variants has since been extended to many other fungal and bacterial pathogens and has been given added impetus with increased use of resistant varieties and the appearance of pesticide-resistant strains, as discussed by several contributors to this volume. Thus the importance of studying the qualitative as well as quantitative aspects of pathogen populations has become widely recognized, although difficulties have arisen in defining and using such terms as subspecies, strain, race, form, variety and biotype (Caten, this volume).

Even greater difficulties have emerged in virology because the species concept has not been widely accepted, and there is no general agreement on the status of the wide range of virus variants that can be isolated from many field populations. Variation is often apparent from differences in such biological features as symptom expression or host range, and in transmission by vectors or through seed. However, methods based on serology or other physicochemical properties of viruses are being used increasingly and some of the available techniques are far quicker and better suited for processing numerous samples than those that involve inoculating whole plants and awaiting the response.

The epidemiological importance of strain variation has long been recognized, as in studies on surgarcane mosaic and sugar beet curly top viruses in the western United States. Giddings (1948) showed that early each growing season the predominant strains of curly top were avirulent to beet and did not kill the overwintering weed hosts of the virus and its leafhopper vector. Later in the season the strains isolated from beet were far more aggressive. They had an advantage over less virulent strains in having such drastic effects on crop growth that stands became sparse and therefore particularly favourable for the growth and reproduction of the vector, which thrives only in exposed habitats (Bennett, 1963). These opposing trends explain the continued occurrence of both types of strain and seasonal shifts in their relative abundance. However, the latest studies indicate that the situation has changed with the occurrence of much more virulent strains following altered land use and the widespread cultivation of beet varieties resistant to curly top (Magyarosy & Duffus, 1977).

Seasonal and other changes in the prevalence of strains have also been recorded in studying the incidence of cucumber mosaic virus in vegetable crops and their weeds in south-east France (Quiot *et al.*, 1979). The strains that predominated in spring-sown crops of tomato, celery and pepper were much more sensitive to heat than those isolated from equivalent summer plantings.

Other complex interactions between the virulence of strains and host response occur with barley stripe mosaic virus, which is transmitted mainly through seed and by direct contact between plants. Sensitive barley varieties infected with virulent strains produce a large proportion of infected seed, although few seeds develop and they tend to be small with poor viability. By contrast, tolerant varieties produce abundant viable seed but few are infected. Thus the virus strains that persist readily for successive generations tend to be of intermediate virulence that infect substantial proportions of seed without drastic effects on fertility or on contact between plants (Timian, 1974). There is a somewhat similar situation with rice stripe virus, as highly virulent strains that soon kill the host plant are less likely to be acquired and transmitted transovarially by the plant-hopper vector than less virulent strains (Kisimoto, 1972).

Variation in other characteristics has been encountered in work on the luteovirus or viruses causing yellow dwarf disease of cereals (Hill, this volume) and there is no agreement on the taxonomic status of the five types of isolate described. Some differ serologically, in their effects on growth and yield and in their pattern of spread in the field. Four are transmitted specifically by different aphid species, whereas the fifth is transmitted non-specifically by two different species. A 20-year study in New York State revealed a marked shift from a specific to a non-specifically transmitted type (Rochow, 1979). Site, regional and seasonal differences have also been reported in the predominant type and these have sometimes been related to changes in the aphid fauna.

These examples illustrate the complex factors influencing the overall epidemiological competence of viruses and virus strains. However, much of the evidence on variation has come from 'taxonomic' or aetiological studies intended to provide definitive descriptions of viruses and to determine their interrelationships. There have been few comprehensive surveys of the prevalence and distribution of the various strains of even the most important viruses. This is a serious limitation in attempts to estimate crop loss, or breed for disease resistance, as there can be great differences between strains in their ability to infect or in their effects, with important interactions between varieties and strains.

One reason why there have been few virus surveys comparable to those of cereal mildew and rust races (Wolfe, this volume; Groth & Roelfs, this volume) is that in work on systemic pathogens it is seldom possible to adopt quick-acting host differentials or convenient detached-leaf techniques. Moreover, less attention has been given to breeding for resistance to viruses compared with equivalent studies on fungal pathogens. Thus it has not often been necessary to determine the range of virus strains against which resistance must be sought and the emergence and spread of resistance-breaking strains has seldom had to be monitored.

Nevertheless, experience with sugarcane mosaic in Louisiana and soybean mosaic in Korea illustrates the rapidity with which resistance-breaking strains can become prevalent following the introduction of resistant varieties (Summers *et al.*, 1948; Cho *et al.*, 1977). Similarly rapid changes in the predominant strain of tomato mosaic virus occurred in England following the widespread use of resistant varieties (Pelham *et al.*, 1970), and later with the introduction of mild strain protection (Fletcher & Butler, 1975). These changes were monitored using isogenic tomato lines, and suggested that the resistance-breaking strain did not compete successfully with the former one in susceptible varieties and declined rapidly when the resistant varieties were replaced.

Similarly, the limited distribution of the resistance-breaking strain of raspberry ringspot virus in Scotland may be due to its limited epidemiological competence, as it is less invasive and less readily seed transmitted than the common strain (Hanada & Harrison, 1977). This and evidence of the apparent inability of resistance-breaking strains of potato virus X to become established in Britain led Harrison (1981) to suggest that the range of biologically effective virus variants may be limited by the small coding capacity of their genomes, and also by the likelihood that many individual gene products have more than one function, so that a mutant gene product must perform all such roles. Harrison also suggested that the dissemination of resistance-breaking strains of virus by vectors may be restricted by the small numbers of vectors that disperse compared with air-borne fungal spores. This comparison is not altogether valid, and overlooks the very large number of virus particles that may be carried by each individual vector and the ability of some vectors to infect many plants and in some instances sustain virus replication throughout their life.

Whatever the reason, the durability of resistance to tobacco mosaic and several other virus diseases suggests that problems due to breakdown will be less than with many diseases caused by fungal pathogens. However, generalizations are premature, as examples of breakdown have occurred and they are all due to virus strains that invade, or damage more severely, cultivars that were previously resistant to or tolerant of infection. It is seldom established whether failure is due to the emergence or introduction of an apparently novel strain or to the increased prevalence of an existing one. Moreover, the type of host resistance adopted and the way it is deployed are not always apparent from publications, and yet this can greatly influence the quantity and type of inoculum available and the opportunity for virulent or resistance-breaking strains to appear and become established. The opportunity and the selection pressures operating can be very different as tolerant cultivars may support considerable multiplication of the usual strain whereas resistant, hypersensitive or immune cultivars do not. The performance and durability of resistant cultivars are also influenced by the degree of exposure to susceptible or tolerant ones and the rapidity with which these are replaced.

There is little evidence with virus diseases of the relative merits of single and multiple resistances, or of major and minor genes and the ease with which these are overcome. Moreover, strains of only three viruses have been considered in terms of the gene-for-gene hypothesis (Drijfhout, 1978; Jones, 1981; Pelham, 1972) and there is no information on the effectiveness of intraspecies mixtures of the type used against some fungal diseases (Wolfe, this volume).

Breeding for resistance has such obvious advantages that it is likely to become increasingly important as a means of avoiding or decreasing the losses due to virus diseases. This is already apparent from much of the current work at the Rice Research and other International Institutes situated in regions where sophisticated control measures are inappropriate and where there are limited opportunities for controlling vectors with pesticides. Thus, virus strain variation merits increased attention in developing the most appropriate strategy for deploying the host resistance genes available in attempts to achieve durable control. Similar studies will also be required if there is to be increased use of 'mild' strains to protect plants from the damaging effects of at least some virulent ones, as practised already against tomato mosaic and citrus tristeza viruses. Moreover, it will eventually be necessary to consider the possible emergence of resistant or tolerant strains in deciding how chemotherapeutants should be used most effectively.

Conclusions

The small size, lack of independent metabolic activity and simple genomic structure of viruses and viroids distinguish them from other pathogens and create difficulties in detection, identification and assay. Nevertheless, their occurrence and spread within plant or animal communities can be studied in ways comparable to those adopted in following the population dynamics of more elaborate pathogens, parasites and free-living organisms.

One of the main aims of comparative epidemiology is to facilitate liaison between those in diverse disciplines by comparing and contrasting the behaviour of different types of pathogen. This paper has sought to show the advantages of such an ecological, holistic approach in a search for unifying concepts of general applicability. The emphasis is placed on the common biological features that influence the evolution, growth, survival, perennation, dispersal and adaptation of all pathogens as these factors tend to be overlooked or treated as separate unrelated phenomena. A sound knowledge of epidemiological principles is essential in developing effective strategies of disease control.

References

Anderson R.M. (1981) Infectious disease agents and cyclic fluctuations in host abundance. In: *The Mathematical Theory of the Dynamics of Biological Populations. II* (Ed. by R.W. Hins & E. Cooke), pp. 47–80. Academic Press, New York.

Bennett C.W. (1963) Highly virulent strains of curly top virus in sugar beet in western United States. *Journal of American Society of Sugar Beet Technology* **12**, 515–20.

Browning J.A. (1981) The agro-ecosystem – natural ecosystem dichotomy and its impact on phytopathological concepts. In: *Pests, Pathogens and Vegetation* (Ed. by J.M. Thresh), pp. 159–72. Pitman, London.

Burdon J.J. & Shattock R.C. (1980) Disease in plant communities. *Applied Biology* **5**, 145–219.

Cho E.K., Chung B.J. & Lee S.H. (1977) Studies on identification and classification of soybean virus diseases in Korea. II. Etiology of a necrotic disease of *Glycine max. Plant Disease Reporter* **61**, 313–17.

Crosse J.E. (1967) Plant pathogenic bacteria in the soil. In: *The Ecology of Soil Bacteria* (Ed. by T.R.G. Gray & D. Parkinson), pp. 552–72. University Press, Liverpool.

Drijfhout E. (1978). Genetic interaction between *Phaseolus vulgaris* and bean common mosaic virus with implications for strain identification and breeding for resistance. *Agricultural Research Report 872*, Pudoc, Wageningen.

Fletcher J.T. & Butler D. (1975) Strain changes in populations of tobacco mosaic virus from tomato crops. *Annals of Applied Biology* **81**, 409–12.

Giddings N.J. (1948) Some studies of curly-top virus in the field. *Proceedings American Society of Sugar Beet Technology* **5**, 531–8.

Hanada K. & Harrison B.D. (1977) Effects of virus genotype and temperature on seed transmission of nepoviruses. *Annals of Applied Biology* **85**, 79–92.

Harper J.L. (1982) After description. In: *The Plant Community as a Working Mechanism* (Ed. by E.I. Newman) pp. 11–25. British Ecological Society, Blackwell Scientific Publications, Oxford.

Harrison B.D. (1981) Plant virus ecology: ingredients, interactions and environmental influences. *Annals of Applied Biology* **99**, 195–209.

Hutchinson G.E. (1978) *An Introduction to Population Ecology*. Yale University Press, New Haven.

Jones R.A.C. (1981) The ecology of viruses infecting wild and cultivated potatoes in the Andean region of South America. In: *Pests, Pathogens and Vegetation* (Ed. by J.M. Thresh), pp. 89–107. Pitman, London.

Kisimoto R. (1972) Biology of rice stripe virus disease. *The Iden* **26**, 34–40 (In Japanese).

Magyarosy A.C. & Duffus J.E. (1977) The occurrence of highly virulent strains of the beet curly top virus in California. *Plant Disease Reporter* **61**, 248–51.

Murant A.F. (1981) The role of wild plants in the ecology of nematode-borne viruses. In: *Pests, Pathogens and Vegetation* (Ed. by J.M. Thresh), pp. 237–48. Pitman, London.

Pelham J. (1972) Strain-genotype interaction of tobacco mosaic virus in tomato. *Annals of Applied Biology* **71**, 219–28.

Pelham J., Fletcher J.T. & Hawkins J.H. (1970) The establishment of a new strain of tobacco mosaic virus resulting from the use of resistant varieties of tomato. *Annals of Applied Biology* **65**, 293–7.

Quiot J.B., Devergne J.C., Cardin L., Verbrugghe M., Marchoux G. & Labonne, G. (1979) Ecologie et épidémiologie du virus de la mosaïque du concombre dans le sud-est de la France. VII. Répartition de deux types de populations virales dans des cultures sensibles. *Annales de Phytopathologie* **11**, 359–73.

Rochow W.F. (1979) Field variants of barley yellow dwarf virus: detection and fluctuation during twenty years. *Phytopathology* **69**, 655–60.

Summers E.M., Brandes E.W. & Rands R.D. (1948) Mosaic of sugarcane in the United States, with special reference to strains of the virus. *Technical Bulletin 955*, USDA, Washington.

Thresh J.M. (1974a) Temporal patterns of virus spread. *Annual Review of Phytopathology* **12**, 111–28.

Thresh J.M. (1974b) Vector relationships and the development of epidemics: the epidemiology of plant viruses. *Phytopathology* **64**, 1050–6.

Thresh J.M. (1978) The epidemiology of plant virus diseases. In: *Plant Disease Epidemiology* (Ed. by P.R. Scott & A. Bainbridge), pp. 79–91. Blackwell Scientific Publications, Oxford.

Thresh J.M. (1980) An ecological approach to the epidemiology of plant virus diseases. In: *Comparative Epidemiology* (Ed. by J. Palti & J. Kranz), pp. 57–70. Pudoc, Wageningen.

Thresh J.M. (1982) Cropping practices and virus spread. *Annual Review of Phytopathology* **20**, 193–218.

Thresh J.M. (1983a) Plant virus epidemiology and control: current trends and future prospects. In: *Plant Virus Epidemiology* (Ed. R.T. Plumb & J.M. Thresh), pp. 349–60. Blackwell Scientific Publications, Oxford.

Thresh J.M. (1983b) Progress curves of plant virus disease. *Advances in Applied Biology* **8**, 1–85.

Thresh, J.M. (1986a) Plant virus disease forecasting. In: *Plant Virus Epidemics: Monitoring, Modelling and Predicting Outbreaks* (Ed. by G.D. MacLean, R.G. Garrett & W.G. Ruesink), pp. 359–86. Academic Press, Sydney.

Thresh J.M. (1986b) Plant virus dispersal. In: *The Movement and Dispersal of Agriculturally Important Biotic Agents* (Ed. by D.R. MacKenzie, C.S. Bradfield, G.G. Kennedy, R.D. Berger & D.J. Taranto), pp. 51–106. Claitors Publishing Division, Baton Rouge.

Timian R.G. (1974) The range of symbiosis of barley and barley stripe mosaic virus. *Phytopathology* **64**, 342–5.

Vanderplank J.E. (1948) The relation between the size of fields and the spread of plant-disease into them. I. Crowd diseases. *Empire Journal of Experimental Agriculture* **16**, 134–42.

Vanderplank J.E. (1949) Vulnerability and resistance to the harmful plant virus. *South African Journal of Science* **46**, 58–66.

Vanderplank J.E. (1963) *Plant Diseases: Epidemics and Control*. Academic Press, London.

Zadoks J.C. & Schein R.D. (1979) *Epidemiology and Plant Disease Management*. Oxford University Press.

12 Cereal virus diseases: contrasting experience

S.A. HILL
Agricultural Development and Advisory Service, Harpenden Laboratory, Harpenden AL5 2BD, UK

Introduction

The population dynamics and genetics of the pathogen are central to the development and evolution of both fungal and viral diseases of plants. However the following studies of two important virus diseases of cereals underline the equal, or greater, importance of the dynamics and genetics of the virus vector. Modelling the response of viral pathogens to control measures is thus more difficult than for fungal pathogens (see also Thresh, this volume).

Barley yellow dwarf virus (BYDV)

General properties

Barley yellow dwarf virus (BYDV) is a luteovirus which occupies the phloem of its host and does not invade other host tissue. It is transmitted by several common aphid species in the persistent manner and infects most genera of the Gramineae. The virus exists as strains which vary in virulence and which are specifically transmitted by different aphid species (Table 12.1); it is not seed-borne.

Table 12.1. BYDV strains in the UK (Plumb, 1974)

Strain	Aphid vector species	Severity of symptom
MAV[a]	*Sitobion avenae*	mild
PAV[a]	*S. avenae* and *Rhopalosiphum padi*	severe
RPV	*R. padi*	severe
MDV	*Metapolophium avenae* and *S. avenae*	mild

[a] PAV and MAV are related serologically.

Yield effects, time of infection and isolate virulence

Watson (1959) showed that, in the field, BYDV affected barley more than wheat, and reduced yield more the earlier that infection occurred in the development of the host. She concluded, however, that the time of symptom expression was more

Wolfe M.S. & Caten C.E. (1987) *Populations of Plant Pathogens: their Dynamics and Genetics.* Blackwell Scientific Publications, Oxford.

closely related to the effect on yield than was the time of infection. It was also noted that two isolates differed in virulence, one causing greater yield loss than the other.

Incidence of virus in wheat and barley

Surveys of foliar diseases by the Agricultural Development and Advisory Service (ADAS) (King, 1977) revealed a significant level of BYDV infection in most spring barley crops in the southern half of the country when they were sampled at late growth stages; wheat was similarly affected. During the summer, however, virus symptoms are rare in spring-sown crops. In contrast, aerial surveys during the summer of many autumn-sown crops of wheat revealed widespread symptoms, evident as discrete patches in 1976 and 1977 (Greaves *et al.*, 1983). Analysis of the survey data showed that BYDV infection was commonest in crops sown before mid-October, and comparatively infrequent in later sown crops (Figure 12.1).

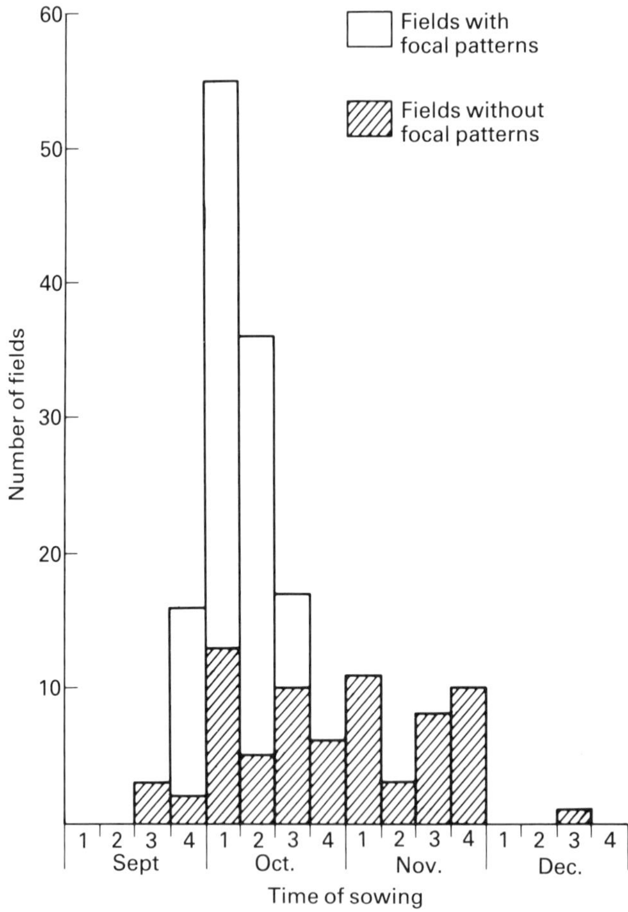

Figure 12.1. Effect of time of sowing on BYDV incidence. Unshaded columns represent fields with clear foci of infection; shaded columns are fields apparently free from infection.

Strain patterns

Aphid transmission tests on samples from field crops show that patches of virus in autumn-sown crops in southern England are caused mostly by severe isolates of the virus, probably spread by the aphid species *Rhopalosiphum padi*. Towards harvest, mild strains of the kind transmitted only by *Sitobion avenae* are also found in autumn-sown crops, and are the most common in spring-sown crops.

Vector control to prevent patch development

As the incidence of damaging outbreaks of BYDV in autumn-sown cereals became more frequent, trials were initiated to investigate control of the vector. Most of the effort was directed towards the definition of optimum times and frequency of application of effective aphicides. Almost without exception, trials have shown that development of field patches of BYDV in the spring can be controlled effectively by one application of aphicide at the end of October or in early November (Table 12.2). The control of spring patch development was directly related to yield increase.

Table 12.2. Yield response from BYDV control – effect of time of application (ADAS and LARS trials)

Host	Sowing date	Spray date	Yield increase (t/ha)[a]
Wheat cv. Armada	21/9/83	6/10/81	0.19
		27/10/81	0.34
		25/11/81	0.51[b]
		16/2/82	0.31
Barley cv. Sonja	15/9/80	10/11/80	1.43[b]
		24/11/80	0.83
		10 and 24/11/80	1.43[b]
Barley cv. Igri	13/9/80	15/10/80	1.9[b]
		30/10/80	2.1[b]
		30/10 and 11/11/80	2.0[b]
		26/3/81	0.3

[a] Over unsprayed control
[b] Significant at $P = 0.05$

BYDV epidemiology: an hypothesis

From these observations and a knowledge of the occurrence and size of the populations of migrating cereal aphids (Taylor *et al.*, 1981), an hypothesis for the epidemiology of BYDV can be proposed (Figure 12.2). The size of the population of alate (winged) aphid vectors of BYDV varies, but the timing of such flights is

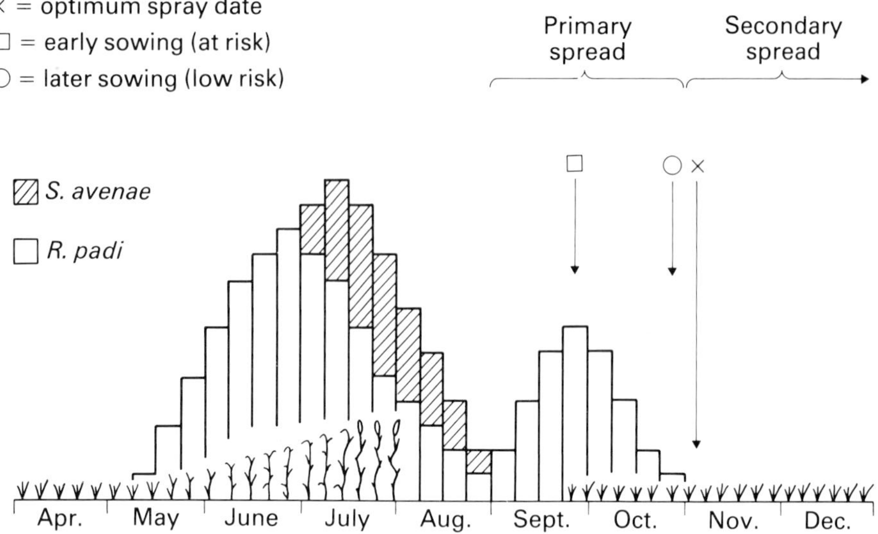

Figure 12.2. BYDV epidemiology diagram relating growth of winter- and spring-sown cereals to virus-vector aphid migrations.

comparatively regular, being largely defined by daylength changes. The expected pattern of vector aphid flights in southern England may thus underlie epidemic development of BYDV. Both BYDV vector species fly during the period from late May to late August, *S. avenae* catches in suction traps being larger than those of *R. padi*. However, *R. padi* has a second period of flight activity during September and October (Figure 12.2). By relating the developmental stages of the cereal crop both to vector aphid catches and to the influence of the timing of infection on final grain yield, it is possible to explain the observed dynamics of virus incidence and strain frequency.

In spring-sown crops, the first vector aphids arrive during mid-May. Probably few of these will be viruliferous, so that infection often does not occur until late May and virus symptoms may not become evident until mid to late June, when cereal plants are well developed (Figure 12.2). Thus the virus infection in spring-sown crops and that which occurs from new infestations of winter crops in the spring is usually too late to be damaging. Only very late-sown spring crops are affected by this phase of the disease and the effects are not severe.

Early-sown winter crops are at much greater risk. *R. padi* migrates from cereals or grasses in September and October and individual aphids have had sufficient opportunity to acquire BYDV. The aphids colonizing early-emerging autumn-sown cereals are thus likely to transmit virus. In most years the virus symptoms may not become evident until March or April, but this is still at a comparatively early stage in the development of the cereal plant. This sequence, together with the severe symptoms of the virus strains transmitted by *R. padi*, explains the damaging nature of BYDV infection in autumn-sown cereals.

Winter crops sown before mid-October emerge during the period of migratory flights of *R. padi*, whilst those sown after mid-October will escape colonization (Figure 12.2). This explains the variation in incidence of BYDV patches in cereal crops revealed by aerial photography, and may also explain the recent increase in incidence of the disease, following the trend to earlier sowing in the autumn, particularly with barley.

The spread of BYDV in autumn-sown cereals can be divided into a primary phase, infection foci created by alighting migrant aphids, and a secondary phase, spread of primary foci by their apterous progeny. The degree of spread depends on the frequency of suitable weather during the winter months. The distinction between the two phases of disease spread explains the effectiveness of a single, properly timed application of aphicide in controlling BYDV. Aphicide application before completion of the primary phase of spread cannot be fully effective, since primary foci of infection may still develop, unless the aphicide is persistent. Application in the spring may be ineffective, since secondary spread of the apterous aphids will have occurred. A single application at the end of the migrant primary phase, but before secondary movement of aphids has started, should kill any aphids present and so prevent further virus spread. The primary foci of virus infection that are left will usually form an insignificant proportion of the crop area.

However, extensive BYDV development even in early-sown autumn cereals is not a regular occurrence in all areas of southern England. Within the last few years BYDV has fluctuated markedly, from severe in 1981 to very little in the following year. Whilst it is possible to explain the disease behaviour within each yearly cycle, the reasons for the variation from one year to the next are less clear, and formulation of advice on BYDV control has been difficult.

Infectivity indices

Plumb *et al.* (1983) have provided a guide for forecasting the risk of severe BYDV infection using an 'Infectivity Index'. The index is the product of the proportion of vector aphids carrying virus and the numbers of aphids caught in the nearest

Table 12.3. Infectivity Indices at Rothamsted and subsequent BYDV severity (Plumb *et al.*, 1982, 1983)

Date	6/9	13/9	20/9	27/9	4/10	11/10	18/10	25/10	Subsequent disease incidence[a]
1980	14	12	81	24	43	11	7	3	Widespread
1981	7	3	0	0	3	0	2	0	Uncommon
1982	2	0	84	7	9	15	12	5	Frequent
1983	0	0	0	0	0	4	0	0	Infrequent

[a] BYDV incidence in autumn sown crops in the following year in the Rothamsted area.

Rothamsted Insect Survey suction trap. The former value is determined by trapping aphids alive and feeding them on oat test plants which produce symptoms if virus is present. Index values in the autumn and the subsequent development of BYDV are positively correlated (Table 12.3); for example, if the cumulative index is less than 50 at the end of October, the subsequent incidence of BYDV is likely to be low. Forecasts using the infectivity index have been accurate for four successive years and have enabled a more positive advisory strategy to be adopted.

Virus-prone areas

The incidence of BYDV varies in different parts of the country. Before the widespread use of insecticides, it was known that in certain areas BYDV occurred regularly in autumn-sown crops, irrespective of infectivity criteria. At the other extreme the incidence of virus in some areas was very low. To extend the BYDV risk assessment based on infectivity indices, a map defining 'virus prone' areas (Figure 12.3) has been developed and a prophylactic spray is advised for high-risk areas.

The high-risk areas are coastal or estuarine areas in the southern half of the country which have a mild maritime climate (Figure 12.3). The most obvious effect of a mild climate would be on the secondary development phase of the aphids. Firstly, the build-up and survival of apterous aphid populations would be greater, secondary aphid feeding would be prolonged, and the consequent spread of virus would be increased. Secondly, plants infected with BYDV would be more susceptible to frost kill, so that in frost-free areas a greater proportion of both primary and secondary inoculum might be expected to survive.

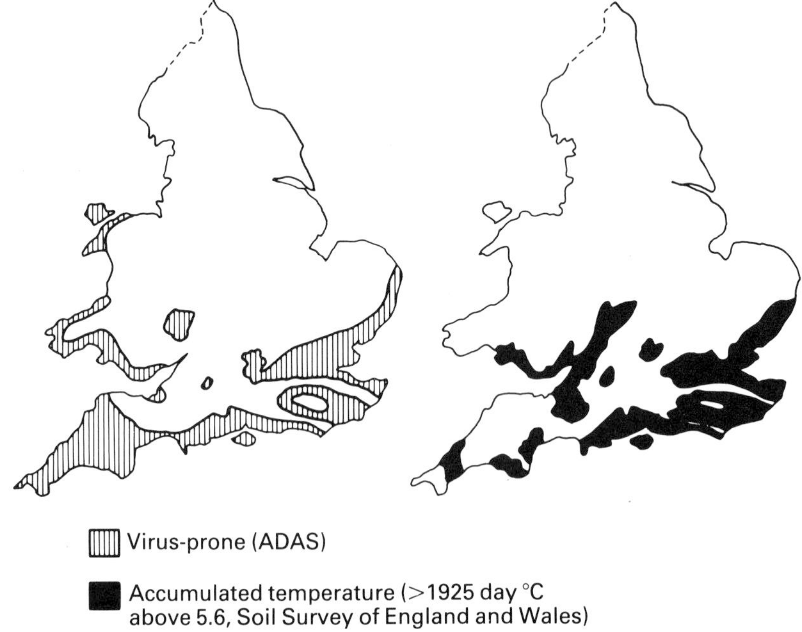

Figure 12.3. BYDV-prone areas and mild areas.

Other phases of the disease spread seem likely to be less affected by climate. The migratory flights of *R. padi* are largely conditioned by daylength changes, and although population size may be related to weather conditions in summer, such variations may not have a consistent geographical distribution. The effect of temperature on populations of *S. avenae*, the other BYDV vector, are probably similar but on a much reduced scale.

Climatic variation which may affect the distribution of severe BYDV may also determine the husbandry of the host crop. For example, if the weather causes a long gap between harvest of susceptible cereal crops and emergence of autumn seedlings, migration of aphid populations will be critical for virus spread. Conversely, with a late harvest followed by early drilling in the autumn, migratory flights may be less important.

Cultivation practice

Traditional cultivation practice in different areas may affect the epidemiology of the disease. For example, in intensive arable areas such as East Anglia, cultivation follows soon after harvest, limiting survival of cereal volunteers or grass weeds that may harbour the virus. In parts of the south-west of England, however, cultivation after cereal harvest may be delayed, allowing profuse generation of volunteers and grass weeds which can be colonized by alate aphids and infected by virus. The relatively short period between cultivation of such fields and emergence of the next cereal crop may allow survival of infective apterous aphids which emerge even after a limited period of burial.

The role of grass in providing inoculum

BYDV infects most wild and cultivated grasses, and surveys have shown that severe isolates of the virus are widespread in most parts of the UK (Doodson, 1967). The most severe BYDV damage often occurs in autumn-sown cereals that follow long-term grass leys. Where the ploughed-out grass is imperfectly buried, BYDV infection by resident apterae is very early and cereal plants may be killed completely leaving bare patches in the field. However, Plumb *et al.* (1982) showed that grass plants may support only low concentrations of BYDV and that acquisition of virus from them by vectors may be less efficient than from infected cereal plants. Thus, the role of grass may be principally as a local source of inoculum; spread over greater distances may require cereal hosts.

The importance of virus strains

The study of the epidemiology of BYDV has been hampered by lack of an efficient diagnostic technique for the occurrence of the virus. Until recently the presence of BYDV could be confirmed only by a lengthy aphid transfer test, which limited the extent of any investigations. However, the test did provide a guide to the virus strains present in a sample, separated according to their vector specificity (Table 12.1). More recently, serological testing has enabled more rapid and easier BYDV diagnosis. However, strain separation is less easy in the serological tests, since the antisera in common use were prepared using virus isolates whose precise character was not accurately defined. More work is needed to define the range of BYDV strains found in the UK, since strain variation may be common. Further, more experience is needed to relate virus levels detected serologically to those that might be aphid transmissable.

BYDV: conclusion

In summary, the hypothesis for the epidemiology of BYDV seems to explain the observations adequately. It depends upon two assumptions: (i) virus is introduced by migrant aphids, the sowing date of the crop being therefore important, as is the infectivity of the migrant aphids; and (ii) secondary spread of BYDV is caused by wingless aphids, which helps to explain the timing and gradation of response to aphicidal control. However, the hypothesis is based on limited information and it is important to determine whether or not further data will refine or reinforce our understanding of the disease.

The hypothesis is not contradicted by the effects of environmental factors discussed above, but these illustrate the complexity of the relationship between BYDV and its vectors. Other factors may also affect the disease development; for example, the biology of the aphid vector species, their relationship with alternate hosts and the roles of the various aphid morphs, all of which may have direct relevance to the geographical distribution of disease. Our lack of knowledge of the effects of such influences, and of many others which may have been omitted, illustrate the comparatively primitive state of our understanding of this disease.

Barley yellow mosaic virus (BaYMV)

General properties

This virus disease was first confirmed in the UK in 1980 (Hill & Evans, 1980), having apparently originated in Japan in 1940; it was recognized in West Germany in 1978. The virus is probably a potyvirus but it has flexuous, filamentous particles of two lengths (300 and 600 nm). The vector is believed to be the fungus *Polymyxa graminis*, which is a weak root parasite. The virus is carried within the motile zoospores of the fungus and survives and is spread passively within the resting spores. BaYMV is not known to be seed-borne and affects only species of the genus *Hordeum*. In the absence of any chemical control, evaluation of the yield effects of the virus have been difficult, but in field comparisons yield losses of up to 42% have been recorded. The virus has become established in most of the cereal growing areas of England.

Distribution of virus in the host

The sporadic recognition of the disease may be due to the variation in sympton expression through the year and in the virus content of affected leaves. Symptoms first become evident in late January and all leaves of infected plants are affected. The typical chlorotic streaks persist until March or April but then disappear as the temperature begins to rise. Controlled environment chamber studies (Friedt, 1983) show that a day/night temperature regime of 10/8°C is required for symptom expression. When symptoms disappear, affected plants appear normal except that they are stunted. Chlorotic streaks may recur if conditions become cooler, and occasional flecks may persist until June. Symptom expression occurs when virus concentration is high in affected leaves. Electron microscope and enzyme-linked immunosorbent assay (ELISA) tests show that virus levels are highest in symptom-bearing leaves and lowest in root tissue.

Virus spread

Local spread of BaYMV by the motile zoospores of *P. graminis* may be restricted to a few centimetres. More effective dispersal is by the passive movement of viruliferous resting spores in soil particles. For example, aerial photography of affected fields often reveals the kite-shaped patches typical of spread by cultivation. *P. graminis* may survive and be distributed in stubble debris, and patches may relate to poor stubble burning, for example, around field boundaries. More difficult to explain is the regular association of BaYMV patches with the sites of hedgerows removed as many as 30 years previously, when no differences can be detected in soils in adjacent areas.

Cultivar reactions to BaYMV

In the absence of chemical controls, efforts have been made to assess the responses of different barley cultivars to BaYMV. A range of reactions has been recorded in replicated trials over three years (Table 12.4). The severity of virus infection within each cultivar was assessed by the percentage of plants affected, since there was no clear variation in symptom intensity. Most if not all cultivars may become infected with BaYMV, but not all produce symptoms. There is a division of response to BaYMV among barley cultivars, the most resistant being of German origin and the most sensitive of UK origin (Table 12.4). Cultivars bred for their malting potential seem more sensitive than feed types. BaYMV resistance seemed also to be positively associated with cold tolerance of cultivars and breeding lines, and was dependent on environmental conditions. For example in the Essex trials, Igri was more affected by virus in the cold winter of 1981/82 than in either the preceding or succeeding years (Table 12.4); the same behaviour was evident though less pronounced in other varieties.

Table 12.4. Reactions[a] of winter barley cultivars to BaYMV

Cultivar	Origin	Trial site			
		Essex			Berks
		1981	1982	1983	1983
Maris Otter	UK	1	1	1	3
Tipper	UK?	2	1	5	–
Halcyon	UK	2	1	1	7
Pirate	France	7	7	–	–
Gerbel	"	8	8	8	1
Igri	Germany	9	3	6	1
Sonja	"	9	9	9	9
Athene	"	9	9	9	8
Birgit	"	9	9	–	–
Hydra	"	9	7	9	2

[a] 9 = resistant, 1 = susceptible

Cultivar-adapted strains of virus or vector?

Three years of trials in Essex apparently defined the reaction of the established winter barley cultivars (Table 12.4), but disturbing variations from this pattern were found in commercial crops in 1982. In a number of fields, crops of previously resistant cultivars were badly affected by BaYMV; all of the fields had been previously cropped with resistant cultivars of feed barley. The trial area, in contrast, had been pre-cropped for many years with a susceptible malting cultivar. A preliminary trial in 1983 on land pre-cropped with a resistant cultivar (Berks) (Table 12.4) supported the field observations and suggested that cultivar-adapted strains of the virus or its vector may occur. No such experiences have been reported from studies of the disease in Germany and France. However, two serologically distinct strains of BaYMV have been reported from Germany (W. Huth, personal communication) which differ only slightly in some characteristics but behave similarly in different barley cultivars. Preliminary observations suggest that isolates of BaYMV from the UK also contain both of the strains identified in Germany.

BaYMV: conclusion

A great deal still needs to be learned about the behaviour of BaYMV. Populations of the virus and of its vector have yet to be analysed, and the epidemiology of the disease may be further complicated by strain variation in virus and vector. Despite the relative immobility of the vector, this virus disease presents a significant threat to the national winter barley crop.

Conclusion

Separate consideration of the population dynamics of virus and vector is impossible;

they interact in the epidemiology of the virus diseases. By comparison with many of the fungal diseases (e.g. Wolfe, this volume) our understanding of the behaviour of virus disease is in its infancy.

References

Doodson J.K. (1967) A survey of barley yellow dwarf virus in S.24 perennial ryegrass in England and Wales, 1966. *Plant Pathology* **16**, 42–5.

Friedt W. (1983) Mechanical transmission of soil-borne barley yellow mosaic virus. *Phytopathologische Zeitschrift* **106**, 16–22.

Greaves D.A., Hooper A.J. & Walpole B.J. (1983) Identification of BYDV and cereal aphid infestations in winter wheat by aerial photography. *Plant Pathology* **32**, 159–72.

Hill S.A. & Evans E.J. (1980) Barley yellow mosaic virus. *Plant Pathology* **29**, 197–9.

King J.E. (1977) Surveys of foliar diseases of spring barley in England and Wales, 1972–75. *Plant Pathology* **26**, 21–9.

Plumb R.T. (1974) Properties and isolates of barley yellow dwarf virus. *Annals of Applied Biology* **77**, 87–91.

Plumb R.T., Lennon E. & Gutteridge R.A. (1982) Barley yellow dwarf virus (BYDV). *Report of Rothamsted Experimental Station for 1981*, pp. 195–8.

Plumb R.T., Lennon E. & Gutteridge R.A. (1983) Barley yellow dwarf virus (BYDV). *Report of Rothamsted Experimental Station for 1982*, pp. 195–6.

Taylor L.R., French R.A., Woiwod, I.P., Dupuch M.J. & Nicklen J. (1981) Synoptic monitoring for migrating insect pests in Great Britain and Western Europe. I. Establishment of expected values for species content, population stability and phenology of aphids and moths. *Report of Rothamsted Experimental Research Station for 1980, part 2*, pp. 41–104.

Watson M.A. (1959) Cereal Viruses in Britain. *NAAS Quarterly Review* **43**, 93–102.

Section 3
Genetic changes in pathogen populations

13 The host population as a selective factor

K.J. LEONARD
US Department of Agriculture, Agricultural Research Service, Department of Plant Pathology, North Carolina State University, Raleigh, NC 27695–7616, USA

Introduction

E.C. Stakman said in 1947 that plant pathogens are shifty enemies. Their populations are genetically variable and they respond to selection pressures. The history of changes in frequencies of pathogenic races of cereal rusts in response to changes in the resistance of cultivars in commercial cereal production is extensively documented in the results of many years of annual surveys of pathogenic races. In spite of that, it is not easy to interpret those race frequency changes in simple terms (Johnson, this volume). The combination of the complex assortments of major genes for resistance in commercial cultivars in different geographical regions of cereal production, the variable roles of sexual and asexual reproduction in the pathogens, and the widespread dissemination of rust urediospores make it difficult to account for specific changes in frequencies of virulence genes in narrowly defined geographical locations and time spans. To better illustrate how host populations exert selective pressures on the populations of their pathogens, I have chosen several simpler examples that may be less well known. These examples will be discussed in terms of some general concepts outlined in the next section.

General concepts

Selection requires genetic variation, and selection of plant pathogens on their host plants requires variation within the pathogen population for virulence on the host. That variation may be qualitative or quantitative.

Qualitative variation for virulence within pathogen populations is well known in the familiar gene-for-gene pattern of the relationship between resistance and virulence in many host–pathogen systems. For the pathogen population, the pattern of variation in virulence on host plants with a particular major gene for resistance may be represented by the solid-line curves in Figure 13.1. The reproductive rate of pathogen genotypes is represented along the x axis, and the frequency of genotypes with various reproductive rates is represented on the y axis.

The solid-line curves in Figure 13.1 represent a pathogen population in which the predominant genotypes lack the gene for virulence necessary to attack the resistant host plants successfully. Consequently, the greatest frequency of genotypes has a

Wolfe M.S. & Caten C.E. (1987) *Populations of Plant Pathogens: their Dynamics and Genetics.* Blackwell Scientific Publications, Oxford.

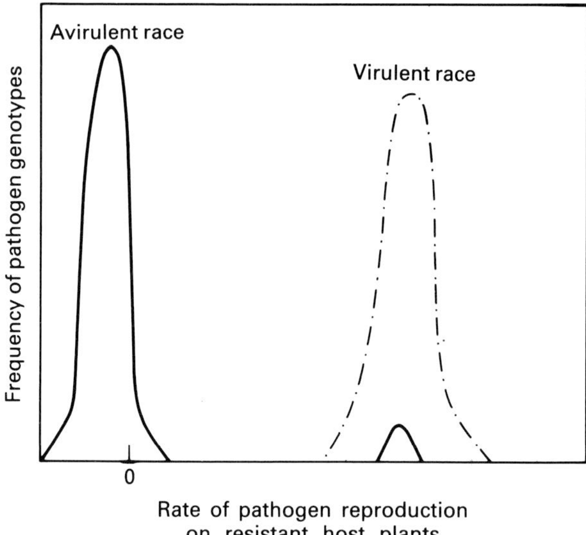

Figure 13.1. Theoretical distribution of frequencies of pathogen genotypes in relation to reproductive capacity on host plants with a major gene for resistance. The solid-line curves represent phenotypic frequencies before selection on the resistant host plants. A small portion of the pathogen population possessing the matching gene for virulence is indicated by the solid-line curve at the right. The dashed-line curve on the right represents the phenotypic frequencies after selection on the resistant plants has eliminated the avirulent genotypes from the population.

negative reproductive rate. If the major gene for resistance is not completely effective, some of the avirulent pathogen genotypes may be able to reproduce enough to just maintain their numbers or to increase them very slowly.

Pathogen genotypes that have the necessary virulence gene are represented in Figure 13.1 by the small group with normal or near-normal reproductive rates. It is easy to see how selection must progress if the host population were made up of just the one resistant genotype. Avirulent pathogen genotypes with negative rates of reproduction would be lost from the population, while the virulent genotypes rapidly increase in frequency. Eventually the pathogen population might exhibit a pattern of variation typical of the normal distribution that is characteristic of many traits in populations of living organisms (dashed-line curve in Figure 13.1). We might be tempted to say that the virulent race had then suppressed the avirulent race, but that could imply a degree of competition greater than has really occurred. As described later, epidemics of highly virulent races may be superimposed on an equilibrium population of races of lesser virulence without drastically affecting the equilibrium population.

The dashed-line curve in Figure 13.1 also resembles the distribution that we might expect to find on a susceptible host with no major genes for resistance. If this kind of variation does exist, and if it has a genetic basis, we would expect to see selection favouring those genotypes with the highest rates of reproduction. That should result in an increase in the mean rate of reproduction for the pathogen population.

With virulence or pathogen reproductive rate, of course, this gain from selection could not go on indefinitely without destroying the host population. Therefore, something must limit the capacity of a pathogen to increase its virulence or its reproductive rate on susceptible plants. If the pathogen population has reached that limit, the dashed-line curve in Figure 13.1 should not be truly representative of genetic variation unless the increased virulence is attained only at a cost to some other part of the pathogen's life cycle. The use of quantitative resistance in the host would reduce the mean rate of reproduction in the pathogen population and might lead to resumption of selection for increased reproductive rate (Figure 13.2). As described later in this chapter, pathogens have been demonstrated to exhibit variation in at least some components of virulence that is similar to a normal distribution (see also Christ *et al.*, this volume), and at least some pathogens have been demonstrated to respond to selection for quantitative variation in virulence under artificial conditions.

In some cases, quantitative variation in virulence to cultivars with quantitative resistance may not be entirely non-specific. Selection of the pathogen population on a single host cultivar may result in the increase in frequency of genes for specific virulence effective against only some resistance genotypes of the host (Figure 13.2). Possible examples of such a phenomenon will be discussed.

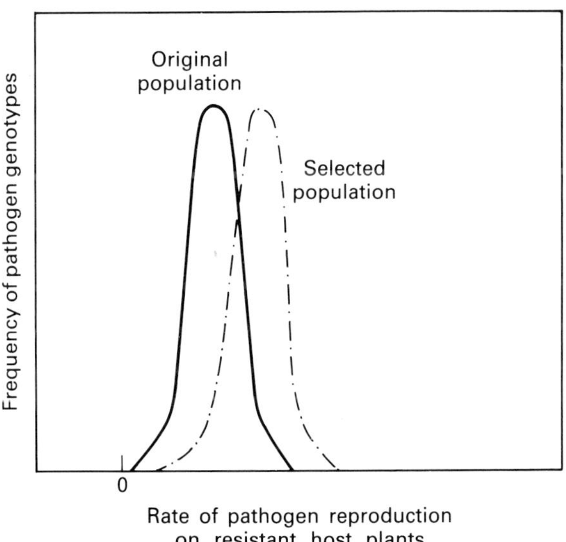

Figure 13.2. Theoretical distribution of frequencies of pathogen genotypes in relation to reproductive capacity on a host with quantitative, partial resistance after selection of those pathogen genotypes with the highest rates of reproduction on the resistant host.

Southern leaf blight of maize

The 1970 epidemic of southern leaf blight of maize in North America is a well-documented example of the selective effect of a host on a pathogen population. The epidemic was caused by a new race of the pathogen, *Bipolaris maydis* (teleomorph,

Figure 13.3. Sites of first-confirmed occurrence of race T of *Bipolaris maydis* in the US (see Foley & Knaphus, 1971; Scheifele, 1971; Ullstrup, 1972). Circles are for three sites in 1968 and triangles are for scattered sites in 1969.

Cochliobolus heterostrophus), which was designated race T to indicate its high level of specific virulence to maize with Texas male sterile cytoplasm (*cms-T*). The old race of *B. maydis*, race O, has a moderate level of non-specific virulence to maize of all cytoplasm types. Race T and race O cause similar symptoms on maize with normal cytoplasm.

In 1970 approximately 85% of the maize crop of the USA contained *cms-T* and was extremely susceptible to race T. *Cms-T* had been introduced into commercial maize hybrids in the 1950s and its frequency in the crop in the USA increased by about 5% each year until 1971. Thus, by the 1960s the maize population in the USA was highly selective for the increase of race T.

The first well-documented occurrence of race T in the USA did not occur, however, until 1968 (Foley & Knaphus, 1971), at three locations in central Iowa (Figure 13.3). This was outside the normal endemic range of race O, which includes the south-eastern part of the USA extending into the Mississippi and Ohio river valleys (Figure 13.4). Within its endemic range, race O was in a more or less stable equilibrium with its host; it appeared in nearly every maize field every year but did not cause extensive damage. In 1969 race T was found in a few scattered fields in Minnesota, Iowa, Illinois, Indiana, Kentucky, and north-eastern Alabama (Figure 13.3) (Moore, 1970; Scheifele, 1971; Ullstrup, 1972). No race T was found among

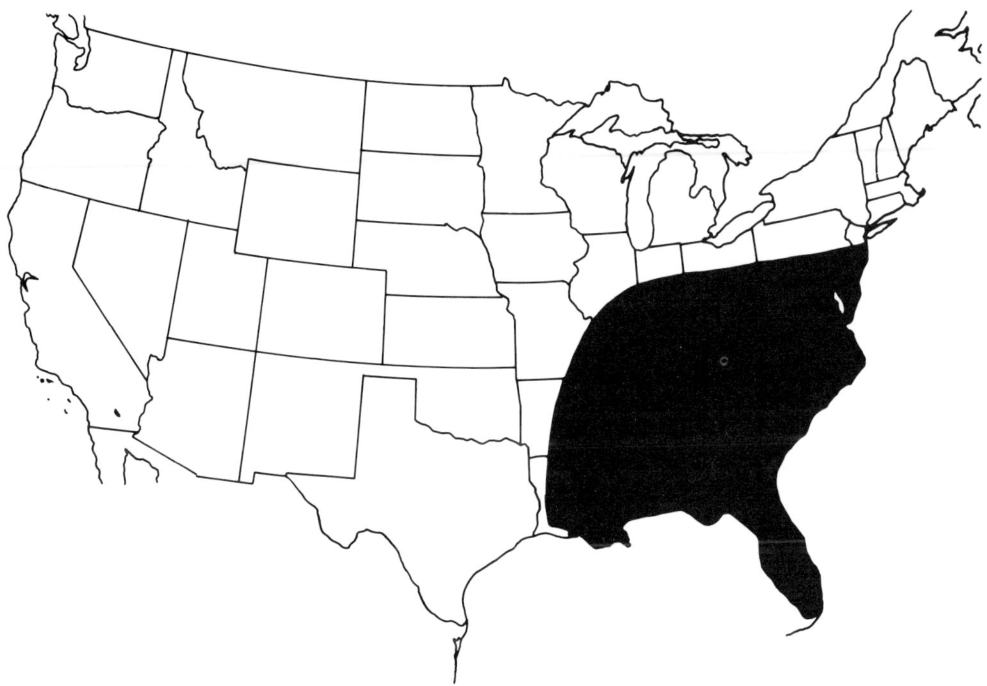

Figure 13.4. Endemic range of race O of *Bipolaris maydis* in the US.

samples collected east of the Appalachian mountains in 1968 and 1969 (Leonard, 1971; Scheifele, 1971).

In February 1970 race T was found in two locations in southern Florida (Moore, 1970; Scheifele, 1971). During the spring and early summer, race T spread rapidly along the Gulf coast and up the Mississippi river valley (Figure 13.5). By mid-August, race T was common throughout most of the Midwest and had invaded the Atlantic coast states as far north as Maryland and Delaware (Moore, 1970; Scheifele, 1971). By the end of the 1970 growing season, race T had colonized virtually every field of maize with *cms-T* in the eastern half of North America (Figure 13.5) (Moore, 1970; Ullstrup, 1972).

The advance of race T during 1970 did not eliminate race O from its endemic area. During 1970 and 1971 race T comprised slightly more than 90% of the isolates of *B. maydis* collected in North Carolina (Figure 13.6) (Leonard, 1971; Leonard, 1973). In 1971 most of the maize planted in North Carolina had normal cytoplasm, and in 1972 only about 1% had *cms-T*. The frequency of race T in the *B. maydis* population dropped dramatically (Figure 13.6) (Leonard, 1974), so that by 1973 race T could be found only occasionally in experimental plantings of *cms-T* maize in the south-eastern US (Leonard, 1977a).

In the mid-western and north-eastern US outside the normal endemic range of race O, the frequency of race T remained high even after maize with *cms-T* had been replaced by maize with normal cytoplasm (Leonard, 1974, 1977b; Nelson, *et al.*,

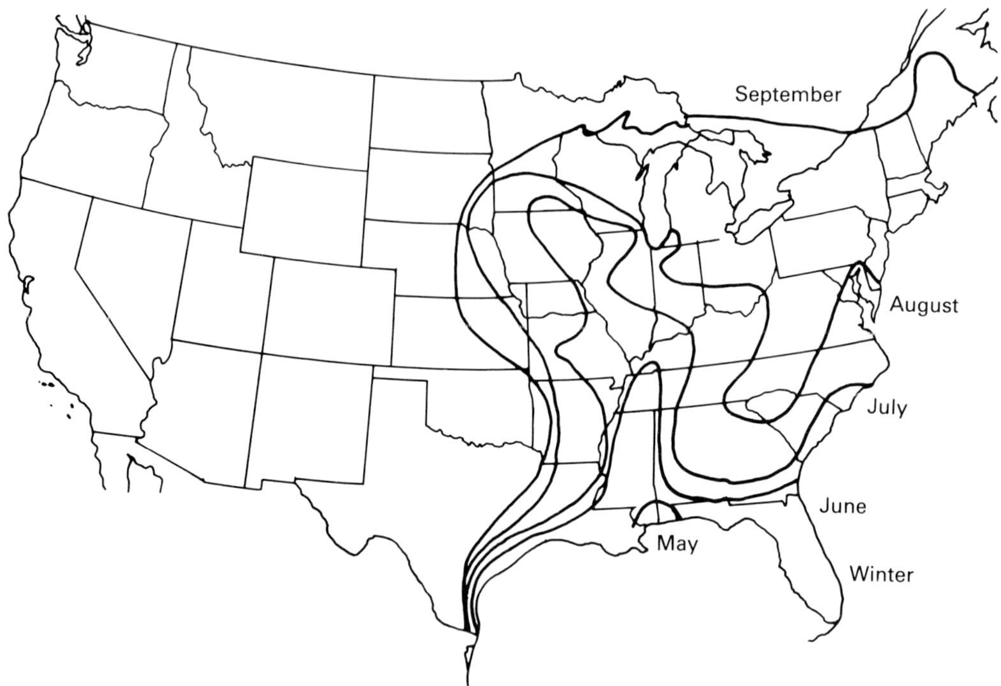

Figure 13.5. Spread of reported damage from race T of *Bipolaris maydis* in the US in 1970. (Data are from Moore, 1970; Scheifele, 1971.)

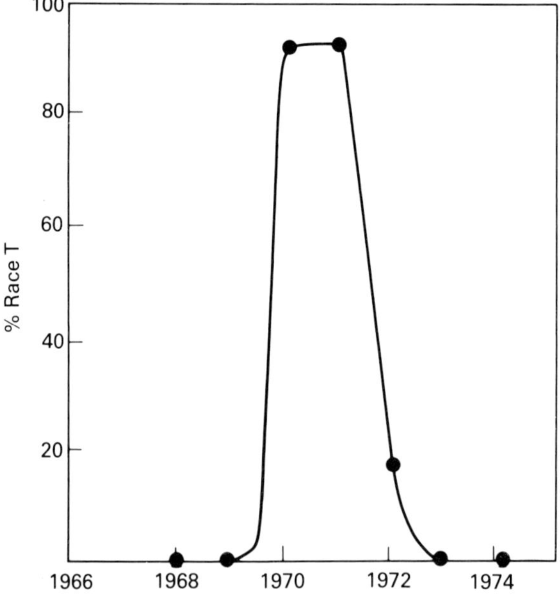

Figure 13.6. Frequency of race T among isolates of *Bipolaris maydis* collected in North Carolina from 1965 through 1974. (Data are from Leonard, 1971, 1973, 1974, 1976, 1977a.)

1973). The continued high frequency of race T in these areas can be accounted for by the lack of a significant residual race O population. Thus, even very small amounts of race T still were reflected as high frequencies.

The southern leaf blight epidemic of maize in the USA illustrates several important features of the selective effect of a host population on a pathogen population. First, the rapid, early increase of race T did not involve direct competition with the predominant, endemic race O. In North Carolina in 1970, for instance, race T swept into the state on winds from the south. In the Midwest, race T expanded outward from the scattered fields where it had occurred in 1969 and where it overwintered (Figure 13.3), but by the time this local expansion was well under way, massive numbers of spores were arriving from the severely blighted fields to the south. Thus, for most of the maize crop in North America in 1970, the increase in frequency of race T can be attributed to a large influx of inoculum relatively late in the growing season, followed by rapid multiplication of race T locally.

A second important feature of the southern leaf blight epidemic was that the endemic race O population in the south-eastern USA was not greatly depressed by the dominance of race T during 1970. Race O was not replaced by race T, but instead, a much larger population of race T was superimposed on the normal race O population. When *cms-T* maize was replaced by maize with normal cytoplasm, the size of the race T population declined very rapidly, while the race O population continued at its usual endemic level. From the rate of decline in frequency of race T, M.W. Grant (personal communication) calculated a selection coefficient of 0.39. This could be regarded as stabilizing selection against virulence that has become ineffective by removal of host genotypes specifically susceptible to it. The selection coefficient calculated by Grant from field data is greater than the value of 0.12 calculated by Leonard (1977b) from studies of selection in mixed populations of isolines of races O and T on normal cytoplasm maize in the glasshouse. In similar studies of selection in a population composed of 50 race O isolates and 50 race T isolates collected from the field, I found a selection coefficient of 0.30 (Leonard, unpublished).

These glasshouse studies were done at relatively low disease intensity, so there was probably very little, if any, interaction among individual infections on the same plant. Thus, the outcome of selection should reflect differences in inherent reproductive rates of the two races. The estimated 12% greater fitness of race O isolines than race T isolines in the glasshouse test corresponded well with measurements of average lesion length. On maize with normal cytoplasm, lesions induced by race O were on average 10% longer than those induced by race T isolines (Leonard, 1977b).

These results suggest that the race T population in the field in North Carolina in 1970 and 1971 was not as fit on maize with normal cytoplasm as it might have been and that even at its best, as represented by the race T isolines, it could not compete successfully with race O on maize with normal cytoplasm. This would explain why race T does not normally occur at detectable frequencies in the *B. maydis* population in the absence of *cms-T*.

We may assume that the rate of mutations from race O to race T is low enough, so that its effect on gene frequencies is easily overwhelmed by selection pressures. Therefore, the equilibrium frequency for race T in the absence of maize with *cms-T* should be approximately equal to the mutation rate divided by the coefficient of selection against race T (Person *et al.*, 1976 see also Gale, this volume; Barrett, this volume). For selection coefficients in the range of 0.4 to 0.1, the equilibrium frequency of race T in the absence of *cms-T* would be approximately 2.5 to 10 times the mutation rate. At low frequencies of *cms-T* in the maize population, race T would not increase in frequency. For instance, if we assume the fitness of race O on maize with either normal or *cms-T* cytoplasm to be 1.0, the fitness of race T on normal cytoplasm maize to be 0.7 to 0.9, and the fitness of race T on *cms-T* maize to be in the range of 3 to 5, the maize population that provided equilibrium between the two races would be in the range of 2 to 13% *cms-T*.

Although race T might have survived in the south-eastern USA at very low frequencies before maize with *cms-T* was widely used, it probably did not survive in the main maize-growing areas of the Midwest before the mid-1960s. *B. maydis* was not commonly found overwintering in those areas before the occurrence of race T there in 1968. Furthermore, all of the race T isolates collected in the Midwest during 1970 and 1971 were of a single mating type, whereas the two mating types of *B. maydis* occur at about equal frequencies in the race O population (Leonard, 1971; Leonard, 1973).

It appears highly likely that a single isolate of race T became established in the Midwest during the mid-1960s, and that it was disseminated through the Midwest in infected seed. The same isolate may have been transported to southern Florida in seed planted in winter breeding nurseries. We know, at least, that the predominant mating type of the isolates of race T found in the south-eastern USA during 1970 was the same as the mating type of the midwestern isolates (Leonard, 1971, 1973.)

Races of *Bipolaris zeicola*

In the closely related pathogen, *B. zeicola* (teleomorph, *C. carbonum*), race 1 produces a host-specific toxin that makes it extremely virulent on maize lines homozygous for a recessive gene for toxin sensitivity at a single chromosomal locus. On maize plants with at least one dominant gene at that locus, race 1 induces only small necrotic spots or flecks that are indistinguishable from those induced by race 2. Race 2, however, is more fit than race 1 on maize plants that are not sensitive to the toxin. With the elimination of toxin-sensitive plants from maize breeding populations and commercial production, race 1 was reduced to undetectable levels in the pathogen population (Leonard, 1978).

Jarvis, an open-pollinated maize variety that has been grown as an experimental population for at least 35 years in North Carolina, contains 7% toxin-sensitive plants. The toxin-producing race 1 has not built up in the Jarvis population, which suggests that the frequency of sensitive plants for equilibrium between race 1 and race 2 is greater than 7% (Leonard, 1978).

With both races O and T of *B. maydis* and races 1 and 2 of *B. zeicola*, resistance in the host and virulence in the pathogen are simply inherited, and the effect of selection is straightforward. If the majority of host plants are sensitive to the toxin, the gene for toxin production confers a large selective advantage to the pathogen, but if toxin-sensitive plants are rare in the host population, the toxin-producing race is at a selective disadvantage. It appears, therefore, that the unregulated production of massive amounts of a toxin reduces the fitness of the fungus if its host is not sensitive to that toxin.

In the following example involving the soybean cyst nematode, neither resistance nor parasitic ability appears to be simply inherited, but the same general principles of selection seem to apply.

Soybean cyst nematode

Resistance of soybean to the cyst nematode (*Heterodera glycines*) is controlled at several loci. The resistance of breeding line P. I. 88788 has been described variously as being due to two dominant and two recessive genes, to one dominant and two recessive genes, or to four or more recessive genes (Luedders, 1983). The resistance of the cultivar Pickett, which is distinct from that of P.I. 88788, is apparently controlled by three recessive genes and one dominant gene (Matson and Williams, 1965).

Field populations of *H. glycines* that have not been exposed to P.I. 88788 or Pickett may show little or no ability to reproduce on these cultivars (Triantaphyllou, 1975). Triantaphyllou showed that a population of *H. glycines* with a low level of parasitic ability on P.I. 88788 could gradually increase its capacity to reproduce on the resistant line. In an original field population of *H. glycines*, the number of females that matured and produced eggs on P.I. 88788 was 16% of that on a fully susceptible soybean cultivar. During selection on P.I. 88788 plants in the glasshouse, the number of females that matured and reproduced gradually increased to more than 60% of that on the susceptible cultivar (Figure 13.7). The gradual increase in parasitic ability on P.I. 88788 without an apparent lag phase suggests that more than a single locus was involved. F_1 and F_2 progeny from mass crosses between the selected and the original nematode populations had intermediate parasitic ability on P.I. 88788, which also suggests complex inheritance of parasitic ability (McCoin, 1977).

Triantaphyllou (1975) subjected another field population of *H. glycines* to selection on both P.I. 88788 and on Pickett, another resistant cultivar. The original field population produced 4.8% as many mature females on P.I. 88788 as on the susceptible cultivar Lee, and 22.1% as many mature females on Pickett as on Lee. After five generations of selection on P.I. 88788 its reproductive capacity on this cultivar increased to 39.4% of that on Lee, but its reproductive capacity on Pickett remained unchanged. After five generations of selection on Pickett, the population increased its reproductive capacity on Pickett to 73% of that on Lee, but its reproductive capacity on P.I. 88788 did not increase. Thus, the ability to parasitize Pickett was

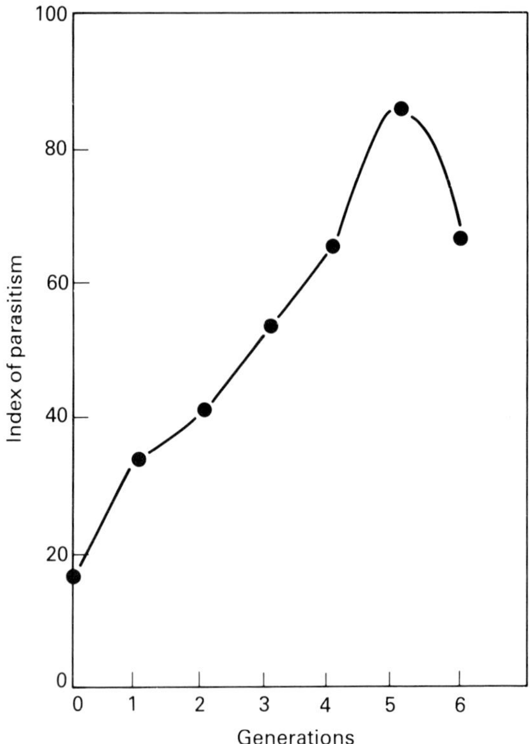

Figure 13.7. Effect of selection of the soybean cyst nematode, *Heterodera glycines*, on the resistant soybean line P.I. 88788 on its ability to parasitize the resistant plants. The index of parasitism is calculated as 100 × the number of mature females found on roots of P.I. 88788 divided by the number on a similarly treated susceptible check plant. (Data are from Triantaphyllou, 1975.)

controlled by different genes from those that controlled the ability to parasitize P.I. 88788 . Selection of this same population for 18 generations on Pickett increased its parasitic ability on Pickett to about 90% of that on the susceptible cultiver (Triantaphyllou, 1975).

After the selected population had been maintained for 22 generations on Pickett, McCoin (1977) mixed the selected population in equal proportion with the original field population that had been maintained on the susceptible cultivar Lee, and he cultured the mixed population on Lee for 10 generations in the glasshouse. All five replicate mixtures showed reduced ability to parasitize Pickett, the resistant cultivar, after selection on the susceptible cultivar (Figure 13.8).

The results of selection and crossing experiments with soybean and with the soybean cyst nematode indicate that the resistant soybean cultivars Pickett and P.I. 88788 each have distinctive sets of several genes that act additively to limit the reproductive ability of the nematode. The nematodes are amphimictic and their populations are, consequently, genetically heterogeneous. Some populations collected from the field contain individuals with a low level of ability to parasitize

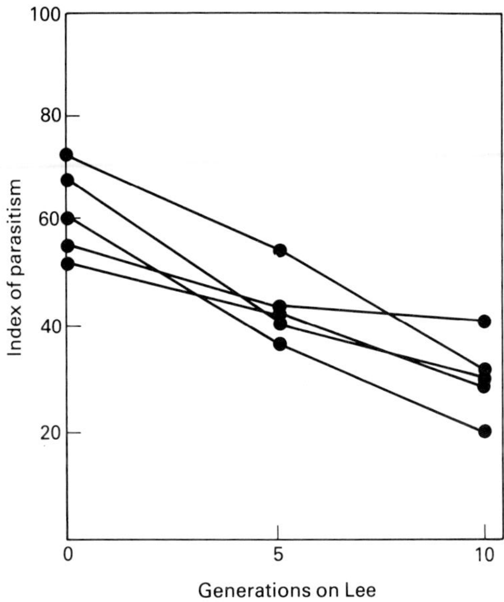

Figure 13.8. Stabilizing selection with respect to index of parasitism of a population of soybean cyst nematode, *Heterodera glycines*, selected for 22 generations on the resistant cultivar Pickett and then grown for 10 generations on the susceptible cultivar Lee. The index of parasitism is calculated as 100 × the number of mature females that developed on Pickett plants divided by the number on similarly treated Lee plants. (Data are from McCoin, 1977.)

either Pickett or P.I. 88788. Selection on the resistant cultivar apparently results in a gradual increase in the frequency of genes for parasitic ability enabling larger proportions of the females to mature and reproduce on the resistant plants. These genes for parasitic ability are specific in their effects; selection for parasitic ability on Pickett did not increase parasitic ability on P.I. 88788, or vice versa.

The ability to parasitize a resistant soybean cultivar appears to be attainable only at a cost in fitness of the nematode on susceptible soybean plants. In McCoin's (1977) experiment continuous culture of a selected population on a susceptible cultivar resulted in a slow decline in the ability of the population to parasitize the resistant cultivar upon which it had originally been selected.

The soybean cyst nematode example illustrates that host–pathogen specificity is not limited to interactions that display a qualitative gene-for-gene relationship between resistance and virulence or parasitic ability. The number of genes controlling resistance to the cyst nematode in soybean cultivars Pickett and P.I. 88788 is small, probably three or four, but they do seem to act more or less additively. The number of genes controlling parasitic ability in the soybean cyst nematode is not known, but the evidence indicates that several genes with additive effects may be involved.

Adaptation in oat stem rust

Leonard (1969) studied the effects of selection in a heterogeneous population of *Puccinia graminis* f. sp. *avenae* established by inoculating the susceptible oat cultivar Craig with aeciospores from more than 300 aecial infections. The aecia were obtained by inoculating barberry plants with basidiospores from overwintered telial-bearing stems of orchard grass collected from the field in New York State.

Urediospores from the first generation on Craig oats were collected and used to inoculate additional sets of Craig and Clintland A oats. Craig has no major genes for stem rust resistance, and Clintland A has *Pg-2*. The Craig and Clintland A subpopulations were carried through seven uredial generations on their respective hosts. After the seventh generation each subpopulation was inoculated on to both Craig and Clintland A plants to test whether the effects of seven generations of selection would be reversed in a single generation on the other host.

After each uredial generation, a portion of the urediospores was used to inoculate a set of seedlings of the standard oat stem rust differential cultivars, and the numbers of resistant- and susceptible-type pustules on each differential were counted. With succeeding uredial generations on the susceptible cultivars, the proportion of avirulent genotypes in the rust population increased, as evidenced by

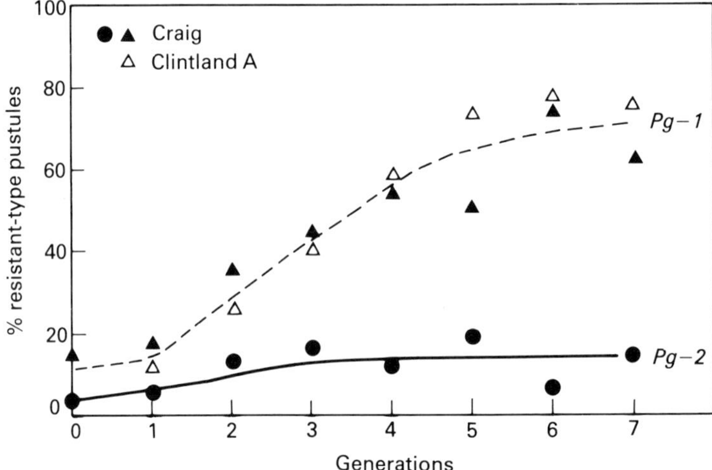

Figure 13.9. Stabilizing selection with respect to frequencies of major genes for virulence in a heterogeneous population of *Puccinia graminis* f. sp. *avenae* cultured through successive uredial generations on Craig or Clintland A oats. Curves show progressive increase of frequencies of genotypes avirulent on differential cultivars with major resistance genes *Pg−1* and *Pg−2*; open symbols are for the subpopulation cultured on Clintland A (*Pg−2*) and solid symbols are for Craig oats (no major genes for resistance). Data for *Pg−1* represent mean values from differential cultivars that share *Pg−1* as well as other genes that did not contribute significantly to the total proportion of resistant type reactions. Data for *Pg−2* are only from the subpopulation on Craig oats; the presence of *Pg−2* in Clintland A prevented the increase of pathogen genotypes without the corresponding gene for virulence. Data are from Leonard (1969).

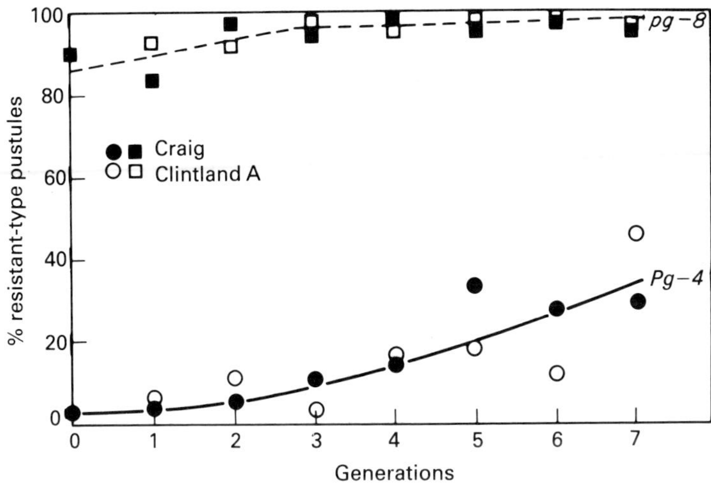

Figure 13.10. Stabilizing selection with respect to virulence of *Puccinia graminis* f. sp. *avenae* on differential cultivars with major resistance genes *Pg−4* and *pg−8*. Data for *pg−8* are mean values from differential cultivars that share *pg−8* as well as having other genes that did not contribute significantly to the total proportion of resistant reactions. See Figure 13.9 for details.

the increasing percentage of resistant-type pustules on the differential cultivars used to test the population (Figures 13.9 & 13.10).

Within the heterogeneous population of *Puccinia graminis* f. sp. *avenae*, selection operated against each of the genes for virulence that could be tested with the set of differential cultivars. The intensity of this selection against unnecessary genes for virulence was surprisingly high; the selection coefficients ranged from 0.14 to 0.39 (Leonard, 1969). This may be due to the high disease intensity on the plants in the test, which would maximize competitive interactions among rust genotypes. Grant & Archer (1983) calculated much lower selection coefficients, in the range of 0.04 to 0.06, for selection against virulent races of wheat stem rust and barley powdery mildew on susceptible cultivars in the field (Editors: but see Johnson, this volume). Under field conditions, competitive interactions among pathogen genotypes would be extremely limited early in epidemics when the incidence of diseased plants is low.

In addition to the change in frequencies of major genes for virulence in Leonard's (1969) heterogeneous population of *P. graminis* f. sp. *avenae*, another more subtle change occurred. The subpopulation of rust cultured on Craig for seven generations had a higher infection efficiency on Craig than on Clintland A oats, whereas the subpopulation cultured on Clintland A for seven generations had approximately equal infection efficiencies on both cultivars (Table 13.1). This adaptation was not reversed by a single generation of culture of the Craig subpopulation on Clintland A, or by a single generation of the Clintland A subpopulation on Craig.

The Clintland A subpopulation produced more urediospores per pustule on both

Table 13.1. Effect of selection in a heterogeneous population of *Puccinia graminis* f. sp. *avenae* on infection efficiency

Selection on host for generations:		
1 to 7	8	Infection efficiency index
Craig	Craig	Clintland A/Craig = 0.74
Craig	Clintland A	Clintland A/Craig = 0.71[†]
		Average = 0.72
Clintland A	Clintland A	Craig/Clintland A = 1.05
Clintland A	Craig	Craig/Clintland A = 0.87
		Average = 0.96

* Infection efficiency index is calculated as the number of pustules induced on the new host divided by the number on the host of selection during generations 1–7.
[†] Significantly different from 1 ($P = 0.01$).
Data are from Leonard (1969).

Table 13.2. Effect of selection in a heterogeneous population of *Puccinia graminis* f. sp. *avenae* on sporulation*

	Sporulation per pustule		
Selection host	Craig	Clintland A	Ratio
Craig	87 000	72 000	0.83
Clintland A	124 000	101 000	0.82

* Data are from Leonard (1969).

Craig and Clintland A than did the Craig subpopulation (Table 13.2). There was no apparent specificity in this adaptive change, however, because the ratio of sporulation on the two cultivars remained the same for both subpopulations. The reasons for this change in sporulation are unclear, but the differences between the two subpopulations illustrate that pathogens as well as hosts exhibit quantitative genetic variation in fitness components (see also Christ *et al.*, this volume). These fitness components may sometimes show host specificity.

Conclusions

The presence of resistance to pathogens in host populations can impose strong selection pressures on pathogen populations only if those populations contain genetic variation upon which selection can act. In the *Bipolaris maydis* population in the USA, the necessary variation in virulence to maize with *cms-T* was apparently not present in sufficient proportions to produce noticeable changes in the pathogen population until 1968. Although Nelson (1973) reported a high frequency of race T of *B. maydis* in maize leaf samples collected from the USA and other countries before 1965, his results are not consistent with other reports of failure to find race T among isolates obtained from maize before 1968 (Leonard, 1973), nor are they consistent with the reports of the spread of race T in 1970, which are well documented.

The increase in frequency of race T in the *B. maydis* population in 1970 did not result from direct competition between race T and race O so much as from a massive influx of inoculum before the local race O populations had had a chance to build up. Most plant pathogens are colonizing species that invade their host plants rapidly during the growing season but survive rather poorly in the absence of their hosts. Consequently, differences in inherent rates of reproduction are important in determining changes in genotype frequencies. Often the initial stages of epidemics occur with populations so small that different pathogen genotypes only rarely occur on the same host plants, so direct competition for the same nutrients would be impossible.

Most glasshouse experiments on selection in populations of plant pathogens are conducted at high inoculum densities, so the effects of selection may be much more intense than would normally occur in the field. While this does not seem to have been the case in the experiments with *B. maydis*, a necrotrophic pathogen, it probably is true of the selection experiments with the biotrophic rust fungi and cyst nematodes. This may be one reason why the results of controlled experiments seem to exaggerate the differences in fitness among pathogen genotypes.

Another important reason why selection in controlled experiments may proceed much faster than it seems to in nature is that natural host populations are usually genetically heterogeneous. Even in agriculture there are few examples of a crop as uniformly susceptible to a single pathogenic race as the North American maize crop was to race T of *B. maydis* in 1970. When a number of different cultivars are grown in a given geographical region and the cultivars differ in their resistance to a pathogen, the forces of selection may be complex and difficult to predict. Selection within individual fields of different cultivars may proceed at different rates and in different directions. In addition, there may be a more or less general reassortment of pathogen genotypes among fields in the dissemination of initial inoculum at the beginning of each growing season, as described by Wolfe & Schwarzbach (1978) for barley powdery mildew.

Each cultivar can be thought of as a unique environment for the pathogen. Just as higher plants exhibit genotype × environment interactions that often involve quantitative inheritance, we should also expect pathogens to show some similar genotype × host environment interactions. Leonard's (1969) selection experiment with oat stem rust showed that a small pathogen genotype × host cultivar interaction for infection efficiency could be obtained by selection for seven generations on separate host cultivars. With soybean cyst nematodes, the adaptation of the pathogen to individual host cultivars was much greater, but it probably involved relatively few genes with greater individual effects.

To suggest that pathogens can adapt to host genotypes through selection from quantitative genetic variation within pathogen populations is not to despair of ever breeding stable disease resistant cultivars of crops. With rare exceptions, selection in natural pathogen populations seems to proceed more slowly than selection in experimental populations. Maintaining genetic diversity in our crops should help delay, or even prevent, the adaptation of pathogen populations to the crops' disease resistance. More research is needed on adaptation of pathogens to host cultivars through quantitative genetic changes. Not only does this topic relate directly to the question of durability of resistance of cultivars in monoculture, but it is also important to the comparison of the relative merits of cultivar mixtures and multiline cultivars.

In typical multiline cultivars, the components differ primarily in major genes for resistance. If the force of stabilizing selection is small, as it seems to be in nature for rusts and powdery mildews, it should be possible for the pathogen to combine virulence to all of the components of a multiline without too great a loss in fitness. Unrelated cultivars, on the other hand, may differ in many characteristics that are polygenically determined. Adaptation to more than one or a few of these might be difficult, if not impossible. Thus, super races of pathogens might be much less likely to arise in cultivar mixtures than in multiline cultivars.

There are a great many pathogens of important crop species that we will probably have to continue to live with. They have had a long history of coevolution with our crops and their wild relatives. It is unreasonable to think that they lack the adaptive capabilities to survive despite our efforts to genetically alter their hosts. However, we can hope to reduce the damage that they cause so that they too will become minor diseases like so many others that appear in our text books but rarely demand our attention.

References

Foley D.C. & Knaphus G. (1971) *Helminthosporium maydis* race T in Iowa in 1968. *Plant Disease Reporter* **55**, 855–7.

Grant M.W. & Archer S.A. (1983) Calculation of selection coefficients against unnecessary genes for virulence from field data. *Phytopathology* **73**, 547–51.

Leonard K.J. (1969) Selection in heterogeneous populations of *Puccinia graminis* f. sp. *avenae*. *Phytopathology* **59**, 1851–7.

Leonard K.J. (1971) Association of virulence and mating type among *Helminthosporium maydis* isolates collected in 1970. *Plant Disease Reporter* **55**, 759–60.

Leonard K.J. (1973) Association of mating type and virulence in *Helminthosporium maydis*, and observations on the origin of the race T population in the United States. *Phytopathology* **63**, 112–15.

Leonard K.J. (1974) *Bipolaris maydis* race and mating type frequencies in North Carolina. *Plant Disease Reporter* **58**, 529–31.

Leonard K.J. (1976) Isolation of *Bipolaris maydis* race O from corn seed harvested in 1965. *Plant Disease Reporter* **60**, 245–7.

Leonard K.J. (1977a) Races of *Bipolaris maydis* in the southeastern U.S. from 1974–1976. *Plant Disease Reporter* **61**, 914–15.

Leonard K.J. (1977b) Virulence, temperature optima, and competitive abilities of isolines of races T and O of *Bipolaris maydis*. *Phytopathology* **67**, 1273–9.

Leonard K.J. (1978) Polymorphisms for lesion type, fungicide tolerance, and mating capacity in *Cochliobolus carbonum* isolates pathogenic to corn. *Canadian Journal of Botany* **56**, 1809–15.
Luedders V.D. (1983) Genetics of the cyst nematode-soybean symbiosis. *Phytopathology* **73**, 944–8.
McCoin G.N. (1977) Studies of the parasitic capability of *Heterodera glycines*. *M.S. Thesis*. North Carolina State University, USA. 29 pp.
Matson A.L. & Williams L.F. (1965) Evidence of a fourth gene for resistance to the soybean cyst nematode. *Crop Science* **5**, 477.
Moore W.F. (1970) Origin and spread of southern corn leaf blight in 1970. *Plant Disease Reporter* **54**, 1104–8.
Nelson R.R. (1973) Further studies on the past occurrence and geographical distribution of isolates of race T of *Helminthosporium maydis*. *Plant Disease Reporter* **57**, 18–19.
Nelson R.R., Ayers J.E., Hill J., Blanco M. Dalmacio S. (1973) Racial composition of *Helminthosporium maydis* on corn hybrids in normal cytoplasm in Pennsylvania in 1972. *Plant Disease Reporter* **57**, 411–12.
Person C., Groth J.V. & Mylyk O.M. (1976) Genetic change in host-parasite systems. *Annual Review of Phytopathology* **14**, 177–88.
Scheifele G.L. (1971) Geographical distribution of *Helminthosporium maydis* race T for 1969 and 1970. *Plant Disease Reporter* **55**, 302–6.
Stakman E.C. (1947) Plant diseases are shifty enemies. *American Scientist* **35**, 321–50.
Triantaphyllou A.C. (1975) Genetic structure of races of *Heterodera glycines* and inheritance of ability to reproduce on resistant soybeans. *Journal of Nematology* **7**, 356–64.
Ullstrup A.J. (1972) The impacts of the southern corn leaf blight epidemics of 1970–1971. *Annual Review of Phytopathology* **10**, 37–50.
Wolfe M.S. & Schwarzbach E. (1978) Patterns of race changes in powdery mildews. *Annual Review of Phytopathology* **16**, 159–80.

tries. All the races carried pathogenicity for several of the differential cultivars and therefore for several resistance genes (McIntosh *et al.*, 1981). Although nothing is known of the resistance genes that may have been present in host wheat cultivars and grasses, the distribution of the wheat cultivars grown at the time appeared to have little effect on the distribution of these races (Watson, 1981). It therefore seems probable that the presence of pathogenicity genes in this population of *P. graminis* was not due to direct selection by matching resistance genes in the host population of Australian wheats and grasses.

Among several parallel examples involving other rust pathogens of cereals are two from Europe. In Britain, commonly occurring races of *P. recondita* carry pathogenicity for many of the identified genes for resistance that have been back-crossed into the cultivar Thatcher (Clifford, 1980; Dyck & Johnson, 1983). Most of these resistance genes are not present in commercial wheat cultivars grown in Britain. Some of them may be present in cultivars grown elsewhere in Europe, but there is no evidence of regular influxes of spores of *P. recondita* to Britain from continental Europe. Similarly, races of *P. hordei* identified in surveys in both Britain and the Netherlands carry pathogenicity for many of the known race-specific resistance genes in barley differential cultivars, although most barley cultivars in commercial use do not carry any of these genes (Clifford, 1974; Parlevliet, 1983).

In all these examples it is probable that the genes for pathogenicity exist not because of continuous selection by resistant cultivars but because they function satisfactorily in controlling the physiology of the pathogen, or at least are not deleterious. The presence of such pathogenicity genes, unnecessary for overcoming resistance genes in the host population, is common in rust pathogen populations, but the actual genes that occur may differ in different parts of the world, as for example in *P. graminis* (Luig, 1983). Of course, their presence at a high frequency in the pathogen population makes it unlikely that the matching resistance genes would be selected in breeding for resistance, except perhaps in combination with other genes.

Dynamics of introduced resistance genes and matching pathogenicity genes

Following the introduction of genes for resistance, rust pathogens have shown diverse responses. Frequently the result has been the selection of previously unknown races carrying pathogenicity for the introduced resistance. Thus the use of resistance has controlled, to some extent, the evolution of pathogenicity (Johnson, 1961).

At the time resistance is introduced, matching pathogenicity must be absent from, or very rare in, the pathogen population. This may indicate that there are forces of selection against that pathogenicity (Leonard, this volume). In the presence of such forces a gene for pathogenicity would be deleterious to the pathogen if it was not necessary to overcome a matching resistance gene. This could be expected to lead to a decline from high frequencies of such unnecessary genes in the population following the withdrawal of a resistance gene. However, it should be noted that rates of change of gene frequencies can be influenced by processes restricting reassortment of genes, random events affecting the survival of genes at low frequencies (stochastic effects) and, in the case of the cereal rusts, by dominance and recessiveness of pathogenicity in the dikaryotic uredial stage of the cereal rust pathogens (Wolfe & Knott, 1982; Gale, this volume; Barrett, this volume; Wolfe, this volume).

The idea that pathogenicity genes declined in frequency in pathogen populations when they were not necessary to overcome resistance was first proposed for *Melampsora lini* on flax by Flor (1956). Subsequently this notion and its implications have been much discussed. Vanderplank (1968) referred to the decline in frequency of unnecessary pathogenicity genes as stabilizing selection, although selection leading to change in a gene frequency is correctly described as directional selection in population genetics. Vanderplank (1968) noted examples where directional selection against unnecessary pathogenicity appeared to occur and others where it did not. He used these observations to classify the quality of the matching resistance genes, as strong where unnecessary pathogenicity declined, and weak where it did not (see Wolfe, this volume). Many models of the action of multilines or cultivar mixtures on pathogen populations incorporate a factor for reduced fitness in members of the pathogen population carrying unnecessary pathogenicity genes (e.g. Barrett, 1978).

Measurement of fitness in rust pathogen populations

Because of its potential importance to disease control it is pertinent to ask by what means any deleterious effect on fitness due to an unnecessary gene for pathogenicity could be measured in rust pathogens.

Competition between races of rust pathogens

Experiments have been designed to compare the competitive ability of races of rust pathogens by growing them together on a host susceptible to both races. Such experiments have often demonstrated the rapid ascendancy of one race over another in a few generations. However, results often do not correspond with observations made in the field. For example, Loegering (1951) indicated that race 56 of *P. graminis* was more common than race 17 in samples collected from Ceres wheat in the USA, although in competition experiments between isolates of the two races on seedlings of Ceres, race 17 predominated. Martens (1973) showed that race C1 of *P. graminis* f. sp. *avenae* predominated over several other races when mixed with them on susceptible seedlings and transferred through several generations. When field plots of the susceptible cultivar Victory were infected with three races, including C1, sampling showed that C1 became rare as the season advanced, corresponding with its rarity in Canadian surveys.

Several factors may lead to differences between laboratory and field experiments and between either type of experiment and the results of surveys of pathogen populations. These could include differences in the viability of inoculum, unsuspected specificity in resistance, a change in specificity of resistance as a host develops, environmental effects on resistance or on races, or the use in experiments of pathogen isolates that are not representative of the naturally occurring population. These observations indicate possible difficulties of using data from experiments to interpret observations made in population surveys, as pointed out by Grant & Archer (1983) (see also Leonard, this volume). They therefore proposed the use of data from population surveys to assess the effects of pathogenicity genes on pathogen fitness.

Assessing pathogen fitness from survey data

Grant & Archer (1983) used data from Australian surveys of *P. graminis* f. sp. *tritici* to assess the strength of directional selection against unnecessary pathogenicity for the resistance gene *Sr6*. Watson & Luig (1963) showed that the frequency of pathogenicity for *Sr6* (*p6*) increased in the population of *P. graminis* following the introduction of a wheat cultivar Eureka, carrying the resistance gene *Sr6*, and subsequently declined when the area of Eureka contracted and cultivars with the resistance gene *Sr11* were introduced. One possible result of the introduction of cultivars with *Sr11* could have been selection against spores possessing *p6* but lacking pathogenicity for *Sr11* (*p11*). Such selection might persist if recombination between genes for pathogenicity was limited due to the asexual mode of reproduction of *P. graminis* in Australia. However, Grant & Archer assumed rates of mutation in genes for pathogenicity and population sizes of *P. graminis* such that *p11* would arise in several hundred thousand spores already possessing *p6* per hectare of an infected crop per day. They further assumed that the part of the host population that did not carry *Sr6* (i.e. cultivars other than Eureka) would be susceptible to all members of the pathogen population, including those with and without *p6*. This enabled them to attribute the decline in frequency of *p6* to a direct effect of reduced fitness in spores carrying *p6* when it was not necessary. They indicated that this example was one of relatively few for which data suitable for such an estimate were available, and that the result suggested a smaller reduction in fitness due to an unnecessary gene for pathogenicity than was commonly applied in theoretical models of pathogen populations.

Though this is an interesting conclusion, some aspects of the data from Australia indicate possible difficulties in attributing declines in frequencies of genes for pathogenicity directly to their supposed deleterious effect when not necessary to overcome resistance. According to Watson & Luig (1968) *p6* was first reported in 1942, four years after the introduction of Eureka as the first stem rust resistant cultivar in northern New South Wales and Queensland. At this time Eureka occupied nine per cent of the total wheat area of the region. It continued to increase up to 18 per cent of the area by 1945 but by this time severe infections of stem rust occurred on it and its popularity quickly declined (Figure 14.1). In 1945, 70 per cent of the

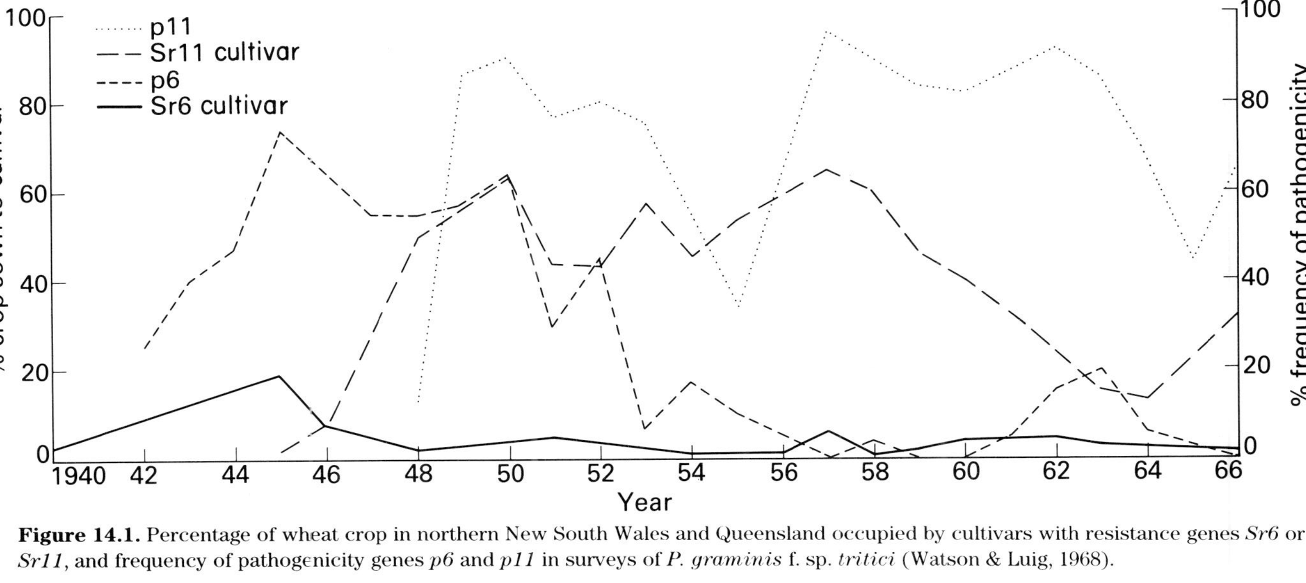

Figure 14.1. Percentage of wheat crop in northern New South Wales and Queensland occupied by cultivars with resistance genes *Sr6* or *Sr11*, and frequency of pathogenicity genes *p6* and *p11* in surveys of *P. graminis* f. sp. *tritici* (Watson & Luig, 1968).

isolates of *P. graminis* collected in surveys possessed *p6*. Starting in 1945 cultivars with the resistance gene *Sr11* were rapidly intoduced. By 1948 these occupied 50 per cent of the total area and, for the first time, stem rust was detected on them. It is likely that, in 1946 and 1947, considerable infection occurred on the declining area of Eureka and that some of the crops were in close proximity to crops of cultivars with *Sr11*. If the rate of mutation to *p11* suggested by Grant & Archer (1983) had occurred it seems most unlikely that infection on the cultivars with *Sr11* would have remained undetected. Thus it seems probable that the effective mutation rate was lower and that there was therefore, for these two years, selection against a high proportion of the spores produced on Eureka, by the cultivars with resistance due to *Sr11*. If such selection occurred it may be questioned why the frequency of *p6* remained at 60 per cent in the surveys when Eureka had declined to between five and 10 per cent of the wheat area. The probable reason was that a high proportion of the samples of *P. graminis* were collected from the small area of infected Eureka while the increasing area of resistant cultivars with *Sr11* gave no samples. Watson (1958) indicated that a high proportion of collections of *P. graminis* from 1941 to 1951 was indeed taken from the cultivar Eureka. It is clearly important to know the cultivars from which survey isolates are collected, and the genes for resistance they contain. Yet this detail is often missing from survey data or ignored in calculations using these data (Wolfe & Knott, 1982). If most isolates are collected from highly susceptible cultivars, these will form a high proportion of the total even when such cultivars occupy only a small proportion of the crop area. The frequency of pathogenicity genes matching resistance genes in such susceptible cultivars will be correspondingly high.

In 1948 the first isolates of *P. graminis* were collected in Australia from cultivars with the gene *Sr11*. These were of two closely related types, one possessing *p11* only, the other possessing both *p6* and *p11*. These coexisted with races possessing *p6* alone, or neither *p6* nor *p11*, as illustrated with data from the Australian survey for 1952 (Table 14.1). Clearly, as cultivars with *Sr11* were widely grown, there was selection against that portion of the population that lacked *p11*, including those with *p6* only. Thus, at least part of the decline of *p6* could be attributed to direct selection against it. However, despite the occurrence of races with combined *p6* and *p11* the frequency of *p6* continued to decline until in 1957 and 1959 it was not detected (Figure 14.1). Could this part of the decline be attributed unequivocally to a deleterious effect of the unnecessary gene *p6*?

Another important occurrence may have influenced events. In 1954 a completely new race of *P. graminis* was identified in Australia. Its many differences from existing races led to the suggestion that it was not derived by mutation from them but may have arrived from outside Australia (Watson, 1981). It was described as highly aggressive, and within eight years the earlier group of races almost completely disappeared from the survey, whether or not they possessed *p6*, *p11*, or both. Apparent derivatives of this new race soon appeared with *p11* and continued to spread rapidly. Thus part of the decline in the frequency of *p6* in the surveys could

Table 14.1. Races of *P. graminis* f. sp. *tritici* in Australia in 1952 in relation to their pathogenicity (+) for *Sr6* and *Sr11**

Race	Pathogenicity for		Number of samples
	Sr6	*Sr11*	
126–0	–	–	49
126–1	+	–	22
222–0	–	–	1
222–2	–	+	10
222–1,2	+	+	4
			Total 86

* From Watson & Luig (1963).

be attributed to spread of the new aggressive races, including those possessing *p11*, but all of which lacked *p6*.

In 1957 the cultivar Eureka was again resistant and was reintroduced but soon became infected with isolates of the new race group now possessing *p6*. After Eureka was again withdrawn the frequency of p6 again declined in surveys up to 1966 in eastern Australia (Figure 14.1). However, in Western Australia races of *P. graminis* with *p6* were increasing and became predominant even though cultivars with *Sr6* occupied only between five and 10 per cent of the wheat area (Luig, 1983). Luig & Watson (1970) believed that, in these races, *p6* had become associated with genes for fitness. In 1969 further new races were discovered in eastern Australia and their many differences from existing races again suggested an origin outside Australia (Watson, 1981). These carried *p6* and rapidly became predominant, although *Sr6* was present in few cultivars, also indicating combination of *p6* with genes for fitness. Of course, genes for fitness could include genes for specific pathogenicity, permitting a wider host range, or other genes affecting survival or reproductive capacity.

These observations on *P. graminis* permit several inferences to be drawn. First, data from surveys in Australia and North America do not indicate random reassortment between genes for pathogenicity in asexually reproducing populations of *P. graminis* (Groth & Roelfs, 1982, and this volume). Deviations from random reassortment in survey data are often a reflection of the frequency with which particular races are isolated with their characteristic combination of pathogenicity genes, as pointed out by Wolfe & Knott (1982) and these deviations are not eliminated by rapid mutational changes.

Because of the restricted gene flow, pathogenicity genes are not independent of their background genotype, including other pathogenicity genes. In these circumstances 'hitch-hiking' of genes must often occur (Gale, this volume). This might explain the recent spread of *p6* in the Australian population of *P. graminis*, when *Sr6* was not common in Australian wheat. Where gene flow is restricted, the

cultivars, had limited the exposure of the *T* gene, and this may have helped to maintain its effectiveness. Populations of *P. graminis* f. sp. *secalis* that could attack barley with the *T* gene were probably sufficiently limited by the small amount of rye grown to prevent severe infection. In addition, Steffenson *et al.* (1985) demonstrated resistance not due to the *T* gene in several barley cultivars.

This interpretation indicates an interesting dynamic interaction involving wheat, rye and barley with the formae speciales *tritici* and *secalis* of *P. graminis* and the epidemiology of stem rust leading to effectiveness of the *T* gene.

Conclusions

These few examples, mainly concerning *P. graminis*, indicate a wide range of effects resulting from the interplay of pathogenicity with the distribution of resistance in the host. Even so, they only illustrate a small part of the dynamic relationships between cereal hosts and their rust pathogens. For example, the interaction of wheat with *P. striiformis* presents an equally wide range of interactions as are shown by wheat with *P. graminis* but with some distinctive features. Rapid matching by *P. striiformis* of resistance in many initially resistant wheat cultivars has curtailed their use and resulted in dynamic changes in specific pathogenicity in the pathogen population. However, numerous cultivars have displayed durable resistance and their use has not led to rapid changes in the pathogen population. The genetic basis of the durable resistance is probably different in different cultivars and much of it remains to be analysed. More details of this complex interaction between a rust pathogen and its host have been outlined elsewhere, and space precludes their inclusion here (Johnson, 1983a, 1984; Sharp, 1983).

An even greater contrast with *P. graminis* on wheat is presented by *P. hordei* on barley and *P. sorghi* on maize. In both examples there is a series of genes giving hypersensitive and usually race-specific resistance which, when effective, is expressed throughout the life of the host plant. These genes interact with specific pathogenicity in the pathogen populations and there is little doubt that their deployment would result in similar types of dynamic changes in the pathogen populations as observed for the rust pathogens of wheat. However, in addition to these genes there is much resistance in commercial cultivars that results in slow disease development, not hypersensitivity, and that shows little race-specificity. *P. sorghi* on maize has remained a disease of minor importance in the USA (Hooker, 1967) and most barley cultivars in recent years in Europe have possessed adequate and durable resistance to *P. hordei*. In both examples the resistance is evidently controlled by the combined effects of several genes, each of relatively small effect individually.

It may be argued that the differences between the several rust diseases of cereals, particularly in the degree of race-specificity of resistance to them, are the result of the way breeders have approached breeding for resistance to them. In my view, it is more likely that pre-existing differences in the biology of the different rust pathogens have contributed largely to the different experiences in attempting to breed for resistance to them. It should not be assumed that experience with one rust disease, such as *P. sorghi* on maize, can serve as a model for breeding for resistance to another rust disease, such as *P. graminis* on wheat. However, similarities between the different rust diseases should be noted and the accumulated knowledge exploited where possible.

Future progress in breeding for resistance to rust diseases of cereals will undoubtedly depend on increasingly precise knowledge of the genetic basis of resistance and of the response of the pathogens to the various components of that resistance. More information on the performance of individual cultivars in relation to their deployment, the effectiveness and durability of their resistance, together with more detailed information about the response of the pathogens, will increase the rate of progress.

Acknowledgement

I thank Dr R.A. McIntosh and his colleagues for providing me with much information about wheat rusts in Australia during my tenure of a Lawrence Pawlett Scholarship at the University of Sydney.

References

Barrett J.A. (1978) A model of epidemic development in variety mixtures. In: *Plant Disease Epidemiology* (Ed. by P.R. Scott & A. Bainbridge), pp. 129–37. Blackwell Scientific Publications, Oxford.

Clifford B.C. (1974) The choice of barley genotypes to differentiate races of *Puccinia hordei* Otth. *Cereal Rusts Bulletin* **2**, 5–6.

Clifford B.C. (1980) Interactions between UK isolates of *Puccinia recondita tritici* and Thatcher wheat lines with single resistance genes. *Cereal Rusts Bulletin* **8**, 10–16.

Dyck P.L. & Johnson R. (1983) Temperature sensitivity of genes for resistance in wheat to *Puccinia recondita*. *Canadian Journal of Plant Pathology* **5**, 229–34.

Flor H.H. (1956) The complementary genic systems in flax and flax rust. *Advances in Genetics* **8**, 29–54.

Grant M.W. & Archer S.A. (1983) Calculation of selection coefficients against unnecessary genes for virulence from field data. *Phytopathology* **73**, 547–51.

Groth J.V. & Roelfs A.P. (1982) The effects of sexual and asexual reproduction on race abundance in cereal rust fungus populations. *Phytopathology* **72**, 1503–7.

Hare R.A. & McIntosh R.A. (1979) Genetic and cytogenetic studies of durable adult–plant resistance in 'Hope' and related cultivars to wheat rusts. *Zeitschrift für Pflanzenzüchtung* **83**, 350–67.

Hooker A.L. (1967) The genetics and expression of resistance in plants to rusts of the genus *Puccinia*. *Annual Review of Phytopathology* **5**, 163–82.

Johnson R. (1983a) Genetic background of durable resistance. In: *Durable Resistance in Crops* (Ed. by F. Lamberti, J.M. Waller & N.A. Van der Graaff), pp. 5–24. Plenum Press, New York.

Johnson R. (1983b) Book review of Host–pathogen Interactions in Plant Disease by J.E. Vanderplank, 1982. *Plant Pathology* **32**, 221–4.

Johnson R. (1984) A critical analysis of durable resistance. *Annual Review of Phytopathology* **32**, 309–30.

Johnson T. (1961) Man–guided evolution in plant rusts. *Science* **133**, 357–62.

Loegering W.Q. (1951) Survival of races of wheat stem rust in mixtures. *Phytopathology* **41**, 55–65.

Luig N.H. (1983) A survey of virulence genes in wheat stem rust, *Puccinia graminis* f. sp. *tritici*. *Advances in Plant Breeding* (Supplement to *Journal of Plant Breeding* **11**). Verlag Paul Parey, Berlin.

Luig N.H. & Watson I.A. (1970) The effect of complex genetic resistance in wheat on the variability of *Puccinia graminis* f. sp. *tritici*. *Proceedings of the Linnaean Society of New South Wales* **95**, 22–45.

McIntosh R.A., Luig N.H., Johnson R. & Hare R.A. (1981) Cytogenetical studies in wheat. X1. *Sr9g* for reaction to *Puccinia graminis tritici*. *Zeitschrift für Pflanzenzüchtung* **87**, 274–89.

Martens J.W. (1973) Competitive ability of oat stem rust races in mixtures. *Canadian Journal of Botany* **51**, 2233–6.

Parlevliet J.E. (1983) Race-specific resistance and cultivar-specific virulence in the barley-leaf rust pathosystem and their consequences for the breeding of leaf rust resistant barley. *Euphytica* **32**, 367–75.

Sharp E.L. (1983) Experience of using durable resistance in the U.S.A. In: *Durable Resistance in Crops* (Ed. by F. Lamberti, J.M. Waller & N.A. Van der Graaff), pp. 385–95. Plenum Press, New York.

Stakman E.C. & Levine M.N. (1922) The determination of biologic forms of *Puccinia graminis* on *Triticum* spp. *University of Minnesota Technical Bulletin* **8**.

Steffenson B.J., Wilcoxson R.D. & Roelfs A.P. (1985) Resistance of barley to *Puccinia graminis* f. sp. *tritici* and *Puccinia graminis* f. sp. *secalis*. *Phytopathology* **75**, 1108–11.

Vanderplank J.E. (1968) *Disease Resistance in Plants*. Academic Press, New York.

Vanderplank J.E. (1982) *Host–pathogen Interactions in Plant Disease*. Academic Press, New York.

Watson I.A. (1958) The present status of breeding disease resistant wheats in Australia. *The Agricultural Gazette of New South Wales* **69**, 630–60.

Watson I.A. (1981) Wheat and its rust parasites in Australia. In: *Wheat Science – Today and Tomorrow* (Ed. by L.T. Evans & W.J. Peacock), pp. 129–147. Cambridge University Press, Cambridge.

Watson I.A. & Luig N.H. (1963) The classification of *Puccinia graminis* var. *tritici*. *Proceedings of the Linnaean Society of New South Wales* **88**, 235–58.

Watson I.A. & Luig N.H. (1968) The ecology and genetics of host–pathogen relationships in wheat rusts in Australia. In: *Proceedings of the 3rd International Wheat Genetics Symposium* (Ed. by K.W. Finlay & K.W. Shepherd), pp. 227–38. Australian Academy of Science, Canberra.

Wolfe M.S. & Knott D.R. (1982) Populations of plant pathogens: some constraints on analysis of variation in pathogenicity. *Plant Pathology* **31**, 79–90.

15 The geographical distribution and frequency of virulence determinants in *Bremia lactucae*: relationships between genetic control and host selection

I.R. CRUTE
National Vegetable Research Station, Wellesbourne, Warwick, UK

Introduction

Lettuce downy mildew caused by the oomycete fungus *Bremia lactucae* Regel is a serious disease of lettuce (*Lactuca sativa* L.) wherever the crop is cultivated (Crute & Dixon, 1981). Resistance to *B. lactucae* has been a major objective of lettuce breeding programmes since the 1920s (Jagger, 1924). However, the identification and deployment of major resistance genes has revealed considerable variation for specific virulence in the pathogen. In this chapter, data on pathogenic variation are examined in relation to information on the structure of the host population and on the genetics of specific virulence.

Disease epidemiology and pathogen life history

During its asexual phase, *B. lactucae* is diploid and coenocytic. It is disseminated by means of air-borne, multinucleate conidiosporangia. The fungus is commonly heterothallic, existing as two sexual compatibility types (Michelmore & Ingram, 1980) and is therefore predominantly outbreeding. Secondary homothallism does, however, occur (Michelmore & Ingram, 1982), and there is some genetic evidence for self-fertilization following stimulation by an isolate of opposite mating type (Norwood & Crute, 1984). The product of the sexual process, the oospore, is a long-lived, resting structure, which may remain dormant until it is specifically stimulated by host exudates (Morgan, 1983). This combination of biological features has importance for a full understanding of the pathogen's population biology.

Downy mildew is not ubiquitous in lettuce crops, and tends to be of seasonal importance occurring mainly after wet conditions which favour infection. Disease outbreaks originate from small, isolated foci which arise from incoming asexual spores or infections from oospores (Crute & Dixon, 1981). In a conducive environment, the disease can spread rapidly, an asexual generation being completed in 7 to 10 days. Sequential sowing of lettuce over a season or even the whole year facilitates crop-to-crop spread. An epidemic may therefore have its origin in a single

Wolfe M.S. & Caten C.E. (1987) *Populations of Plant Pathogens: their Dynamics and Genetics.* Blackwell Scientific Publications, Oxford.

primary infection or in several, and this has obvious implications for the structure of the pathogen population.

The lettuce crop

Lettuce is a crop of considerable diversity both in the genetic sense and in terms of the scale and means of production. Different morphological and physiological types of lettuce (e.g. cos, crisphead or butterhead; long-day or day-neutral) are grown in different geographical regions at different times of year to satisfy market demands. At the two extremes, lettuce may be grown on a small 'market-garden' scale for local consumption or may occupy extensive areas (e.g. in the Salinas Valley of California) from which the product is transported considerable distances both within and between countries. Often, production is in the hands of specialist producers concentrated in particular regions of a country (e.g. the Westlandse region of The Netherlands and the Hesketh Bank area of Lancashire in the UK).

The crop may be seeded directly or be grown from transplants, sometimes purchased from specialist raisers often outside a potential production area. Under optimum growing conditions, some types of lettuce may be marketable in as little as 60 days after sowing and under intensive production systems, ten crops in a year is not exceptional. All the above factors have implications for the population biology of *B. lactucae*, a pathogen intimately associated with its host.

Downy mildew assumes importance in lettuce production because relatively minor attacks can result in reduced quality or even unmarketability as a consequence of the need to remove diseased leaves prior to packing. For this reason, the objective of any control procedure is total absence of disease. As a result, plant breeders have concentrated their efforts on the utilization of major resistance genes which, when effective, provide the high degree of control demanded.

The history of breeding for downy mildew resistance in lettuce and the genetic relationship between host and pathogen

Breeding for resistance to downy mildew in lettuce commenced in 1924 in the USA (Jagger, 1924) and almost 60 years later continues to be actively pursued in that country. Downy mildew resistance is also a major objective of breeding programmes in The Netherlands, UK, France, Sweden, Czechoslovakia, Israel, Japan and Australia.

The sources of resistance used and the resulting combinations represented in commercially relevant lettuce cultivars are best examined by reference to breeding programmes conducted in the USA, The Netherlands, and the UK. In the USA, the concentration has been almost exclusively on crisphead types, whilst initially, butterhead types for protected cropping were the major concern of Dutch programmes. In the UK, breeding concentrated on butterhead types for field cropping. Figure 15.1 illustrates the parallel development and interrelations between these programmes and gives dates for the release of important cultivars. Further details and relevant references are given in the review by Crute & Dixon (1981).

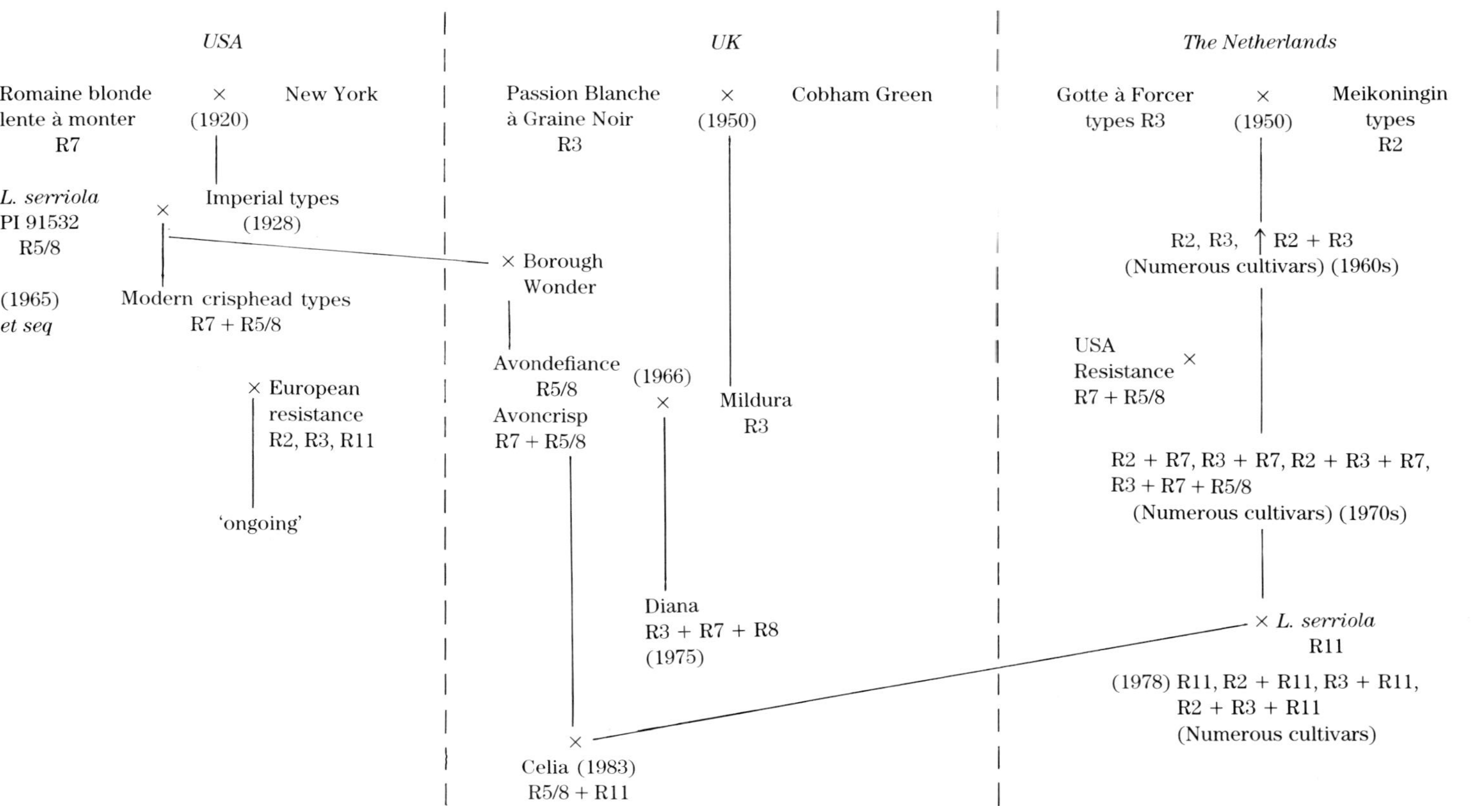

Figure 15.1. A summary of the main stages in the history of lettuce breeding for downy mildew resistance carried out in USA, UK and The Netherlands, identifying the utilization of specific resistance factors R2, R3, R5/8, R7 and R11.

Simultaneously with the exploitation of resistance sources, variation for specific virulence became apparent and began to be investigated. For 50 years, these studies resulted in the taxonomic categorization of numerous physiologic races conventionally identified by their range of virulence on different sets of differential cultivars (Crute & Dixon, 1981; Caten, this volume).

Interpretation of host variation for specific resistance and pathogen variation for specific virulence in terms of a hypothetical gene-for-gene relationship was subsequently made (Crute & Johnson, 1976a), As a result, 16 race-specific resistance factors (R-factors 1–16) have now been identified in lettuce cultivars (Johnson *et al.*, 1977, 1978; Crute & Lebeda, 1981, 1983; Yuen & Lorbeer, 1983). This does not include other race-specific factors identified in related wild *Lactuca* species (Norwood *et al.*, 1981). These 16 factors exist in more than 40 different combinations (Table 15.1). Where genetic studies have been conducted, R-factors appear to be inherited as single dominant genes (*Dm* genes) and both linkage and allelism have been reported (Norwood & Crute, 1980; Bannerot & Boulidard, 1976).

Of these 16 R-factors, only five (R2, R3, R7, R5/8 and R11) have been deliberately used to any great extent in cultivar development, and it is on these and the matching virulence factors in *B. lactucae* that this chapter concentrates.

Table 15.1. Putative combinations of 16 specific resistance factors to *B. lactucae* located in lettuce cultivars

R-factor combination	Example cultivar/line
0	Cobham Green
1	Blondine
2	
3	Dandie
4	R4/T57/E
5/8[a]	Valmaine/Valverde[a]
6	Sabine
7	
9	
10	
11	Mondian
12	Hilde
13[b]	Vanguard 75[b]
14	Lednicky
15[b]	Pennlake[b]
16	Saffier
1 2	Cristallo R2[c]

1		3										Mildura
1						7						Bremex
1									11			Kitty
	2		4									Plus
	2					7						Ravel
	2								11			Topass
	2									14		Liba
		3	4									Kwiek
		3		5/8								Polo
		3				7						Plevanos
		3							11			Juliet
			4								15[b]	Gelber Winterkönig
				5/8	6							Avondefiance
				5/8		7						Calmar
				5/8			9					Bourguignonne[d]
				5/8				10				Sucrine
					6	7						Karma
					6				11			Mussette
						7					15[b]	Mesa 659
1	2		4									Ancora
1	2					7						Portato
1					6	7						Ardente
	2	3	4									Type 57
	2	3				7						Pascal
	2	3							11			Tannex
1	2								11			Hamlet
		3		5/8		7						Diana
				5/8	6	7						Avoncrisp
				5/8	6				11			Celia

Notes: [a] R5 and R8 are now thought to be identical (see text).
[b] Nomenclature of R13 as detailed by Yuen & Lorbeer (1983).
'R13' defined by Crute & Lebeda (1983) becomes R15.
[c] NVRS selection of Cristallo.
[d] Stocks of Bourguignonne may differ in specific resistance.

The structure of the host population

It is evident that R-factor deployment varies from area to area between and within countries, between seasons, and between crop types. Sometimes the result is the homogeneous exposure of a particular R-factor combination. For example, the cultivar Calmar (R7 + R5/8) was grown almost exclusively during the 1960s in parts of California and the same is now true of the cultivar Salinas (also R7 + R5/8). In contrast, a mosaic of R-factors may also result. Table 15.2 shows the number of samples of *B. lactucae* obtained from cultivars carrying different combinations of R-factors R2, R3, R7, R5/8 and R11 during a survey conducted jointly by the Agricultural Advisory and Development Service (ADAS), National Institute of Agricultural Botany (NIAB) and National Vegetable Research Station (NVRS) in the UK between 1975 and 1978. Three hundred and five samples were obtained from commercial lettuce crops over this period and their virulence phenotypes were tested at NIAB by Mr I. Wright. The data in Table 15.2 are inevitably biased by the fact that these were the cultivars most attacked by the fungus, however the 'mosaic' effect is clear and it is also obvious that the mosaic is uneven, with some R-factor combinations being more frequently represented than others.

Differences which may exist from season to season in R-factor deployment are also seen from Table 15.2 in which are listed the combinations of R-factors R2, R3, R7, R5/8 and R11 present in commercial cultivars for outdoor summer (F) and

Table 15.2. Number of samples of *B. lactucae* received from cultivars carrying different combinations of R2, R3, R5/8, R7 and R11 during a survey conducted in the UK between 1976 and 1978

R-factor combination	Cultivar type[a]	No. of samples	% of total
0	F	71	23
2	P	43	14
3	P	21	7
5/8	F	14	5
7	F	7	2
11	F	2	1
2 + 7	P	41	13
2 + 11	P	1	<1
3 + 7	F & P	69	23
5/8 + 7	F	16	5
2 + 3 + 7	P	4	1
3 + 7 + 5/8	F	16	5
		305	

[a] F = used in summer field crops; P = used in winter protected crops.

protected winter (P) use in the UK. It is apparent that R2 and R5/8 are found exclusively in protected and field cultivars respectively.

In summary, the structure of the host population is characterized by differences between countries and seasons, the existence of numerous R-factor combinations and the fact that these have been in commercial use over different time periods in different regions.

The genetics of specific virulence

Since the discovery that *B. lactucae* is normally heterothallic, and exists as two sexual compatibility types (B1 and B2) (Michelmore & Ingram, 1980), it has been possible to conduct matings between isolates differing in specific virulence and observe the segregation in resulting progeny. Details of the methods used to produce, recover and test sexual progeny have been described elsewhere (Michelmore & Ingram, 1981; Norwood *et al.*, 1983; Michelmore & Crute, 1982; Norwood & Crute, 1983; Michelmore *et al.*, 1984), and substantial information on the genetic control of specific virulence has accumulated (Norwood *et al.*, 1983; Michelmore *et al.*, 1984; Norwood & Crute, 1984). Some of the findings have an important bearing on the interpretation of population studies.

In common with many other parasites (Christ *et al.*, this volume), specific avirulence (*A*) in *B. lactucae* is dominant to virulence (*a*) and these characters may be controlled by alleles at a different locus for each R-factor. In the case of the R11/A11 combination, segregation ratios imply that two matching gene pairs control the relationship. There is also evidence that loci controlling virulence for R2 and R11 are linked, but no other indications of linkage between virulence loci have been obtained. In no case is there evidence for duplicate genes controlling virulence for the same R-factor.

In crosses where virulence for R1 and R3 segregated together, there was an apparent epistatic relationship between avirulence for the R1 differential (Blondine) and for the R3 differential (Mildura). Isolates virulent on Blondine and avirulent on Mildura were found, but not *vice versa*, and all isolates virulent on Mildura were also virulent on Blondine. Further investigation suggested that rather than carrying only R3, Mildura also carried R1, which was previously undetected. Detection of two genes in Mildura was complicated by the fact that they are probably linked.

A similar situation was discovered in the relationship between differentials thought to carry R5 (Valmaine), R8 (Valverde), R9 (Bourguignonne) and R10 (Sucrine). Michelmore *et al.* (1984) drew attention to the close association between virulence for R5, R8 and R10 in both segregating populations and also isolate surveys. In addition, an apparent epistatic relationship between avirulence for R9 and that for R5, R8 and R10 was also evident in that Bourguignonne was not recorded as susceptible when differentials for R5, R8 and R10 were resistant. Apparent associations between virulence for these four R-factors have been reported in various studies (Dixon & Wright, 1978; Wellving & Crute, 1978; Lebeda,

1981; Yuen & Lorbeer, 1982; Gustafsson *et al.*, 1983; Trimboli & Crute, 1983) and linkage together with a 'hitch-hiking' effect have been suggested to explain the findings. As a result of further genetic studies, it now seems likely that R5 and R8 are in fact identical, that Sucrine (R10) carries R5/8 in addition to R10 and that Bourguignonne carries R5/8 in addition to R9 (Norwood & Crute, 1984). Some batches of Bourguignonne may also carry R1 (Norwood & Crute, 1985).

Evidence for a genuine non-allelic interaction in the expression of specific virulence has been found for three loci. Crosses between some isolates resulted in segregations for virulence on R1, R4 and R5/8 which were best accommodated by implicating a second locus at which a dominant allele (*I*) 'inhibited' the expression of avirulence. Hence, while avirulence may result from the genotypes *Aaii* or *AAii*, all of the following genotypes result in virulence: *AAIi*, *AAII*, *AaIi*, *AaII*, *aaIi*, *aaII* or *aaii*. The best evidence for the existence of the I allele is when an avirulent isolate (*AAii*) is crossed with two different virulent isolates (*aaii*, *aaII*) and yields F1 progeny which are all avirulent in the former case and all virulent in the latter. Alternatively, crosses between virulent and avirulent parents may give 7:1, 13:3 and 3:1 ratios for virulent: avirulent progeny.

So far, genotypes have been established for 13 parent *B. lactucae* isolates with respect to 11 R-factors and evidence for non-allelic modification by inhibitor loci exists for three A loci. Since such inhibitor loci have been identified in other host-parasite systems (Lawrence *et al.*, 1981; Christ *et al.*, this volume) it is not unreasonable to suggest that it could be a general phenomenon indicative of a further tier of genetic control above the well-established gene-for-gene relationship. These findings have considerable implications for studies aimed at understanding the molecular basis of gene-for-gene specificity. The existence of the *I* allele is consistent with the view that incompatibility results from the direct interaction between the products of the dominant *R* and *A* alleles (Ellingboe, 1981). It can be suggested that the product of *I* interferes with this association and thus restores compatibility.

The fact that several different genotypes can result in a similar phenotype also has bearing on the interpretation of pathogen population surveys. It is not possible to assume that the frequency of an observed virulence necessarily equates with the frequency of a particular allele in the population. Furthermore, different genotypes classed together as conditioning virulence may in fact express this virulence to quantitatively different degrees which has implications for the interpretation of studies on pathogen evolution.

Of the 13 isolates of *B. lactucae* for which genotypes have been established, all but one are heterozygous at least at one locus involved in the control of specific virulence. This finding may be expected for an organism which is outbreeding under most circumstances. One possible implication of this is that it could theoretically allow for the production of virulent segregants as a result of somatic recombination during the asexual phase. This has been discussed previously (Crute & Norwood,

1980) and in a different context, somatic recombination is known to occur in oomycete fungi (Sansome, 1980). However, no clear evidence for somatic recombinants expressing additional virulence has yet been obtained. The multinucleate asexual spore may mean that recombinant nuclei are rarely expressed.

Another important implication of heterozygosity relates to phenotypic expression. The number of copies of the *A* allele may, for some genes and in some environments, affect how strongly incompatibility is expressed. Evidence has been obtained that gene-dosage effects in both the fungus and the host can influence the extent of colonization by the parasite (Crute & Norwood, 1986). This phenomenon, together with the observed variation for general fitness determinants or specific modifying genes, may explain the occurrence of reaction types intermediate between complete virulence/susceptibility and complete avirulence/resistance in both segregating sexual progeny and isolate surveys.

The distribution and frequency of specific virulence determinants

Sampling and testing methods

The sporadic and unpredictable occurrence of downy mildew infections in lettuce crops has inevitably meant that virulence surveys have tended to be unstructured and non-random. Surveys have relied upon opportunist sampling and testing of isolates. Furthermore, studies have generally been made of temporally and geographically discrete populations which complicates interpretation of the data and makes some forms of analysis inappropriate (Wolfe & Knott, 1982). Since lettuce is a minor crop, the resources allocated to population sampling and testing have been more restricted than is the case with major crops. The data discussed below may not therefore describe accurately pathogenic variation in *B. lactucae* and such shortcomings in technique are acknowledged. Future work should pay more attention to survey design and sampling techniques (Groth & Roelfs, this volume).

A 'sample' of *B. lactucae* in surveys so far conducted may comprise one diseased leaf, several leaves from one or a few plants, or many leaves collected randomly from a crop. Whichever is the case, it is usual to prepare a spore suspension from the diseased leaves either immediately after collection or after incubation at high humidity to encourage further sporulation. The resulting suspension is either used to inoculate a 'universally susceptible' cultivar in order to produce a new crop of spores for testing, or may be directly applied to a set of appropriate differential lettuce cultivars.

Various methods, which differ only in detail, have been described for inoculating and assessing the reaction of differential cultivars (Crute & Dickinson, 1976; Wellving & Crute, 1978; Dixon & Wright, 1978; Lebeda, 1981; Yuen & Lorbeer, 1982) but the procedures recently described by Michelmore & Crute (1982) have proved less labour intensive than those previously employed.

Data from surveys of specific virulence

Surveys of the frequency of occurrence of *B. lactucae* virulence phenotypes within different countries have been reported for Denmark (K. Thinggard, personal communication), the Federal Republic of Germany (Zinkernagel, 1975; Crüger, 1976; Handke & Bandze, 1976; Handke *et al.*, 1979), Sweden (Wellving & Crute, 1978; Gustafsson *et al.*, 1983), Israel (Netzer, 1973, and personal communication), Finland (Osara & Crute, 1981), Czechoslovakia (Lebeda, 1981, 1982), Australia (Trimboli & Crute, 1983), USA (Zink *et al.*, 1978; Yuen & Lorbeer, 1982) and The Netherlands (Blok & Van Bakel, 1976). In the UK, data collected between 1973 and 1975 have been reported by Dixon & Wright (1978) and the data collected during the joint survey conducted by ADAS, NIAB and NVRS between 1976 and 1978 were summarized by Dixon (1978). In this Chapter, the data collected during this survey have been re-examined, intepreted in the light of recent developments, and are discussed in more detail.

One feature common to all the above studies is the frequent occurrence of apparently 'unnecessary' virulence. Isolates of the fungus commonly appear to carry specific virulence factors which are not required to facilitate pathogenicity on either the cultivar of origin or cultivars commonly grown in the area. However, the suggestion (see above) that some cultivars carrying R3 may also carry R1, and that R5 and R8 are identical, means that the interpretation that virulence is 'unnecessary' may in some cases be incorrect.

Crute & Lebeda (1981, 1983) identified three R-factors for which matching virulence occurs so frequently in most populations as to be almost fixed. However, the detection of these R-factors in the presence of those which are more commonly effective is not possible with available isolates. It is therefore possible that the frequent occurrence of matching virulence does in fact reflect the location of these

Table 15.3. Proportion of *B. lactucae* isolates virulent on eight lines of *L. serriola* carrying seedling resistance not attributable to R1–16

L. serriola line	% of virulent isolates[a]
01	8
02	17
03	46
04	33
05	35
07	11
08	4
09	0

[a] Out of 76 isolates examined.

Table 15.4. A summary of the results of surveys conducted in various countries on the occurrence and frequency of specific virulence to *B. lactucae*

Country: Years:	DK 1979–81	S 1971–82	SF 1972–76	CS 1977–79	D 1974–77	NL 1976	UK 1973–75	UK 1976–78	IL 1973 + 1979	AUS 1979–81	US–CA 1974–78
Virulence factor											
1	0.91	0.73	0.75	0.89	1.00	nd	1.00	1.00	nd	L	0.22
2 + 4	0.60	0.42	0.50	1.00	0.55	0.26	0.76	0.81	0.52	L	0.22
3	0.86	0.67	0.67	0.39	0.66	0.81	0.74	0.94	0.03	L	0.13
5/8	0.33	0.92	0.83	0.65	0.16	nd	0.81	0.39	0.08	H	1.00
6	0.49	0.89	0.42	0.42	nd	nd	nd	0.98	nd	H	1.00
7	1.00	0.97	0.92	0.25	0.57	0.81	0.77	0.88	1.00	H	1.00
11	0.12	0.31	0.08	0.15	0.02	nd	nd	0.05	H	L	0.28
Sample size:	57	468	11	449	137	310	90	305	53	32	32

DK = Denmark
S = Sweden
SF = Finland
CS = Czechoslovakia
D = W. Germany
NL = Netherlands
UK = United Kingdom
IL = Israel
AUS = Australia
US–CA = USA, California

nd = no data

H = present at high frequency
L = absent or present at low frequency

R-factors undetected in commonly deployed cultivars. Crute & Lebeda (1981) have discussed other implications resulting from the discovery of these genes.

The occurrence of apparently superfluous specific virulence is also illustrated by the finding that a substantial proportion of isolates from several different countries proved to be virulent on some members of a set of *L. serriola* lines known to carry seedling resistance not attributable to R1–16 (Table 15.3).

The results of the surveys referred to above are summarized in Table 15.4. Some clear contrasts between virulence factor frequencies in different countries are apparent. Sometimes differences can be readily attributed to the history of R-factor usage while in other cases there is no obvious explanation. One of the most obvious contrasts is between European countries on the one hand and the USA and Australia on the other with respect to matching virulence for R1, R2, R3 and R4. These R-factors originate in early European 'long-day' or 'forcing' butterhead cultivars and have been deployed extensively in cultivars grown in the region for more than a decade. These R-factors are not represented in the 'crisphead' cultivars used almost exclusively in the USA and Australia and, as a consequence, matching virulence, although recorded, is comparatively rare. Within Europe there are some local differences with respect to these virulence factors and these are less easy to explain. For example, in Scandinavia, the frequency of virulence to match R2 + R4 is lower than elsewhere in Europe, while in Czechoslovakia this virulence factor combination appears to be fixed.

Another contrast worthy of note concerns the frequency of virulence for R7. In Czechoslovakia, where American crisphead and modern Dutch-bred cultivars carrying this factor have been less used, matching virulence occurs less frequently than in the rest of Europe. Interestingly, in Israel, virulence for R7 occurred in all isolates, despite the fact that at the time of the survey, cultivars carrying R7 were not commonly grown. Because R7 is present in most American crisphead cultivars developed since the 1920s, it is not at all surprising that in the USA virulence for R7 is apparently fixed.

The other R-factor deployed extensively in the USA is R5/8. The cultivar Calmar (R7 + R5/8) remained free of disease in California for 9 years from its release in 1965 (Welch *et al.*, 1965), although isolates carrying virulence for R5/8 occurred elsewhere in the USA (Sleeth & Leeper, 1966). In contrast, cultivars carrying R5/8 have never proved to be resistant in Australia (D.S. Trimboli, personal communication) and in the UK, virulence to match R5/8 was detected before cultivars developed from American breeding material and carrying this R-factor (Avoncrisp and Avon-defiance) were grown commercially (Channon *et al.*, 1965; Crute & Johnson, 1976b). Virulence for R5/8 is now frequent in the USA and as a result, breeders are concentrating their efforts on European sources of resistance (R1, R2, R3 and R4).

The most recent R-factor to be exposed commercially is R11, and in most countries virulence to match it is generally found at a lower frequency than is virulence to match R-factors which have been in use for longer. An inexplicable exception to this trend is in Israel, where virulence for R11 was detected at a high frequency despite the fact that no cultivars carrying R11 were grown at this time.

Within countries, geographical differences in the occurrence of virulence phenotypes may also occur. Lebeda (1982) has examined this in detail in Czechoslovakia. In the UK, Dixon & Wright (1978) highlighted an example where, in 1974 in the Evesham production area, a phenotype lacking virulence for R2 predominated (Crute & Davis, 1976) although this was not the case elsewhere in the country. It appears that in any particular disease outbreak, a particular virulence phenotype can tend to dominate a region, suggesting a clonal effect from a single initial, source (Gale, this volume). This supposition is strongly supported by more recent findings. Numerous samples of *B. lactucae* taken from sites in the Hesketh Bank area of Lancashire during a severe outbreak of downy mildew at the end of 1983 and early 1984 have been shown to be a single phenotype. All were metalaxyl-resistant (see below), carried virulence for R1–10 but not R11 or R16, and were of B2 sexual compatibility type. Despite these findings, there is also evidence for heterogeneity in the fungus population within a single crop and these data are discussed below.

The UK B. lactucae *virulence survey 1976–1978*

Between 1976 and 1978 tests for specific virulence were conducted at NIAB on 305 isolates of *B. lactucae* collected from commercial lettuce crops in the UK. The original data have been reappraised for the purposes of this paper and will be discussed solely in relation to virulence to match R-factors R2, R3, R7, R5/8 and R11. These are the R-factors of greatest commercial relevance and are those for which, under certain circumstances, matching virulence may be rare (see above).

On the basis of five R-factors it is possible to identify 2^5 (= 32) different virulence phenotypes. Of these 32 possible virulence phenotypes, 18 were identified during the study (Table 15.5). The majority of isolates carried virulence for three out of the five R-factors, while very few carried a single virulence factor, or all five (Table 15.6). The population was dominated by one virulence phenotype which

Table 15.5. Virulence phenotypes of *B. lactucae* identified on the basis of virulence for R2, R3, R5/8, R7 and R11 and their occurrence during a survey of the pathogen population in the UK between 1976 and 1978

Virulence phenotype[a]					Number of times recorded	%
2	3	5/8	7	11		
+	–	–	–	–	10	3
–	–	+	–	–	1	<1
–	–	–	+	–	1	<1
+	+	–	–	–	11	4
+	–	+	–	–	3	1
–	+	–	+	–	10	3
–	–	+	+	–	6	2
–	–	+	+	+	3	1
–	+	+	+	–	37	12
+	–	+	–	+	1	<1
+	–	+	+	–	2	1
+	+	–	+	–	148	49
+	+	+	–	–	10	3
+	–	+	+	+	2	1
+	+	–	+	+	4	1
+	+	+	–	+	2	1
+	+	+	+	–	51	17
+	+	+	+	+	3	1
					Total 305	

[a] The remaining 14 possible virulence phenotypes were not recorded.

Table 15.6. Proportion of *B. lactucae* isolates carrying virulence for different numbers of R-factors R2, R3, R5/8, R7 and R11 found during a survey in the UK between 1976 and 1978

No. of virulence factors	Percentage of isolates (total = 305)
0	0
1	4
2	10
3	66
4	19
5	1

comprised almost half of the total sample (Table 15.5). Isolates of intermediate complexity therefore appeared to predominate, but this could be artifactual.

The frequency of occurrence of virulence to match these R-factors was significantly correlated ($P > 0.01 < 0.05$) with the frequency of the R-factor in the host population sampled. Unfortunately, there is no way of knowing whether the sampled host population reflected accurately the actual host population. No correlation exists when virulence to match other uncommonly utilized R-factors is also considered, which confirms findings reported by Lebeda (1981). For some virulence factors, therefore, frequency does appear to be affected directly by host selection (Leonard, this volume) but this would seem not to be the case for others.

A further finding of interest, which has been discussed in a different context elsewhere (Crute, 1984), was the highly significant disassociation ($P < 0.001$) between virulence to match R2 and R5/8. This appears to result from disruptive selection. Factors R2 and R5/8 are exclusively located in cultivars used in different seasons (Table 15.2). The cultivars are therefore rarely exposed to infection at the same locality and the same time. Consequently, directional selection for combined virulence has not been imposed.

The population structure within crops

Little attention in studies on variation for specific virulence in *B. lactucae* has been focused on the population existing within a lettuce crop or within narrow geographical boundaries. What limited data are available is discussed here.

In the UK survey outlined above, samples were obtained from five sites in both 1976 and 1977, and on each occasion several samples were collected from cultivars carrying different R-factors. The data obtained with respect to virulence for R-factors R2, R3, R5/8, R7 and R11 are given in Table 15.7. Several points of interest emerged from this exercise. It was clear that at a single site, more than one virulence phenotype may occur at any time and that the phenotype detected to some extent depended upon the cultivar of origin. For example, it was generally true

Table 15.7. Variation for specific virulence detected in samples of *B. lactucae* collected from several different cultivars at five sites in the UK in 1976 and 1977

Site	Year	R-factors in cultivar	Virulence phenotype					No. of of samples
			2	3	5/8	7	11	
1	1976	0	−	−	+	+	−	1
		3 + 7	+	+	−	+	−	2
		3 + 7 + 5/8	−	+	+	+	−	1
	1977	0	+	−	+	+	−	1
			+	+	−	+	−	2
		5/8	−	−	+	+	+	1
		3	+	+	−	+	−	1
		2	+	+	−	+	−	2
		3 + 7	+	+	−	+	−	4
2	1976	3 + 7	+	+	−	+		1
		2 + 7	+	+	+	+	−	1
		2	+	+	−	−	−	1
		3	+	+	−	+	−	1
	1977	3 + 7	+	+	−	+	−	3
		0	+	+	+	+	−	1
3	1976	3 + 7	+	+	−	+	−	1
		2	+	+	+	+	−	1
	1977	3 + 7	+	+	−	+	−	2
		2 + 7	+	+	−	+	−	1
		3 + 7 + 5/8	−	+	+	+	−	1
4	1976	7 + 5/8	+	+	+	+	−	2
		0	+	+	+	+	−	1
		3 + 7	+	+	−	+	−	1
	1977	7 + 5/8	−	+	+	+	−	1
		3 + 7	+	+	−	+	−	1
		3 + 7 + 5/8	−	+	+	+	−	1
5	1977	0	+	+	−	+	−	9
			+	+	−	+	+	2
		3	+	+	−	+	+	1
		7 + 5/8	+	+	+	+	+	1
		5/8	+	+	+	+	−	1
		3 + 7 + 5/8	−	+	+	+	−	1
		7	+	+	−	+	−	2

that virulence for R5/8 was confined to isolates from cultivars carrying R5/8, or those without R-factors. It was also clear that more than one virulence phenotype occurred on a single cultivar. Differences and similarities occurred between the virulence phenotypes detected in the successive years and no conclusion on the epidemiological continuity from year to year, either through the sexual or asexual phase, is justified.

Data collected by I. Gustavsson (personal communication) in Sweden illustrate a similar situation. Table 15.8 shows the results of sampling a single cultivar (Great Lakes III, probably carrying R7) at one site on 15 occasions over a 2-year period. The number of samples tested on each occasion varied, but up to seven different virulence phenotypes were detected, with six being found at a single sampling date. In a second series of tests (Table 15.9), isolates taken from different individual plants at the same location on two separate occasions also exhibited different virulence phenotypes.

Table 15.8. Variation for specific virulence detected in samples of *B. lactucae* taken on 15 occasions over a 2-year period from a single cultivar grown at one site in Sweden (data collected by I. Gustavsson)

Year	Sample time	Virulence phenotype					Samples taken
		12	3	5/8	7	11	
1981	1	−	+	+	+	−	4
	2	−	+	+	+	−	1
	3	+	+	+	+	+	3
	4	+	+	+	+	+	2
1982	1	+	+	+	+	−	5
	2	+	+	+	+	+	10
		−	+	+	+	+	
		−	+	+	+	−	
	3	+	+	+	+	+	11
		−	+	+	+	+	
	4	+	+	+	+	+	14
		−	+	+	+	−	
		−	+	+	+	+	
		−	−	+	+	−	
	5	+	+	+	+	+	2
	6	−	−	+	+	+	1
	7	−	−	+	+	+	10
		−	+	+	+	−	
		+	−	+	+	−	
		+	+	+	+	−	
		+	+	+	+	+	
		−	−	+	+	−	
	8	+	+	+	+	+	10
		+	+	+	+	−	
	9	+	+	+	+	+	10
		+	+	+	+	−	
	10	+	+	+	+	+	10
		+	+	+	+	−	
	11	+	+	+	+	−	6
		−	+	+	+	−	
		−	−	+	+	+	

Table 15.9. Variation for specific virulence detected in samples of *B. lactucae* from individual plants of a single cultivar taken from the same site in Sweden at two different dates (data collected by I. Gustavsson)

Sample date	Plant	Virulence phenotype				
		2	3	5/8	7	11
1	1	−	+	+	+	+
	2	−	+	+	+	+
	3	−	−	+	+	+
	4	−	+	+	+	+
2	1	−	+	+	+	+
	2	−	+	+	+	+
	3	−	−	+	+	−
	4	−	−	−	+	−

Insensitivity to metalaxyl

One reason why the UK survey of specific virulence in *B. lactucae* stopped in 1978 was due to the widespread use of the fungicide metalaxyl for downy mildew control. The disease ceased to cause major problems for growers using regular prophylactic treatments and there was no longer a ready supply of isolates from commercial crops.

In a limited survey of *B. lactucae* isolates conducted between 1978 and 1982, Wynn & Crute (1981) found no evidence for variation in sensitivity to metalaxyl in the fungus which would threaten commercial control. Using a specially developed bioassay (Wynn & Crute, 1983), it was determined that following recommended application rates, control failure would only be experienced if an isolate insensitive to 0.1−1.0 μg/ml emerged. All normally sensitive isolates tested were completely inhibited at 0.1 μg/ml or less in a standard laboratory test. However, in October 1983, isolates insensitive to $>$ 100 μg metalaxyl per ml were obtained from a severe outbreak of the disease on protected lettuce crops in the Hesketh Bank area of Lancashire.

There may be several reasons why it took 5 years of intensive usage before metalaxyl resistance was detected in *B. lactucae* whereas with some other host−pathogen combinations, insensitivity was detected soon after commercial use commenced (Skylakakis, Georgopoulos, this volume). Firstly, because the adoption of metalaxyl for lettuce downy mildew control was so extensive, the size of the pathogen population probably declined to a low level very rapidly. Hence the size of the fungus population from which selection could occur was reduced. Secondly, the crop area involved in the UK is relatively small and as a result, the number of successful host penetrations on which selection could operate may have been too low to select a rare mutant (Gale, this volume). Finally, the fact that most lettuce cultivars carry race-specific resistance may have resulted in an unconscious integration of control measures. It may have been particularly beneficial for control that the introduction of metalaxyl coincided with the more extensive utilization of cultivars carrying R11 (Crute, 1984).

Although it has serious consequences for disease control, the occurrence of a high level of insensitivity to metalaxyl in a single-virulence phenotype (see above), confined to one production area, presents an excellent opportunity to study in detail the process of gene-flow within the *B. lactucae* population.

Acknowledgement

I am grateful to Ingrid Gustavsson for permission to quote the results of unpublished work.

References

Bannerot H. & Boulidard L. (1976) Contribution to the study of inheritance of resistance to *Bremia lactucae. Proceedings Eucarpia Meeting on Leafy Vegetables, Wageningen* pp. 86–7. Institute for Horticultural Plant Breeding, Wageningen, The Netherlands.

Blok I. & Van Bakel J.J.M. (1976) Wit in sla. *Groenten en Fruit* No. **31638**, 172.

Channon A.G., Webb M.J.W. & Watts L.E. (1965) Studies on two races of *Bremia lactucae* Regel. *Annals of Applied Biology* **56**, 389–97.

Crüger G. (1976) Zur verbreitung von pathotypen des erregers des falschen mehltaus (*Bremia lactucae*) in salatbau. *Gemüse* **12**, 350–1.

Crute I.R. (1984) The integrated use of genetic and chemical methods for control of lettuce downy mildew (*Bremia lactucae* Regel). *Crop Protection* **3**, 223–41.

Crute I.R. & Davis A.A. (1976) New virulence gene combinations in British isolates of *Bremia lactucae. Annals of Applied Biology* **83**, 173–5.

Crute I.R. & Dickinson C.H. (1976) The behaviour of *Bremia lactucae* on cultivars of *Lactuca sativa* and on other composites. *Annals of Applied Biology* **82**, 433–50.

Crute I.R. & Dixon G.R. (1981) Downy mildew diseases caused by the genus *Bremia* Regel. In: *The Downy Mildews* (Ed. by D.M. Spencer), pp. 423–60. Academic Press, London.

Crute I.R. & Johnson A.G. (1976a) The genetic relationship between races of *Bremia lactucae* and cultivars of *Lactuca sativa. Annals of Applied Biology* **83**, 125–37.

Crute I.R. & Johnson A.G. (1976b) Breeding for resistance to lettuce downy mildew, *Bremia lactucae. Annals of Applied Biology* **84**, 287–90.

Crute I.R. & Lebeda A. (1981) Evidence for a race-specific resistance factor in some lettuce (*Lactuca sativa* L.) cultivars previously considered to be universally susceptible to *Bremia lactucae* Regel. *Theoretical and Applied Genetics* **60**, 185–9.

Crute I.R. & Lebeda A. (1983) Two new resistance factors to *Bremia lactucae* identified in cultivars of lettuce. *Tests of Agrochemicals and Cultivars* No.4 (*Annals of Applied Biology* **102**, Supplement), pp. 128–9.

Crute I.R. & Norwood J.M. (1980) Inter-isolate variation for virulence in *Bremia lactucae. Annals of Applied Biology* **94**, 273–310.

Crute I.R. & Norwood J.M. (1986) Gene-dosage effects on the relationship between *Bremia lactucae* (downy mildew) and *Lactuca sativa* (lettuce): the relevance to a mechanistic understanding of host parasite specificity. *Physiological Plant Pathology*. (In press).

Dixon G.R. (1978) Monitoring vegetable diseases. In: *Plant Disease Epidemiology* (Ed. by P.R. Scott & A. Bainbridge), pp. 71–8. Blackwell Scientific Publications, Oxford.

Dixon G.R. & Wright I.R. (1978) Frequency and geographical distribution of specific virulence factors in *Bremia lactucae* populations in England from 1973 to 1975. *Annals of Applied Biology* **88**, 187–294.

Ellingboe A.H. (1981) Changing concepts in host–pathogen genetics. *Annual Review of Phytopathology* **19**, 125–43.

Gustafsson M., Arhammer M. & Gustavsson I. (1983) Linkage between virulence genes, compatibility types and sexual recombination in the Swedish population of *Bremia lactucae. Phytopathologische Zeitschrift* **108**, 341–54.

Handke S. & Bandze E. (1976) Mehltaurassen im salatanbau in den jahren 1974 und 1975. *Rheinische Monatszeitschrift für Gemüse* **64**, 341.

Handke S., Bandze E. & Radies M. (1979) Mehltaurassen im salatanbau in den jahren 1976 und 1977. *Gemüse* **15**, 186–8.

Jagger I.C. (1924) Immunity to mildew (*Bremia lactucae* Regel) and its inheritence in lettuce. *Phytopathology* **14**, 122.

Johnson A.G., Crute I.R. & Gordon P.L. (1977) The genetics of race specific resistance in lettuce (*Lactuca sativa*) to downy mildew (*Bremia lactucae*). *Annals of Applied Biology* **86**, 87–103.

Johnson A.G., Laxton S.A., Crute I.R., Gordon P.L. & Norwood J.M. (1978) Further work on the genetics of race specific resistance in lettuce (*Lactuca sativa*) to downy mildew (*Bremia lactucae*). *Annals of Applied Biology* **89**, 257–64.

Lawrence G.J., Mayo G.M.E. & Shepherd K.W. (1981) Interactions between genes controlling pathogenicity in the flax rust fungus. *Phytopathology* **71**, 12–19.

Lebeda A. (1981) Population genetics of lettuce downy mildew (*Bremia lactucae*). *Phytopathologische Zeitschrift* **101**, 228–30.

Lebeda A. (1982) Geographic distribution of virulence factors in the Czechoslovakian population of *Bremia lactucae* Regel. *Acta Phytopathologica Academiae Scientiarum Hungaricae* **17**, 65–79.

Michelmore R.W. & Crute I.R. (1982) A method for determining the virulence phenotype of isolates of *Bremia lactucae. Transactions of the British Mycological Society* **79**, 542–6.

Michelmore R.W. & Ingram D.S. (1980) Heterothallism in *Bremia lactucae. Transactions of the British Mycological Society* **75**, 47–56.

Michelmore R.W. & Ingram D.S. (1981) The recovery of sexual progeny following sexual reproduction of *Bremia lactucae* Regel. *Transactions of the British Mycological Society* **77**, 131–7.

Michelmore R.W. & Ingram D.S. (1982) Secondary homothallism in *Bremia lactucae. Transactions of the British Mycological Society* **78**, 1–9.

Michelmore R.W., Norwood J.M., Ingram D.S., Crute I.R. & Nicholson P. (1984) The inheritance of virulence in *Bremia lactucae* to match resistance factors 3, 4, 5, 8, 9, 10 and 11 in lettuce (*Lactuca sativa*). *Plant Pathology* **33**, 301–15.

Morgan W.M. (1983) Viability of *Bremia lactucae* oospores and stimulation of their germination by lettuce seedlings. *Transactions of the British Mycological Society* **80**, 403–8.

Netzer D. (1973) Physiologic races of *Bremia lactucae* in Israel. *Transactions of the British Mycological Society* **61**, 375–8.

Norwood J.M. & Crute I.R. (1980) Linkage between genes for resistance to downy mildew (*Bremia lactucae*) in lettuce. *Annals of Applied Biology* **94**, 127–35.

Norwood J.M. & Crute I.R. (1983) Infection of lettuce by oospores of *Bremia lactucae. Transactions of the British Mycological Society* **81**, 144–7.

Norwood J.M. & Crute I.R. (1984) The genetic control and expression of specificity in *Bremia lactucae* (lettuce downy mildew). *Plant Pathology* **33**, 385–400.

Norwood J.M. & Crute I.R. (1985) Race specific resistance to lettuce downy mildew (*Bremia lactucae*) in the lettuce cultivar Bourguignonne Grosse Blonde d'Hiver. *Annals of Applied Biology* **106**, 595–9.

Norwood J.M., Crute I.R. & Lebeda A. (1981) The location and characteristics of novel sources of resistance to *Bremia lactucae* Regel (downy mildew) in wild *Lactuca* L. species. *Euphytica* **30**, 659–68.

Norwood J.M., Michelmore R.W., Crute I.R. & Ingram D.S. (1983) The inheritance of specific virulence in *Bremia lactucae* (downy mildew) to match resistance factors 1, 2, 4, 6 and 11 in *Lactuca sativa* (lettuce). *Plant Pathology* **32**, 177–86.

Osara K. & Crute I.R. (1981) Variation for specific virulence in the Finnish *Bremia lactucae* population. *Annales Agriculturae Fenniae* **20**, 198–209.

Sansome E. (1980) Reciprocal translocation heterozygosity in heterothallic species of *Phytophthora* and its significance. *Transactions of the British Mycological Society* **74**, 175–85.

Sleeth B. & Leeper P.W. (1966) Mildew resistant lettuce susceptible to a new physiologic race of *Bremia lactucae* in south Texas. *Plant Disease Reporter* **50**, 460.

Trimboli D.S. & Crute I.R. (1983) The specific virulence characteristics of *Bremia lactucae* (lettuce downy mildew) in Australia. *Australasian Plant Pathology* **12**, 58–60.

Welch J.E., Grogan R.G., Zink F.W., Kihara G.M. & Kimble K.A. (1965) Calmar – a new lettuce variety resistant to downy mildew. *California Agriculture* **19**, 3–4.

Wellving A. & Crute I.R. (1978) The virulence characteristics of *Bremia lactucae* populations present in Sweden from 1971–1976. *Annals of Applied Biology* **89**, 251–6.

Wolfe M.S. & Knott D.R. (1982) Populations of plant pathogens: some constraints on analysis of variation in pathogenicity. *Plant Pathology* **31**, 79–90.

Wynn E.C. & Crute I.R. (1981) Variation for sensitivity to metalaxyl in *Bremia lactucae* (lettuce downy mildew). *Netherlands Journal of Plant Pathology* **87**, 246.

Wynn E.C. & Crute I.R. (1983) Bioassay of metalaxyl in plant tissue. *Annals of Applied Biology* **102**, 117–21.

Yuen J.E. & Lorbeer J.W. (1982) Virulence factors of *Bremia lactucae* in New York. *Phytopathology* **72**, 1363–7.

Yuen J.E. & Lorbeer J.W. (1983) A new gene for resistance to *Bremia lactucae*. *Phytopathology* **73**, 159–62.

Zink F.W., Grogan R.G. & Kimble K.A. (1978) Search for resistance to downy mildew and the nature of inheritance of resistance. *5th Annual Report of the Californian Iceberg Lettuce Research Program, Salinas, California, USA*, pp. 23–7.

Zinkernagel V. (1975) Das auftreten physiologischer rassen von *Bremia lactucae* Regel, dem erreger des falschen mehltaus bei salat. *Nachrichtenblatt des Deutschen Pflanzenschutzdienstes* **27**, 185–8.

16 Recent genetic changes in the *Ophiostoma ulmi* population: the threat to the future of the elm

C.M. BRASIER
Forest Research Station, Alice Holt Lodge, Farnham, Surrey, UK

Introduction

A genus of some 40 species, the elm is one of the most useful trees in the Northern Hemisphere. Owing to its combination of beauty and sometimes quite remarkable hardiness against wind, salt, drought and cold it has been increasingly planted in recent centuries to provide shade trees in towns and cities, shelterbelts in exposed areas, and to furnish rural landscapes (Richens 1984). Since the arrival of Dutch elm disease, however, the elm's position both as an obvious planting choice and as a principal component of the landscape has been increasingly in question. Dutch elm disease is a vascular wilt disease caused by a combination of the fungus *Ophiostoma* (*Ceratocystis*) *ulmi* and beetle vectors of the family Scolytidae. Its first recorded appearance was in north-west Europe at the end of the First World War, at the onset of what would now, with hindsight, be considered the first epidemic of the disease. This epidemic spread both rapidly and widely causing quite serious losses in many areas (see Peace, 1960; Gibbs, 1978) but subsequently declined in severity in most of its locations.

More recently, major changes have occurred in the structure of the *O. ulmi* population which have led to new and catastrophic losses among the elm populations of North America, Europe and south west Asia. These new epidemics stimulated studies aimed at understanding the changes in the pathogen population. Some of these changes will be outlined in the following account, together with evidence as to their likely outcome and their significance for the future of the elm in the affected areas.

The sub-groups of *O. ulmi*

An early fact to emerge from investigations into the causes of the present epidemics was that *O. ulmi* occurred in nature not, as had been previously supposed, as a continuum of variation within one population, but as three discrete subpopulations or subgroups, each with its own characteristics and range of variation (Gibbs & Brasier, 1973; Brasier & Gibbs, 1973; Brasier, 1979, 1981, 1982b). These have been termed the non-aggressive strain and the Eurasian (EAN) and North American (NAN) races of the aggressive strain (Figure 16.1).

Wolfe M.S. & Caten C.E. (1987) *Populations of Plant Pathogens: their Dynamics and Genetics.* Blackwell Scientific Publications, Oxford.

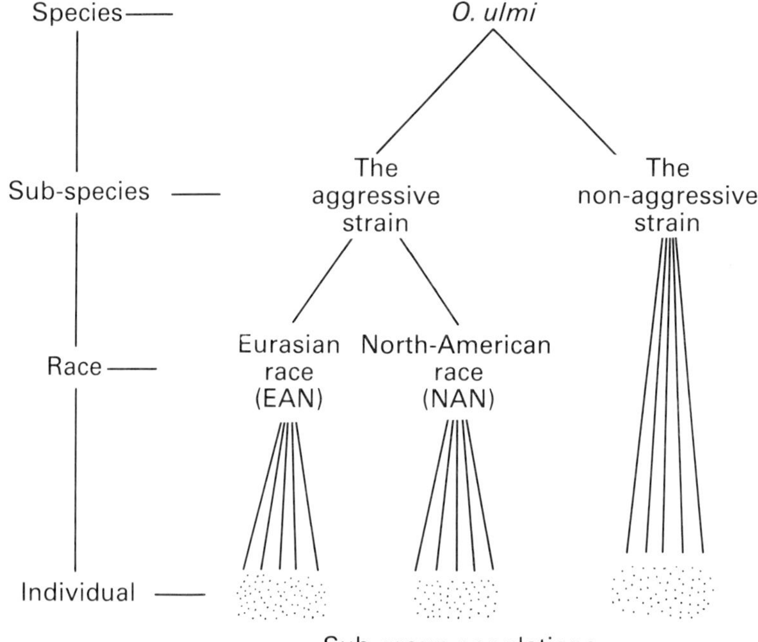

Figure 16.1. The division of *O. ulmi* into three subgroups or subpopulations, each with its own characteristics and range of variation. Adapted from Brasier (1983a).

An important distinction between the two major subgroups, the aggressive and non-aggressive strains, is that the former is a strong and the latter a relatively weak pathogen (Gibbs & Brasier 1973; Brasier 1982a,b). These two subgroups differ widely in many other physiological respects,. including colony morphology, growth rate, optimum temperature for growth, toxin and enzyme production, soluble protein patterns and mating type frequency (e.g. Brasier, 1981, 1982a, b, 1984; Brasier *et al.*, 1981; Bernier *et al.*, 1983; Jeng & Hubbes, 1983; Scheffer & Elgersma, 1982; Svaldi & Elgersma, 1982; Takai, 1974, 1980). Since they also show evidence of considerable reproductive isolation and genome incompatibility (Brasier, 1977, 1978, 1982a, 1984), they are now considered equivalent to subspecies (Brasier, 1982a). The physiological differences between the EAN and NAN races of the aggressive strain, on the other hand, roughly balance their similarities and only a moderate degree of reproductive isolation exists between them (Brasier, 1979, 1981, 1982a, b, 1984). They are presently considered to be equivalent to races in the broadest sense of the term, i.e. as classically applied in population and evolutionary biology (e.g. *sensu* Stebbins, 1966) rather than in the narrow sense of 'physiological race' (Caten, this volume).

The changes in the *O. ulmi* population

The non-aggressive strain is now thought to have been responsible for the first rather milder epidemics of Dutch elm disease which began in north-west Europe around 1910–20, later spreading westwards to North America around 1927 and eastwards into central and eastern Europe and south-west Asia, reaching central Asia around 1939 (Gibbs & Brasier, 1973; Gibbs, 1978). The present much more severe 'second' epidemics are due to the recent spread of the two races of the aggressive strain into areas previously occupied by the non-aggressive. The origin of the aggressive and non-aggressive strains is not the concern of the present paper, but it seems likely that the NAN aggressive may have arrived in the mid-west of North America around 1940–50 (McNabb, 1974), whence it has spread widely (Gibbs *et al.*, 1979; Brasier 1983a), and that the EAN aggressive probably arrived in Romania at about the same time (Brasier, 1983a). In Europe, this has led to two separate migratory episodes: (i) the importation of the NAN race from North America into Britain in the mid-1960s (Brasier & Gibbs, 1973) and its subsequent spread into neighbouring countries of north-west Europe and beyond (Brasier, 1979, 1983a); (ii) the westward migration of the EAN race from central and eastern Europe or from further east (Brasier, 1979, 1983a).

Rather unusually, therefore, there has been a simultaneous development of two massive but independent epidemics in Europe which are now overlapping to the extent that the EAN and NAN races can be found side by side in several European countries (Figure 16.2). The situation is also changing rapidly as the EAN race spreads westwards and the NAN eastwards. In some areas which represent early centres of the current epidemics such as the American mid-west, Britain and Romania, extremely heavy disease losses have already occurred. In Britain alone some 20 million elms are estimated to have died between 1970 and 1980. In countries where EAN or NAN have only recently arrived, such as Spain and Sweden,

Figure 16.2. Summary of the known positions of the EAN and NAN races of the aggressive strain of *O. ulmi* in Europe and south-west Asia in 1983. Based on > 1500 samples collected and tested by the author. T = Tashkent. The distribution of the non-aggressive strain is not shown. Adapted from Brasier (1983a).

equivalent losses are only just beginning. The probability is that as a result of the second epidemics a majority of the mature field elms will be destroyed from the Rocky Mountains to the east coast of North America and throughout Europe to central Asia.

The outcome of the changes

An important question is what the future holds for the next generation of elms, the young seedlings and root suckers which are now developing, sometimes in great numbers, in areas where the mature elms have recently died. In particular, will the disease return and destroy these young elms when they are large enough to support a population of breeding beetles? The answer to this question depends upon the future behaviour of the aggressive strain: whether it will decline or die out in the face of the collapsing host and beetle population, or whether it will survive. At present there appear to be three ways in which a decline of the aggressive strain might come about: (i) through hybridization with the non-aggressive strain, (ii) through its replacement by the non-aggressive strain, (iii) through a reduction in the pathogenicity of the aggressive strain itself, i.e. via internal genetic mechanisms. These possibilities will now be considered in the light of current evidence.

Hybridization of the aggressive and non-aggressive strains

In epidemic outbreak areas sexually compatible isolates of the aggressive and non-aggressive strains can often be obtained from the same piece of diseased elm bark (e.g. Brasier & Gibbs, 1976). Therefore there is a possibility of hybridization occurring, especially during the earlier stages of an epidemic (see later). However, four lines of evidence suggest that such hybridization is unlikely to be an important factor in the decline of the aggressive strain.

Firstly, when crosses are made between the aggressive and non-aggressive strains in the laboratory, the resulting progeny are quite unlike either parental type, being extremely unusual in cultural characteristics and, most importantly, generally rather weak pathogens. Although their production might be expected to contribute to a decline in pathogenicity of the *O. ulmi* population, their various properties also indicate that they are likely to be generally unfit and unlikely to survive (Brasier & Gibbs, 1976; Brasier, 1977, 1982a). Secondly, although such crosses can be forced in the laboratory, other laboratory experiments have shown that a reproductive barrier exists which largely prevents the aggressive strain from being fertilized by the non-aggressive strain in aggressive ♀ × non-aggressive ♂ pairings (Brasier, 1977, 1984). Thirdly, experiments using mites as fertilizing agents in an attempt to reproduce field conditions indicate that in nature the barrier to hybridization between the strains would be virtually total (Brasier, 1978, 1984). The fourth line of evidence comes directly from nature. Although many thousands of fresh isolates of *O. ulmi* have been examined by the author from localities in North America, Europe and south-west Asia where both strains occur side by side, very few have been seen which cannot be confidently assigned to either the non-aggressive or one or other race of the aggressive strain. These few isolates have usually conformed to mutant types of the normal aggressive or non-aggressive wild types, and have not had the characteristics of hybrids. Thus, on present evidence it appears that hybridization between the aggressive and non-aggressive strains is likely to be a rare event, and even if it does occur, any resulting progeny are unlikely to survive. It therefore seems reasonable to suggest that such hybridization is unlikely to be involved in any future decline in pathogenicity either of the aggressive strain or of the *O. ulmi* population as a whole.

Replacement of the aggressive by the non-aggressive strain

It is necessary to look in more detail at the possibility that the aggressive strain may eventually be replaced by the non-aggressive strain as the elm population declines. We begin by considering the relationship between the two strains from the moment that the aggressive strain arrives in territory previously occupied only by the non-aggressive strain. The likely course of events as the epidemic progresses through a locality, both in terms of changes in the frequency of the aggressive and non-aggressive strains and in sizes of the elm and beetle populations, is summarized in Figure 16.3.

When the aggressive strain is first introduced into an area (Figure 16.3B), only the non-aggressive strain is being carried by the local beetle population. The first tree(s) killed by the aggressive strain will therefore be used as breeding material mainly by local beetles carrying the non-aggressive strain. A large number of beetles will therefore leave this tree carrying the non-aggressive strain, resulting in an increase in the number of infections caused by the non-aggressive strain to above its previous (unspecified) level (Figure 16.3C). Thus the non-aggressive strain may initially benefit from the arrival of the aggressive strain. As the aggressive strain gradually increases in frequency through killing more trees and providing more beetle breeding material the number of infections initiated by the non-aggressive strain is overtaken by those initiated by the aggressive strain (Figure 16.3D).

Evidence for this sequence of events comes from sample data which show a relatively high proportion of non-aggressive strain infections in the early to mid-term stages of recent disease outbreaks. Thus, in two of the three main disease outbreak areas in Britain during 1971, the non-aggressive strain accounted for over 22% of infections (Gibbs & Brasier, 1973). The number of infections was probably well above that which would have occurred if the non-aggressive strain had been present on its own. Similar high concentrations of the non-aggressive strain can be found in other EAN or NAN/non-aggressive epidemic front areas, for example in central Turkey (C.M. Brasier, 1981 survey, unpublished).

Data are also available showing the change in the proportion of the aggressive and non-aggressive strains as outbreaks have progressed. The most comprehensive

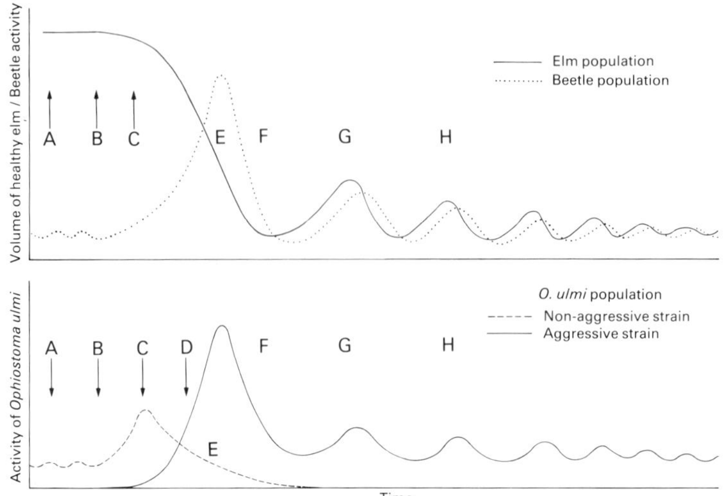

Figure 16.3. Proposed course of events in the current Dutch elm disease epidemics in Europe and their projected outcome. Upper graph, changes in the field elm and beetle vector populations. Lower graph, changes in the *O. ulmi* population.

A Prior to the arrival of the aggressive strain, the elm population is subject to periodic flare-ups of the non-aggressive strain.

B The aggressive strain arrives in a locality. The first tree killed by the aggressive strain is used as breeding material by local beetles carrying the non-aggressive strain. A large number of beetles leave this tree also carrying the non-aggressive strain, and some the aggressive strain.

C This process results in an increase in the number of infections caused by the non-aggressive strain to well above its previous level, and a gradual increase in the number of trees killed by the aggressive strain.

D Through killing more trees and providing a greater volume of beetle breeding material, infections initiated by the aggressive strain increase, rapidly overtaking those initiated by the non-aggressive strain.

E The non-aggressive strain declines.

F Most of the accessible large elms are killed, resulting in a collapse of the beetle population and that of the aggressive strain. The non-aggressive strain is by now virtually eliminated in the main epidemic areas. It survives in a few isolated pockets of elm untouched by the aggressive strain.

G Elm seedlings and root suckers regenerate in large numbers. When large enough to support beetle breeding, they are attacked by the aggressive strain. Due to the marked reduction in the size of available breeding material, the principal beetle vector *Scolytus scolytus* declines, and smaller beetles such as *S. multistriatus*, *S. kershi* and *S. ensifer* become of greater importance in disease transmission.

H The cycle F–G is repeated. Field elms are largely reduced to a scrub or understorey population. Most surviving mature trees are escapes in woodlands, on islands and in upland valleys. Adapted from Brasier (1983a).

data come from the Netherlands where samples have been taken in the twelve individual provinces over the period 1974–1980. These show clearly that there has been a steady and dramatic decline in the frequency of the non-aggressive strain in all 12 provinces as the epidemic has progressed (Figure 16.4).

In Britain, where the present epidemic began earlier than in the Netherlands, changes in the frequency of the two strains have been monitored since 1971 at the three original main disease outbreak areas. These data (Table 16.1) show that by

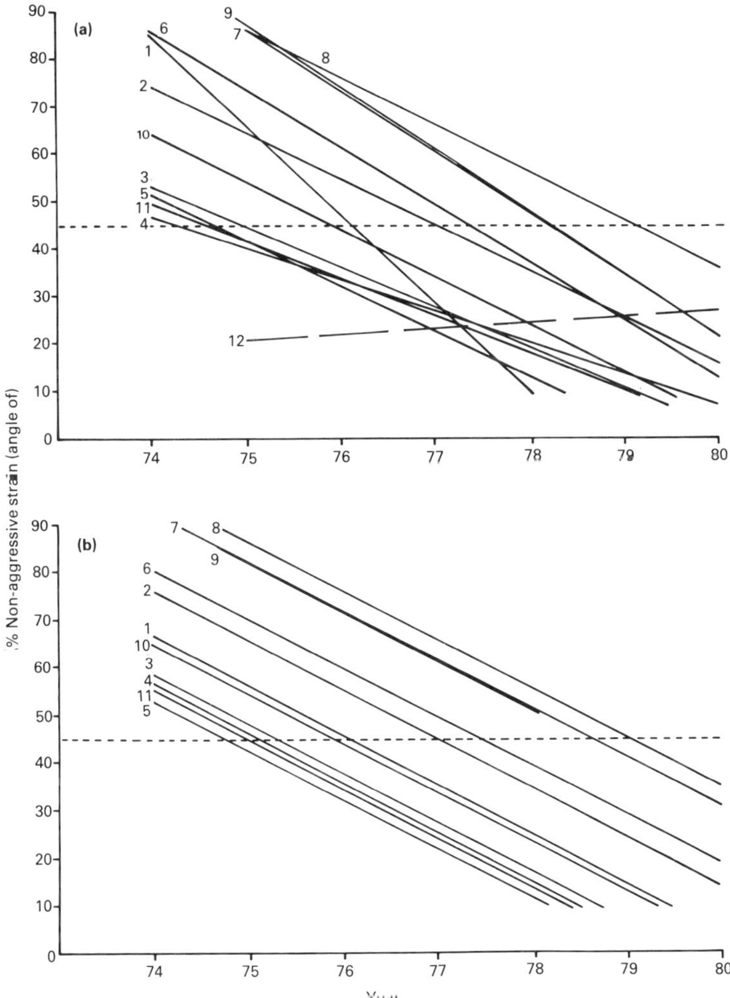

Figure 16.4a. Regression slopes showing the decline in the frequency of the non-aggressive strain of *O. ulmi* in the twelve provinces of The Netherlands between 1974 and 1980. Based on *c.* 1300 samples (Gremmen *et al.*, 1976; C.M. Brasier & H.M. Heybroek, unpublished data). The percentages were transformed to angles for the analysis. The slopes show significant interactions ($P = <0.05$). However, when the 1975 figure for Province 12 is omitted, the latter shows the same trend as the other Provinces, and there are no differences between slopes. A logit transformation of the data gives similar results.

Figure 16.4b. The regression slopes of 11 Provinces (no. 12 omitted) plotted according to the common slope. Note the rapid decline in the proportion of the non-aggressive strain. Note also that the four most eastern provinces (3, 4, 5, 11) reach the 50% level (dashed line) several years ahead of the three most western provinces (7, 8, 9). Adapted from Brasier (1983a).

1971 the non-aggressive strain in Britain had already declined to a level not reached in many Dutch provinces until 1980. The decline continued until, by 1978, the non-aggressive strain was probably less than 1% of the *O. ulmi* population at all three sites. It was not recovered in a sample of over 400 isolates in 1983.

Similar evidence comes from North America, where in the context of the prob-

Table 16.1. Decline of the non-aggressive strain as a proportion of the *O. ulmi* population in southern Britain* since 1971

Year	No. of isolates in sample		
	Non-aggressive	NAN aggressive	% Non-aggressive
1971	23	110	17.3
1972	10	173	5.5
1974	2	86	2.3
1975	4	79	4.8
1978	0	232	0.0
1983	0	413	0.0

* Samples of xylem origin collected around the three original British epidemic outbreak areas of Tewkesbury, Gloucestershire; Chichester, Sussex and Orsett, Essex (see Gibbs & Brasier 1973). In 1971, 1972 and 1974, samples were also collected at Ilchester, Somerset.

able eastward spread of the NAN aggressive from the mid-west in recent decades (McNabb, 1974; Gibbs *et al.*, 1979), the non-aggressive strain has declined as a proportion of the *O. ulmi* population from *c.* 49% in 1977 to *c.* 5% in 1983 at sites across Vermont, and from 85% (1977) to *c* 22% (1983) at Millinocket, Maine (Gibbs *et al.*, 1979; Young & Houston, 1982; Houston, 1985). Likewise, around Dublin in Ireland the non-aggressive has declined relative to the aggressive from *c.* 91% to *c.* 37% between 1977 and 1979 (Mangan & Walsh, 1980 and A. Mangan, personal communication).

While these data indicate that we can expect a marked reduction in the frequency of the non-aggressive strain during the course of an epidemic (Figure 16.3F), they do not answer the question of whether the non-aggressive strain will die out completely, or whether it could still return to replace the aggressive strain. An indication of possible future developments may be found in Romania. Although many of Romania's elms survived the first epidemic in the 1930s, a second more severe epidemic occurred as early as the 1950s (Petrescu *et al.*, 1963), beginning in the Moldavian plain region around Iasi, and killed most of the large elms, including those in mixed stands with oak. A sample survey of *O. ulmi* in Romania conducted by the author in 1980 showed that the EAN race of the aggressive strain was present, consistent with the postulated westward spread of this race across central and eastern Europe in recent years (Brasier, 1979). It was therefore concluded that the second wave of the disease in Romania was due to the arrival of the EAN race and that the original outbreak was caused by the non-aggressive strain (Brasier, 1983a).

On this assumption, Romania currently represents a situation some 30 years after the arrival of the aggressive strain, i.e. some 10–15 years ahead of the present situation in much of north west Europe, and can be taken as something of a pointer to the future. The following observations made during the 1980 survey are therefore particularly pertinent: (i) in the Moldavian region of Romania the elms (mostly *Ulmus carpinifolia*) are largely reduced to bushes and saplings at the margins of forests and in clearings; (ii) the disease is still remarkably heavy, even among relatively small saplings; (iii) all 120 samples of the fungus collected in the area were the EAN race of the aggressive strain.

The evidence from Romania, together with that from the other locations cited above, therefore suggests that in the aggressive/non-aggressive confrontation the non-aggressive strain will not only decline but may well be heading for virtual extinction (Brasier, 1983a).

Reduction in pathogenicity of the aggressive strain through internal genetic mechanisms

If, as the evidence suggests, young elms continue to be attacked by one or other race of the aggressive strain in the post-epidemic period (Figure 16.3F–H) any future decline in the pathogenicity of the *O. ulmi* population may come about only through genetic changes within the aggressive strain itself. Present knowledge is insufficient to assess the likelihood of such attenuation, but from the information available it is possible to suggest ways in which it might occur and to consider some consequences.

The aggressive strain should not be considered as a single unit for this purpose, because there are a number of potentially important differences between the EAN and NAN races, including differences in pathogenicity. Thus, the mean pathogenicity of EAN isolates is typically a little less than that of NAN isolates, there is a greater range of pathogenic ability within the EAN race than within the NAN (Brasier, 1982a, b) and EAN isolates are less likely to cause disease recurrence in a second season (Brasier, 1982a). In the post-epidemic period therefore, the EAN race might show greater flexibility in response to selection imposed by the collapse of the host and beetle populations, and hence be better equipped to survive in direct competition with the NAN race where the two occur together, as for example in Denmark, Ireland, The Netherlands and Italy (Figure 16.2). Mutation, and recombination within or between the EAN and NAN races, are other factors which could contribute to attenuation, assuming that more weakly pathogenic mutants or recombinants are at a selective advantage. A sterility barrier similar to the aggressive/non-aggressive barrier is operated by the EAN race against the NAN but this EAN/NAN barrier is only partial and is more likely to reduce than to prevent hybridization in nature (Brasier, 1979, 1984).

Although pathogenicity in *O. ulmi* is thought to be largely under the control of nuclear genes (Brasier, 1977, 1982a), attenuation might also come about through the increased influence of cytoplasmic factors. A prime candidate must be the *d*-factor, a cytoplasmically transmitted disease of *O. ulmi* believed to be spread mainly between adjacent mycelia during the overwintering saprophytic phase of the fungus in diseased elm bark (Brasier, 1983b). The *d*-factor significantly reduces the growth and reproductive vigour of cultures of *O. ulmi* and the viability of their asexual spores. The impact of such a disease on *O. ulmi* could well be greater in the post-epidemic period when the *O. ulmi* population is small and subject to the constraints of smaller elm and beetle populations (Figure 16.3F–H) than during an epidemic (Figure 16.3C–E) (Brasier, 1983b). By critically reducing the general growth and reproductive fitness of the fungus in the post-epidemic period, therefore, the *d*-factor could conceivably bring about a decline in the effectiveness of the aggressive strain comparable to that produced by the *h*-factor in *Endothia parasitica*, which has brought about a natural remission of Chestnut blight epidemics in Italy since the 1950s (Grente & Sauret, 1969; Anagnostakis, 1982). The natural spread of *d*-factors is, however, subject to a number of constraints, including restriction of transmission between adjacent mycelia by the vegetative incompatibility system of the fungus and probable non-transmission via ascospores (Brasier, 1983b, 1984, 1986). The potential influence of *d*-factors on attentuation will therefore depend

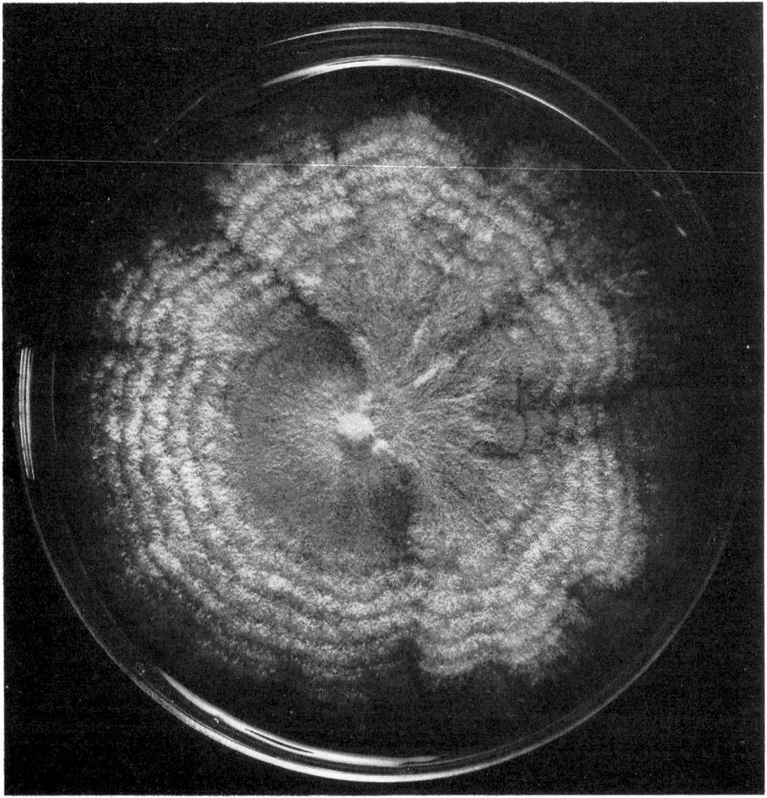

Figure 16.5. The *up-mut* (left) and wild-type (right) phenotypes expressed in a single colony of an EAN aggressive isolate of *O. ulmi*.

upon the effects of these constraining or regulatory factors in relation to the structure of the *O. ulmi* population (e.g. the number of vegetative incompatibility groups, the intensity of ascospore production etc.) during the post-epidemic period.

Another factor which might influence growth and pathogenic fitness in the aggressive strain is that regulating the *up-mut* or uniform powdery mutant (Figure 16.5), an unstable colony dimorphism found in the EAN race (and also in the non-aggressive strain) but not in the NAN race. The dimorphism is a prominent feature of the cultural behaviour of the EAN and seems likely to have an important though as yet unidentified physiological function.

Developmentally expressed factors such as *d* and *up-mut* could conceivably bring about a rapid reduction in pathogenicity of the aggressive strain population, but since they may also tend to affect other fitness characters, such as growth and reproductive vigour, they are likely to be superseded by nuclear gene modifications to the pathogenicity, growth and reproductive systems via mutation and recombination. Ultimately, attenuation might lead to the emergence of a more moderately pathogenic form of the fungus in better balance with the host population. It would probably also be better adapted to the post-epidemic period in a number of other ways, and would hence be unlike the aggressive or non-aggressive strains as we now know them, i.e. it would form another distinct subgroup of *O. ulmi*. It might even first appear as a successful variant in a 'hybrid zone' where the EAN and NAN races overlap. Whatever the possibilities, however, both the probability of and the likely time-scale for attenuation are unknown.

Discussion

If the above conclusions regarding the likely structure of the *O. ulmi* population in the post-epidemic period are broadly correct, then it appears that in the epidemic areas the elm is likely to be reduced largely to a scrub or marginal population under constant attack by one or another form of the aggressive strain. The fate of the elm, therefore, may parallel the fate of the American chestnut in eastern North America which, in the wake of Chestnut blight, now survives mainly as coppice shoots under persistent attack from the disease. With Dutch elm disease, however, we are witnessing not just a regional continental event involving a single species, but a massive pandemic extending across North America and Europe to at least as far east as Iran (Brasier & Afsharpour, 1979) and Tashkent (Brasier, 1983c) and involving the destruction of many species of elm in many habitats: a colossal natural disaster, which will impoverish a diverse array of human cultures.

It also appears that in the epidemic areas the elm/*O. ulmi* relationship may remain seriously out of balance, unless the aggressive strain attenuates in some way. Clearly this does not augur well for the control of the disease in the near future. Where there are specific requirements for elm, as for example for shade trees in the towns of Italy or the USA, or for shelterbelts in the polders of Holland or the steppes of the USSR, breeding for resistance by incorporating an element of resistant Asiatic parentage may be one way of redressing the balance (see e.g. Lester & Smalley, 1972; Smalley & Lester, 1973; Heybroek, 1976, 1983; Brasier 1982a). This does not, however, answer the problem of the loss of our field elms and it is in the countryside that the battle for a natural balance between host and pathogen will always be fought. Certainly in Europe and North America resistant elms are unlikely to be planted, at least for a very long time to come, in sufficient numbers relative to the size of the local susceptible field elm population (even if the latter is largely a scrub population) to exert strong directional selection on the pathogen. Hence, it is from local field elms that any *O. ulmi* inoculum attacking locally planted resistant elms is likely to originate.

Considerable attention must continue to be paid therefore, to the disease situation in field elms, especially with a view to monitoring any attentuation in the aggressive strain, to assessing the threat to plantings of resistant elm and to saving the field elm itself. Regarding the latter, serious consideration should be given to ways of improving the balance of the field elm and pathogen populations, including for example, the artificial spread of *d*-factors (Brasier, 1983b) or the dissemination of pollen of ecologically similar but resistant elms of Asiatic origin in order to raise the baseline of disease resistance among field-elm seedlings (Brasier, 1983a).

Further understanding of ways in which a return to a more satisfactory host–pathogen balance might be achieved could come from studying the causes of the decline of the non-aggressive strain during the current epidemics. It is difficult to account for this decline on the basis of lower pathogenicity alone, since even on the highly susceptible North American elm population (mainly *U. americana*), on which the non-aggressive strain generally causes much higher levels of disease than on European elms (Gibbs *et al.*, 1975), it appears to be declining as rapidly as it is in Europe (Young & Houston, 1982; Houston, 1985). Part of the cause of its decline may lie not in the pathogenic phase at all, but in the long, overwintering, saprophytic phase associated with beetle breeding galleries in the bark of diseased elms. The latter phase accounts for much the larger part of the annual *O. ulmi* cycle, and indeed may often bypass the pathogenic phase altogether (Webber & Brasier, 1984). Factors which might contribute to the decline of the non-aggressive relative to the aggressive strain during the saprophytic phase include differences in growth rate and reproductive physiology, competitive vegetative incompatibility interactions, the 'penetration effect' (Brasier, 1984), and *d*-factor transfer between the subgroups (Brasier, 1983b). The possible impact of such factors on the decline of the non-aggressive strain require investigation since, if major factors in addition to pathogenic ability can be identified, new ways of manipulating the pathogen population might become apparent (Brasier, 1986).

The geographical source or centre of origin of Dutch elm disease remains unidentified. On the basis of the greater resistance of Asiatic elms the disease is popularly supposed, like Chestnut blight, to have an origin in eastern Asia (Heybroek, 1966). So long as doubt as to its origin persists, however, any approach to disease control in Europe and North America, whether involving host or pathogen manipulation, must remain under threat from the arrival of further pre-existing forms or subgroups of *O. ulmi* from the putative source, in much the same way that elm breeding programmes were thrown off balance by the NAN and EAN races of the aggressive strain (Heybroek, 1976; Brasier, 1982a). A salutary and perhaps timely warning of this problem comes from recent evidence that a single isolate of *O. ulmi* from Kashmir in the Himalayas may be of an entirely different subgroup from either the EAN, NAN or the non-aggressive forms (Brasier 1983c). Surveys of Dutch elm disease in the Himalayas and China are needed to resolve these vital questions.

References

Anagnostakis S.L. (1982) Biological control of Chestnut blight. *Science* **215**, 466–71.

Bernier L., Jeng R.S. & Hubbes M. (1983) Differentiation of aggressive and non-aggressive strains of *Ceratocystis ulmi* by polyacrilamide gel electrophoresis of intramycelial enzymes. *Mycotaxon* **17**, 456–72.

Brasier C.M. (1977) Inheritance of pathogenicity and cultural characters in *Ceratocystis ulmi*. Hybridisation of protoperithecial and non-aggressive strains. *Transactions of the British Mycological Society* **68**, 45–52.

Brasier C.M. (1978) Mites and reproduction in *Ceratocystis ulmi* and other fungi. *Transactions of the British Mycological Society* **70**, 81–9.

Brasier C.M. (1979) Dual origin of recent Dutch elm disease outbreaks in Europe. *Nature* **281**, 78–9.

Brasier C.M. (1981) Laboratory investigations of *Ceratocystis ulmi*. In: *Compendium of Elm Diseases* (Ed. by R.J. Stipes & R.J. Campana), pp. 76–9. American Phytopathological Society, Society, St Paul, Minnesota.

Brasier C.M. (1982a) Genetics of pathogenicity in *Ceratocystis ulmi* and its significance for elm breeding. In: *Resistance to Diseases and Pests in Forest Trees* (Ed. by H.M. Heybroek, B.R. Stephan & K. von Weissenberg), pp. 224–35. Pudoc, Wageningen, The Netherlands.

Brasier C.M. (1982b) Occurrence of three sub-groups within *Ceratocystis ulmi*. In: *Proceedings of the Dutch Elm Disease Symposium and Workshop*, Winnipeg, Manitoba, 5–9 October 1981 (Ed. by E.S. Kondo, Y. Hiratsuka & W.B.G. Denyer), pp. 298–321. Manitoba Department of Natural Resources, Manitoba, Canada.

Brasier C.M. (1983a) The future of Dutch elm disease in Europe. In: *Research on Dutch Elm Disease in Europe* (Ed. by D.A. Burdekin). *Forestry Commission Bulletin* **60**, 96–104. HMSO, London.

Brasier C.M. (1983b) A cytoplasmically transmitted disease of *Ceratocystis ulmi*. *Nature* **305**, 220–3.

Brasier C.M. (1983c) Dutch elm disease. In: *Report on Forest Research*, p. 32 HMSO, London.

Brasier C.M. (1984) Inter-mycelial interactions in *Ceratocystis ulmi*: their physiological properties and ecological importance. In: *The Ecology and Physiology of the Fungal Mycelium* (Ed. by D. Jennings & A.D.M. Rayner), pp. 451–97. Cambridge University Press, Cambridge.

Brasier C.M. (1986) The population biology of Dutch elm disease. *Advances in Plant Pathology* **5**, 53–118.

Brasier C.M. & Afsharpour F. (1979) The aggressive and non-aggressive strains of *Ceratocystis ulmi* in Iran. *European Journal of Forest Pathology* **9**, 113–22.

Brasier C.M. & Gibbs J.N. (1973) Origin of the Dutch elm disease epidemic in Britain. *Nature* **242**, 607–9.

Brasier C.M. & Gibbs J.N. (1976) Inheritance of pathogenicity and cultural characters in *Ceratocystis ulmi*. I. Hybridisation of aggressive and non-aggressive strains. *Annals of Applied Biology* **83**, 31–7.

Brasier C.M., Lea J. & Rawlings M.K. (1981) The aggressive and non-aggressive strains of *Ceratocystis ulmi* have different temperature optima for growth. *Transactions of the British Mycological Society* **76**, 213–18.

Gibbs J.N. (1978) Intercontinental epidemiology of Dutch elm disease. *Annual Review of Phytopathology* **16**, 287–307.

Gibbs J.N. & Brasier C.M. (1973) Correlation between cultural characters and pathogenicity in *Ceratocystis ulmi* from Europe and North America. *Nature* **241**, 381–3.

Gibbs J.N., Brasier C.M. Heybroek H.M. & McNabb H.M. (1975) Further studies on the pathogenicity of *Ceratocystis ulmi. European Journal of Forest Pathology* **5**, 161–74.

Gibbs J.N. Houston D.R. & Smalley E.B. (1979) Aggressive and non-aggressive strains of *Ceratocystis ulmi* in North America. *Phytopathology* **69** 1215–19.

Gremmen J., Heybroek H.M. & De Kam M. (1976) The aggressive strain of *Ceratocystis ulmi* in the Netherlands. *Nederlands Bosbouw Tidschrift* **48**, 137–43.

Grente J. & Sauret S. (1969) L''hypovirulence exclusive' est-elle controlée par des determinants cytoplasmiques? *Comptes Rendus Hebdomadaires des Séances de l'Académie des Sciences*, Serie D, **268**, 3173–6.

Heybroek H.M. (1966) Dutch elm disease abroad. *American Forestry* **72**, 26–9.

Heybroek H.M. (1976) Chapters on the genetic improvement of elms. *Proceedings of the Symposium of Better Metropolitan Landscapes.* USDA Forest Service General Technical Report NE-22. pp. 203–13.

Heybroek H.M. (1983) Resistant elms for Europe. In: *Research on Dutch Elm Disease in Europe* (Ed. by D.A. Burdekin). *Forestry Commission Bulletin* **60**, 108–13. HMSO, London.

Houston D.R. (1985) Spread and increase of *Ceratocystis ulmi* with cultural characteristics of the aggressive strain in northeastern North America. *Plant Disease* **69**, 677–680.

Jeng R.S. & Hubbes M (1983) Identification of aggressive and non-aggressive strains of *Ceratocystis ulmi* by polyacrilamide gradient gel electrophoresis of intramycelial proteins. *Mycotaxon* **17**, 445–55.

Lester D.T. & Smalley E.B. (1972) Response of backcross hybrids and three species combinations of *Ulmus pumila, U. japonica* and *U. rubra* to inoculation with *Ceratocystis ulmi. Phytopathology* **62**, 845–8.

McNabb H.S. (1974) Further speculation on the aggressive strain of *Ceratocystis ulmi*. Abstract., Paper No. 12, *86th Session of the Iowa Academy of Sciences, April 19–20, 1974.*

Mangan A. & Walsh P.F. (1980) Some observations on the spread of Dutch elm disease in Ireland, 1978–79. *Irish Journal of Agricultural Research* **19**, 133–40.

Peace T.R. (1960) The status and development of elm disease in Britain. *Forestry Commission Bulletin* **33**, 1–44. HMSO, London.

Petrescu M., Ditu I., Poleac E. & Cucuianu E. (1963) Elm disease in Romania. *Studii se Cercetari. Institutul de Cercetari Forestiere, Bucharest* **23**, 135–50.

Richens R.F. (1984) *Elm.* Cambridge University Press, Cambridge.

Scheffer R.J. & Elgersma D.M. (1982) A scanning electron microscopical study of cell wall degradation in elm wood by aggressive and non-aggressive isolates of *Ophiostoma ulmi. European Journal of Forest Pathology* **12**, 25–8.

Smalley E.B. & Lester D.T. (1973) 'Sapporo Autumn Gold' elm. *Hortscience* **8**, 514–15.

Stebbins G.L. (1966) *Processes of Organic Evolution.* Prentice Hall, Englewood Cliffs, New Jersey.

Svaldi R. & Elgersma D.M. (1982) Further studies on the activity of cell wall degrading enzymes of aggressive and non-aggressive isolates of *Ophiostoma ulmi. European Journal of Forest Pathology* **12**, 29–36.

Takai S. (1974) Pathogenicity and cerato-ulmin production in *Ceratocystis ulmi. Nature* **252**, 124–6.

Takai S. (1980) Relationship of the production of the toxin cerato-ulmin to synnemata formation, pathogenicity, mycelial habit and growth of *Ceratocystis ulmi* isolates. *Canadian Journal of Botany* **58**, 658–62.

Young K.R. & Houston D.R. (1982) Increase in frequency of the aggressive strain of *Ceratocystis ulmi* in New England. *Phytopathology* **72**, 264 (abstract).

Webber J.F. & Brasier C.M. (1984) The transmission of Dutch elm disease: a study of the process involved. In: *Invertebrate-microbial Interactions* (Ed. by J.M. Anderson, A.D.M. Rayner & D. Walton), pp. 271–306. Cambridge University Press, Cambridge.

17 Changes in the composition of pathogen populations caused by resistance to fungicides

G. SKYLAKAKIS
Eli Lilly International Corporation, Lilly House, Hanover Sqare, London W1, UK

Introduction

Only 20 years ago, Georgopoulos & Zaracovitis (1967) wrote 'The reported cases of tolerance to agricultural fungicides are very few and the knowledge accumulated hardly justifies a review'. Since then, however, the situation has changed drastically; substantial changes in the populations of several major plant pathogens in terms of their sensitivity to agricultural fungicides have been observed, leading frequently to significant crop damage and to discontinuation or substantial modification in the use of important products. It is the objective of this chapter to review our understanding of how such changes have occurred and to discuss the methods available to manage them.

Basic assumptions

There seems to be general agreement on the following basic assumptions.

1. Fungicide-resistant mutants exist as subpopulations at some low but finite frequency before the exposure of the total population to a new fungicide (Skylakakis, 1982b; see also Gale, this volume).

2. The existence of such mutants at a low frequency suggests that, in the absence of the fungicide, they are less fit than the more common sensitive individuals in the population. If they were not, they would have been prevalent prior to the introduction of the fungicide; the new product would not have provided satisfactory disease control and would not have been released for manufacture and use (Skylakakis, 1982a; Wolfe, 1982; see also Gale, this volume).

3. Large-scale application of the new fungicide provides a selective advantage for resistant mutants, which may then increase.

Let us examine some of the practical implications of these assumptions. Firstly, resistant mutants exist at a low frequency (perhaps 10^{-6} to 10^{-8}) prior to the introduction of the fungicide. This frequency, determined by the mutation rates(s) at the site(s) that cause resistance and the fitness of the resistant mutants, defines the level from which selection proceeds (see also Barrett, this volume; Gale, this volume). Other factors being equal, the higher the initial frequency of the resistant

Wolfe M.S. & Caten C.E. (1987) *Populations of Plant Pathogens: their Dynamics and Genetics.* Blackwell Scientific Publications, Oxford.

mutants, the sooner will the degree of disease control provided by the new fungicide be affected. Secondly, because they are initially so scarce, resistant mutants, if they exist, can only be detected by very extensive sampling. Initial testing for resistance is a 'numbers' game. Thirdly, since economic disease control in the field is affected only by relatively high frequencies of resistant mutants (10^{-2} to 10^{-1}), the bulk of their population increase occurs before any decrease in disease control is observed.

The two forms of selection

Selection caused by introduction of the new fungicide may be either quantitative or qualitative. The quantitative response occurs when there is single, wide, unimodal distribution in the pathogen population in terms of fungicide sensitivity. As selection operates, it may be observed that the whole distribution moves towards decreased sensitivity and the shift can be quantified by measuring any suitable location parameter of the distribution, for example the LD_{50}. Disease control in the field may or may not be affected, depending on the extent of the shift in relation to the initial position of the practical application rate of the fungicide.

In contrast, a qualitative response occurs when there are two or more non-overlapping distributions in terms of fungicide sensitivity, i.e. distinct sensitive and resistant subpopulations can be recognized. If the application rate of the fungicide is such that the one population is sensitive while the other is resistant, selection will change the relative frequency of these distributions and disease control in the field will be lost.

Although in theory both forms of response could occur simultaneously or sequentially for a particular fungicide/pathogen interaction, in practice either the one or the other seems to dominate.

The process of selection

It is useful to distinguish three major components of the process of selection. First, there is a component of magnitude. One can usefully define this difference in relative fitness of the resistant and sensitive subpopulations by their difference in absolute rates of increase (MacKenzie, 1978; Groth & Barrett, 1980). Its sign will define the direction of change, and its magnitude will define the rate at which change will occur. The second component of selection is temporal. Duration of selection can be defined as the length of time during which selection acts. Depending on whether the overall population change can best be described by discrete steps or by continuous change, either pathogen generations or physical time units (days, weeks etc.) respectively, can be used to measure the duration. The third component of selection is spatial and is defined by the proportion of the total pathogen population upon which selection acts.

Magnitude of selection

All of these components can be affected by factors unique to the biology and ecology of each pathogen. For example, in relation to the magnitude of selection, the less fit the resistant subpopulations are in the absence of the fungicide, the slower is their increase in its presence likely to be. On average it is likely that resistance involving several genetic changes will impose a greater burden on pathogen fitness than resistance which is achieved by a single mutation. In passing, attention should be drawn to the methods used to measure fitness. Fitness is often measured by comparing sporulation and lesion size or carrying out replacement studies under ideal conditions (Horsten, 1979). It is quite conceivable, however, that in nature, fitness is determined primarily by the capacity of a genotype to survive and multiply under suboptimal or even extreme conditions. Greater emphasis on fitness evaluations under a range of environments rather than a limited set of near optimal conditions is needed (see also Barrett, this volume).

The effect of fungicide efficacy is straightforward (see Georgopoulos, this volume). It is the products that are inherently more efficient and used in a programme designed to achieve a high degree of disease control that carry the maximum risk. Within the limits set by practical application rates, the higher the degree of resistance, the greater the intensity of selection. Once total immunity to the practical application rate is reached, further increases in resistance are unlikely to have any effect.

Finally, the rate of increase of the resistant subpopulation is reduced if it has to compete with the sensitive one for the occupation of a limited number of available susceptible host sites (Skylakakis, 1980; Wolfe, 1982). In practical terms, the potential for faster build-up of resistance is greater in the early rather than the late stage of an epidemic.

Duration of selection

Concerning the duration of selection, and with all other factors being equal, the shorter the generation time of the pathogen, the faster will be the build-up of resistance. There is an inherently greater risk of rapid development of resistance in pathogens with generation times that are as short as those that cause potato late blight or cucumber downy mildew than from those that cause cereal foot rots or cereal smut (Skylakakis, 1982a). If follows that whatever slows down the progress of an epidemic without selectively favouring the resistant subpopulation, also increases the time taken for resistance to build up.

Pathogens that affect high-value crops, e.g. ornamentals or vegetables, attack directly a marketed product, e.g. fruit scab, or that reproduce very rapidly, e.g. *Phytophthora infestans*, are often controlled by repeated fungicide applications. For such pathogens the duration of selection is longer than is the case for relatively slowly reproducing pathogens attacking the non-marketed part of lower value crops that are usually controlled by a limited number of applications.

Pathogens that spend most of their life-cycle on treated crops create a higher risk than those that spend a substantial part of their life-cycle surviving or multiplying as saprophytes. Conversely, with a given programme of fungicide application, the greater the persistence of fungicide, the longer the duration of selection.

The spatial component of selection

Finally, the spatial component of selection is perhaps the most elusive. The first and major problem occurs when we attempt to define the extent of the population that is exposed to selection. The extremes are obvious: one isolated field or one closed glasshouse versus a province or a whole country. At present we have no set of rules that would allow us to choose rationally between these two extremes. There are only two intuitive generalizations that it is safe to make: (i) in terms of population size, the greater the mobility of the target pathogen the greater the population size on which prevalence should be measured; (ii) in terms of risk, the greater the popularity of the new product(s) the greater the risk.

The shifting unimodal distribution

Ethirimol seed treatment for control of powdery mildew on barley

Ethirimol was introduced as a seed treatment for control of powdery mildew on barley in 1970. Reduced sensitivity of *Erysiphe graminis* f. sp. *hordei* to ethirimol was first reported by Wolfe (1971). The results of systematic surveys have been summarized by Brent (1982) and data from this summary will be used in this chapter (see also Georgopoulos, this volume).

Within 3 years after the introduction of ethirimol, a clear shift towards less sensitive forms had occurred in East Anglia, assuming that the situation in Scotland in the same year represented the baseline sensitivity (Figure 17.1). In his recent review, however, Brent (1982) points to two differences between East Anglia and Scotland in that not only was ethirimol widely introduced one year earlier in East Anglia, but there was also more winter barley in that area. Four years later the sensitivity of the pathogen to ethirimol was virtually identical in both East Anglia and Scotland and the overall variation in sensitivity had been reduced (Figure 17.2). Clearly, the population had adapted to a new optimum defined by the selection exercised by ethirimol seed treatment on spring barley only (see also Wolfe, this volume).

In spite of the existence of individual strains of *E. graminis* differing in sensitivity from an ED_{50} of 0.019 to one of 3.35 ppm (Hollomon, 1978), and a shift in the population so that strains of intermediate sensitivity became predominant, the performance of ethirimol was not significantly affected; indeed the product is still used today.

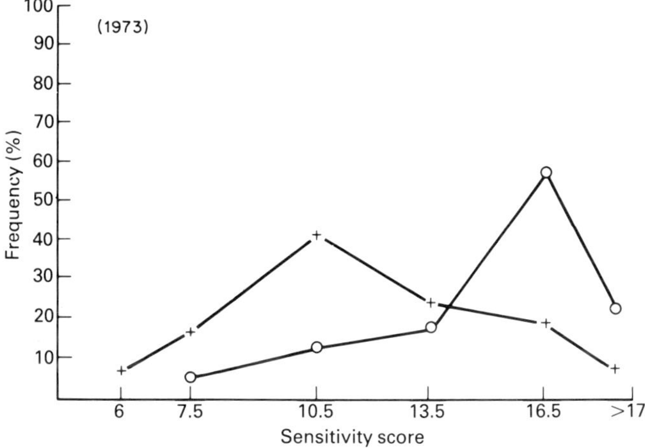

Figure 17.1. Barley powdery mildew: distribution of sensitivity to ethirimol in Scotland and East Anglia in 1973. + = East Anglia; ○ = Scotland. Mean scores: 0 = insensitive; 20 = very sensitive. (After Brent, 1982.)

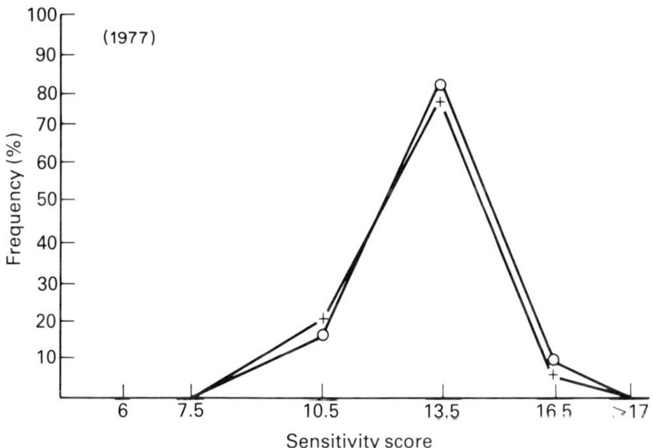

Figure 17.2. As Fig. 17.1, only in 1977.

Although the genetic basis of reduced sensitivity to ethirimol has not yet been clarified, the pattern of the changes observed would seem to indicate the additive effect of several genetic factors. If such were the case, both the initial wide, continuous, variation and the apparently greater fitness of strains of intermediate sensitivity over those highly insensitive would be explained.

Ergosterol biosynthesis inhibitors (EBIs) and cereal powdery mildew

There is a great degree of similarity between the pattern of response of *Erysiphe graminis* f. sp. *hordei* to ethirimol in the early 1970s and to ergosterol biosynthesis inhibitors (EBIs) in the early 1980s. Data from the Newcastle area indicate that a significant shift in sensitivity to triadimenol took place between 1980 and 1981/82 (Figure 17.3) (Fletcher & Wolfe, 1981; Fletcher, 1983).

It is generally accepted that within this magnitude of shift disease control has not yet been adversely affected in the field. In this context, the important question is whether or not the EBIs will follow the ethirimol pattern and continue to give effective control. A closer look into the Newcastle situation indicates that although there was no shift in the population between 1981 and 1982, the overall variation was still great (Figure 17.4). The narrowing down towards intermediate forms clearly identified with ethirimol (Figure 17.2) has not been observed here. As long as this process is not observed, the risk of a decrease in the field performance of EBIs on cereals cannot be excluded. The question remains whether intermediate forms will prevail or whether the distribution is going to be shifted further towards resistance (see also Wolfe, this volume).

It is also interesting to compare the influence of ethirimol and EBIs on *Erysiphe graminis* on the one hand and that of dimethirimol and EBIs on *Sphaerotheca fuliginea* on cucurbits on the other, where disease control has been adversely affected (Bent *et al.*, 1971; Huggenberger *et al.*, 1984). There seems to be no evidence of a different nature of resistance or of a greater inherent variability (Bent *et al.* recorded LD_{50}s of from 0.32 to 5 ppm for dimethirimol in *Sphaerotheca fuliginea*). What seems to have made the difference was the intensity, duration and prevalence of selection rather than the inherent properties of the fungicides involved.

It is interesting to note the common factors in the cases reviewed up to now in this chapter. We have dealt with apparently shifting unimodal distributions of sensitivity to the fungicide, where the genetic control of reduced sensitivity was either not well understood, i.e. with ethirimol and dimethirimol (Brent, 1982), or, possibly, controlled by several additive genes and modifiers, i.e. with the EBIs (van

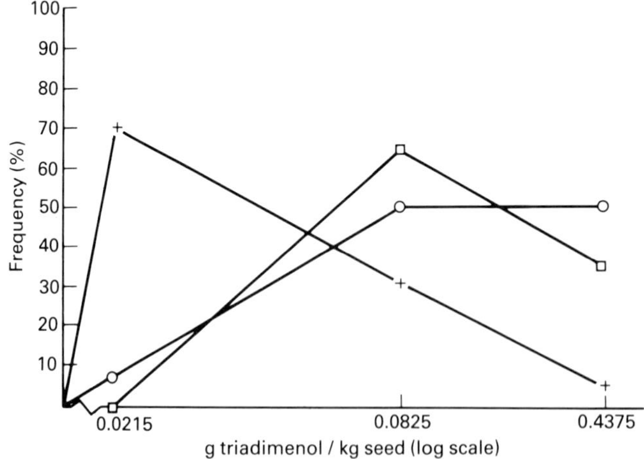

Figure 17.3. Barley powdery mildew: distribution of sensitivity to triadimenol in the Newcastle area. + = 1980; ⊙ = 1981; □ = 1982. (After Fletcher, 1983.)

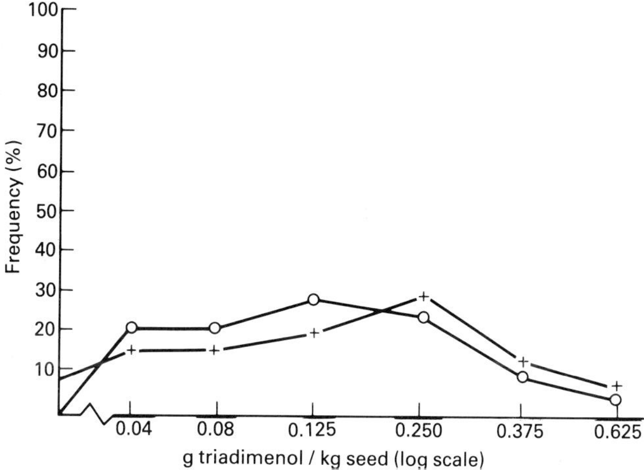

Figure 17.4. As Fig. 17.3.

Tuyl, 1977; Waard *et al.*, 1982). When we face this kind of situation, it seems that we can rationalize a posteriori how differences in the components of selection can account for the difference in the end result (i.e. loss or preservation of disease control), but we cannot predict.

Non-overlapping sensitivity distributions

When we are dealing with changing proportions of two or more discrete non-overlapping distributions of pathogen response to fungicides, namely less than and greater than the fungicide application rate, we are in a better position, at least as far as predictions are concerned. Mathematical models capable in *some degree* of describing the events, *usually by oversimplification*, have been constructed.

The mathematics supporting these models have been recently reviewed (Barrett, 1983; Skylakakis, 1982b, 1983). In Table 17.1, modified after Skylakakis (1982b), the fit between predictions from the models and observed facts has been summarized for a few cases of non-overlapping sensitivity distributions. It is interesting to note

Table 17.1. Predicted and observed period before in the field loss of disease control

Pathogen	Chemical	Standard selection time (days)*	Period to loss of control	
			Predicted	Observed
Phytophthora infestans	Metalaxyl	3.7– 3.8	51–70 days	1–2 seasons
Cercospora beticola	Benomyl	9.5–14.3	130–263 days	140–200 days
Ustilago spp	Carboxin	158	5–7 years	not observed

* Time taken for the ratio of the two genotypes to increase by *e*.

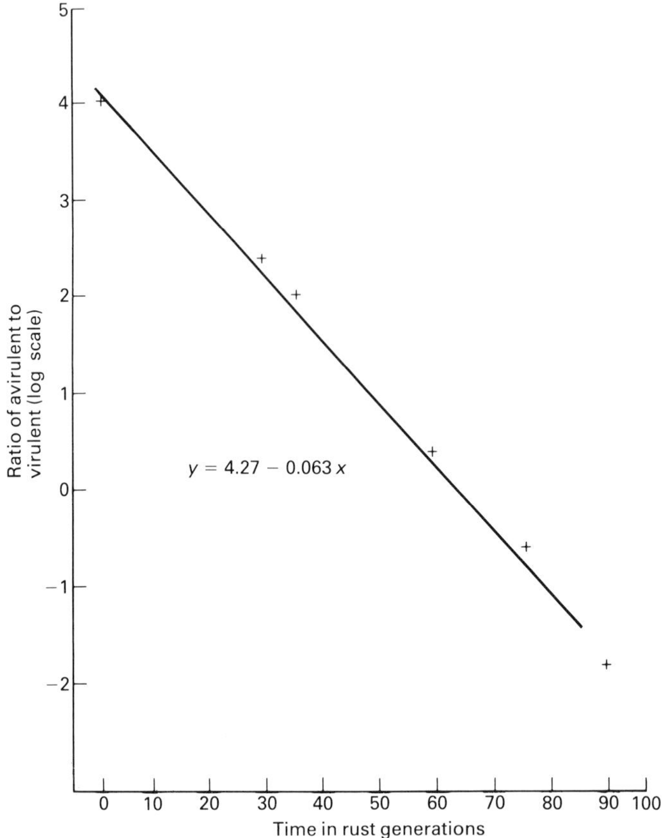

Figure 17.5. Change in relative frequency of rust isolates, avirulent and virulent, on oat cultivar TAM 0-312. After Simons & McDaniel (1983).

the clear capacity of the models to differentiate between pathogens with a potential for very fast, moderate, and very slow build-up of resistance, and the reasonable coincidence between predictions and observations.

Indirect evidence in support of this quantitative approach can be obtained from a different field. Simons & McDaniel (1983) monitored the increasing frequency of crown rust virulent on oat cultivar TAM 0−312 carrying the resistance gene *Pc-59*. Their data seem to indicate two non-overlapping distributions caused by one major mutation from avirulence to virulence, a situation very similar to the development of resistance to the benzimidazoles. If the exponential model is a good approximation, one would expect the log of the ratio of the frequency of the avirulent to the virulent race to decrease linearly with time. In Figure 17.5 a linear regression has been fitted to the data assuming 15 generations of crown rust per calendar year; the regression (b = 0.063 ± 0.0046) is very highly significant, indicating that the exponential model could, in this case, be adequately used to predict the race frequency changes.

The same quantitative considerations that have been applied in predicting the

Table 17.2. The effect of fitness of the resistant subpopulation on the rate of selection of the sensitive subpopulation in the absence of the fungicide

	Standard selection time (days)*			
	Selection coefficient			
Pathogen	0.01	0.05	0.1	0.5
Phytophthora infestans	497.5	97.5	47.5	7.2
Erysiphe graminis	796.0	156.0	75.9	11.5
Cercospora beticola	1194.0	233.9	113.9	17.3
Cereal foot rot pathogens	2985.1	584.9	284.7	43.3

* In favour of the sensitive subpopulation.

rate of build-up of resistance can also be applied to two other questions of considerable interest. Firstly, once the selection exerted by the fungicide is removed, will the situation revert back to that prevailing before the introduction of the fungicide, and if so, at what rate? Secondly, what means are available to delay or stop the build-up of resistance?

We have estimates of neither selection coefficients nor standard selection times (i.e. time necessary for the ratio of two genotypes to increase by e) for fungicide resistance genes in the absence of the toxicant. Recently however, Grant & Archer (1983) estimated selection coefficients for unnecessary virulence genes from field data for *Puccinia graminis* in Australia (1945–1955, s = 0.0524 to 0.04) and *Erysiphe graminis* in the UK (1969–1979, s = 0.0608 to 0.0543) (see also Barrett, this volume; Leonard, this volume). The calculations in Table 17.2 were made using these values together with the values 0.1 and 0.5 quoted by Barrett (1978). The implications of Table 17.2 are clear; for any given pathogen, the faster the development of resistance (corresponding with small selection coefficients), the longer will it take for the population to revert to the *status quo ante*. Taking a pathogen like sugar beet leafspot and a country like Greece (140–160 days of epidemic development per year), it will take 7–8 years of total absence of benzimidazoles for the frequency of common resistant genotypes (coefficient of selection 0.01) to decrease from 90 to 75%. It is only for resistant strains that are highly unfit in the absence of fungicide that the speed at which the initial balance is restored is comparable to the speed at which it was originally changed by an efficient fungicide.

Strategies to delay the build-up of resistance

Among the factors affecting the intensity of selection there are some that can be effectively manipulated. Any factor other than the fungicide at risk that slows down an epidemic also slows down the rate of build-up of resistance. For example, the use of a companion fungicide in a tank mix or ready-to-mix prepack has attracted a lot of attention, as it can reduce the intensity of selection by slowing down the overall speed of the epidemic and reducing the degree of resistance of the mutants (see also Wolfe, this volume). This delaying effect, although theoretically supported, had not been detected experimentally until the recent results of Staub & Sozzi (1983). It is interesting to assess how their result fits with the theory. From their Figure 1, one can assess the delay in increase of the resistant subpopulation. It took 61 days for the proportion of the resistant subpopulation on plants treated with a fungicide mixture (metalaxyl plus mancozeb) to reach the same level as it had reached on plants treated with metalaxyl alone in 39 days, an increase of 56%. Since the overall level of attack was kept relatively low on the treated plants, the exponential model would be expected to provide a reasonable approximation. Indeed the model (Skylakakis, 1981, Table 1, line 3) provides a standard selection time of 3.5 days in the presence of metalaxyl alone, compared to 5.4 days in the presence of a mixture such as metalaxyl plus mancozeb, i.e. a delay of 54%.

The action of the other approach which is often suggested, i.e. alternation of application of a high-risk and a low-risk fungicide, exercises its effect through the duration of selection. It can be argued that it delays the build-up of resistance in physical time units but does not increase the total number of applications with the fungicide at risk before increase of resistant genotypes in the population affects disease control.

Skylakakis (1984) has recently analysed the available experimental evidence on the delaying effect of fungicide mixtures and alternation. He observed that most of the available experimental evidence did not support this delaying effect. This discrepancy between circumstantial evidence and theory on the one hand, and direct experimental evidence on the other, can only be resolved by further research. He suggested that such work should be carried out on inherently slow pathogens with a low starting frequency of resistant strains, should utilize experimental designs which minimize cross-contamination between plots, and involve assessments carried out when overall disease severity in the experiment is still low.

Conclusions

As more changes in the composition of pathogen populations caused by fungicide resistance have been observed, so our understanding of the forces responsible for such changes and the ways to influence them has increased. Substantially more work is needed, however, before this increased knowledge can be reflected in generally agreed recommendations dealing with the assessment of resistance risk in the early phases of development of new products, and the minimization of such a risk, once it has been recognized, after a new product has been commercially introduced.

References

Barrett J.A. (1978) A model of epidemic development in variety mixtures. In: *Plant Disease Epidemiology* (Ed. by P.R. Scott & A. Bainbridge), pp. 129–37. Blackwell Scientific Publications, Oxford.

Barrett J.A. (1983) Estimating relative fitness in plant parasites: Some general problems. *Phytopathology* **73**, 510–12.

Bent K.J., Cole A.M., Turner J.A.W. & Woolner M. (1971) Resistance of cucumber powdery mildew to dimethirimol. *Proceedings of the 6th British Insecticide and Fungicide Conference.* British Crop Protection Council, pp. 274–82.

Brent K.J. (1982) Case study 4: Powdery mildews of barley and cucumber. In: *Fungicide Resistance in Crop Protection* (Ed. by J. Dekker and S.G. Georgopoulos), pp. 219–30. Pudoc, Wageningen.

Fletcher J.T. (1983) Report of a survey of *Erysiphe graminis* sensitivity to triadimenol 1982. *ADAS Plant Pathologists Technical Conference*, Malvern 1983, 9 pp.

Fletcher J.T. & Wolfe M.S. (1981) Insensitivity of *Erysiphe graminis* f. sp. *hordei* to triadimefon, triadimenol and other fungicides. *Proceedings of the 1981 British Crop Protection Conference — Pests and Diseases*, pp. 633–40.

Georgopoulos S.G. & Zaracovitis C. (1967) Tolerance of fungi to organic fungicides. *Annual Review of Phytopathology* **5**, 109–30.

Grant M.W. & Archer S.A. (1983) Calculation of selection coefficients against unnecessary genes for virulence from field data. *Phytopathology* **73**, 547–51.

Groth J.V. & Barrett J.A. (1980) Estimating parasitic fitness: a reply. *Phytopathology* **70**, 840–2.

Hollomon D.W. (1978) Competitive ability and ethirimol sensitivity in strains of barley powdery mildew. *Annals of Applied Biology* **90**, 195–204.

Horsten J.A.H.M. (1979) Acquired resistance to systemic fungicides of *Septoria nodorum* and *Cercosporella herpotrichoides* in cereals. *PhD thesis*, University of Wageningen, The Netherlands, 107 pp.

Huggenberger F., Collins M.A. & Skylakakis G. (1984) Decreased sensitivity of *Sphaerotheca fuliginea* to fenarimol and other ergosterol-biosynthesis inhibitors. *Crop Protection* **3**, 137–49.

MacKenzie D.R. (1978) Estimating parasitic fitness. *Phytopathology* **68**, 9–13.

Simons M.D. & McDaniel M.E. (1983) Gradual evolution of virulence of *Puccinia coronata* on oats. *Phytopathology* **73**, 1203–5.

Skylakakis G. (1980) Estimating parasitic fitness of plant pathogenic fungi: a theoretical contribution. *Phytopathology* **70**, 696–8.

Skylakakis G. (1981) Effects of alternating and mixing pesticides on the build up of fungal resistance. *Phytopathology* **71**, 1119–21.

Skylakakis G. (1982a) Epidemiologic factors affecting the rate of selection of biocide resistant plant pathogenic fungi. *Phytopathology* **72**, 271–3.

Skylakakis G. (1982b) The development and use of models describing outbreaks of resistance to fungicides. *Crop Protection* **1**, 249–62.

Skylakakis G. (1983) Theory and strategy of chemical control. *Annual Review of Phytopathology* **21**, 117–35.

Skylakakis G. (1984) Quantitative evaluation of strategies to delay fungicide resistance. *Proceedings of the 1984 British Crop Protection Conference — Pests and Diseases*, pp. 565–72.

Staub T. & Sozzi D. (1983) Recent practical experiences with fungicide resistance. *Proceedings of the 10th International Congress of Plant Protection*, pp. 591–8.

Tuyl J.M. van (1977) Genetic aspects of resistance to imazalil in *Aspergillus nidulans. Netherlands Journal of Plant Pathology* **83** (Suppl. 1) 169–76.

Waard M.A. de, Groeneweg H. & van Nistelrooy J.G.M. (1982) Laboratory resistance to fungicides which inhibit ergosterol biosynthesis in *Penicillium italicum. Netherlands Journal of Plant Pathology* **88**, 99–112.

Wolfe M.S. (1971) Fungicides and the fungus population problem. *Proceedings of the 6th British Insecticide and Fungicide Conference.* British Crop Protection Council, pp. 724–34.

Wolfe M.S. (1982) Dynamics of the pathogen population in relation to fungicide resistance. In: *Fungicide Resistance in Crop Protection* (Ed. by J. Dekker & S.G. Georgopoulos), pp. 139–48. Pudoc, Wageningen.

18 The development of fungicide resistance

S.G. GEORGOPOULOS
Athens College of Agricultural Sciences, Votanikos GR.- 118 55 Athens, Greece

Introduction

Our efforts to keep populations of fungal pathogens acceptably low on our crops often depend on the use of antifungal chemicals. For the control of diseases on high-value crops the use of fungicides is indispensable and frequent applications of the same chemical during a growing season are common. This provides target fungi with an opportunity to evolve resistance, but the outcome may vary from a high degree of success to an apparently complete failure of the pathogen populations to overcome the effects of the fungicides.

Evidence now suggests that the occurrence and the magnitude of a resistance problem depends mainly on the fungicide, although the type of disease and the method(s) of fungicide application are also of importance. Thus, populations of most fungi seem able to overcome completely the effects of some fungicides. While at the other extreme not a single species has been shown to develop resistance to other fungicides, in spite of extensive and often exclusive use. It is now being realized that a number of antifungal compounds lie between these two extremes, allowing for intermediate population responses and usually only partial and/or temporary loss of effectiveness. This chapter examines the reasons for these differences and attempts a classification of available fungicides according to the potential risk of resistance.

The criteria for the development of fungicide resistance

The major factors which determine whether the population of a target species will be successful in overcoming the effects of a biocide are the availability of appropriate genetic variability and the fitness of sensitive and resistant forms. It is first important to know whether genes which can modify sensitivity are present in the target organism and what is the effect of each gene on the phenotype. The second criterion is the ability of less sensitive forms to survive in the absence of the chemical and to multiply at a rate comparable to that of the rest of the population. If this ability is seriously affected by mutation to fungicide resistance, the frequency of mutants will increase slowly, if at all, during exposure to the fungicide, and will decline when the fungicide is absent (Wolfe, 1982; Skylakakis, this volume).

In other words, if 1 and $1 - s$ are the fitnesses of sensitive and resistant

Wolfe M.S. & Caten C.E. (1987) *Populations of Plant Pathogens: their Dynamics and Genetics*. Blackwell Scientific Publications, Oxford.

genotypes on untreated crops and $1 - t_1$ and $1 - t_2$ the corresponding values for crops treated with a fungicide, a high resistance risk is involved in cases where the selection coefficient, s, is small and the sensitivity of the resistant mutants, t_2, is much less than that of the wild type, t_1. Conversely, the risk is low if substantial differences in sensitivity do not occur or are small and costly in terms of fitness. In intermediate situations, the risk of resistance is moderate (Georgopoulos, 1984).

Both the degree of resistance and the relative fitnesses are determined by comparing rates of reproduction of sensitive and resistant strains on treated and untreated crops. An estimate of the differences in sensitivity can be obtained in the laboratory. The magnitude of these differences should be considered in relation to the rates of fungicide which can be applied. In cases where the rate cannot be increased substantially above the dose required to suppress the wild type population, because of phytotoxicity or other reasons, even a low degree of resistance may be important in practice.

Differences in fitness which are recognized by comparison of only a few isolates, particularly of those obtained in the laboratory, may not reflect the situation in nature. Whenever possible, field isolates or, better, fungicide-sensitive and -resistant fractions of field populations, should be compared. Some difference in fitness between sensitive and resistant genotypes in the absence of the fungicide must be assumed if the latter were infrequent in the population before applications began, for example, if the resistant mutants were maintained by the mutation–selection equilibrium (see Barrett, this volume). If this difference is very small, resistant forms will increase rapidly if part of the crop is treated, and decline only slowly if the treatment is removed or on neighbouring untreated crops (Wolfe, 1982).

Whether appropriate variability is available, and of what kind, has been investigated for several important fungus–fungicide combinations. Measurements of the degree of resistance have been made in many cases and indications of fitness differences have been obtained. From evidence summarized below, it appears that for a particular chemical, target fungi do not differ substantially in their response, while for a particular fungus the degree of resistance attainable and its physiological burden may differ greatly depending on the fungicide. Both the mutational change in sensitivity and the fitness of the mutants are, of course, determined by the mechanism of resistance. The similarity of response suggests that different fungi have equally effective or, perhaps, the same mechanisms at their disposal in order to overcome toxicity of the same type of fungicide. A classification of the available fungicides on this basis can, therefore, be attempted.

Major-gene differences in fungicide sensitivity

High-level resistance to some fungicides is acquired in a single step and has been shown or appears to be controlled by genes of major effect. In this situation, the sensitivity distribution is discontinuous. Resistance alleles usually exist at very low frequencies in untreated populations, so that a small sample is unlikely to contain resistant strains. Following exposure to the fungicide, however, distinct non-overlapping subpopulations can be distinguished. The level of resistance is often high enough for mutants to be unaffected by the chemical at field application rates, i.e. $t_2 \leqslant s$. Depending on whether or not mutant alleles have serious adverse effects on fitness, fungicides allowing for such one-step, major changes in sensitivity can be divided into two groups.

Fitness not seriously affected: high-risk fungicides

On existing evidence the fungicides in this group are the benzimidazoles, the acylalanines, the polyoxins, and perhaps also the carboxanilides. Mutants resistant to these fungicides are not seriously inhibited by the chemical and they are able to compete successfully with sensitive strains on untreated crops. With fungicide application, the resistant part of the population can increase rapidly, depending upon the proportion of crop treated, the persistence of the fungicide, and the frequency of application (Skylakakis, this volume). In most cases, effectiveness is suddenly and completely lost and the possibility for re-use of the fungicide after such a failure is remote.

(a) Benzimidazoles
The use of benzimidazole derivatives has undoubtedly provided the best examples of major gene fungicide resistance without serious adverse effects on fitness (Georgopoulos, 1977). In all cases where genetic analysis was performed, genes with major effects on benzimidazole sensitivity were recognized (Borck & Braymer, 1974; Katan *et al.*, 1983; Shabi & Katan, 1979; van Tuyl, 1977). In addition, benzimidazole resistance is generally very stable in the field, indicating a low coefficient of selection (s) in the absence of the fungicide. Figure 18.1A shows the benomyl sensitivity distribution of field isolates of the sugar-beet leaf spot pathogen, *Cercospora beticola*. These isolates were obtained 8–10 years after the failure of benzimidazoles in Greece, with practically no subsequent use of these fungicides against leaf spot. The distribution is bimodal with two distinct subpopulations, since resistant isolates are unaffected by a concentration of benomyl causing 100% inhibition of sensitive isolates. Such a distinction is also easy in the field, where sensitive isolates are controlled by foliar applications of benomyl at the rate of 150 g per hectare, while resistant isolates produce as much disease on treated as on untreated plots, even at levels of application greater than 150 g per hectare. The general fitness of the resistant isolates, on the other hand, is evident from their continuing predominance despite several years without selection by the use of the benzimidazoles (Figure 18.1A).

For some fungi, a biochemical explanation of the high degree of resistance and fitness of benzimidazole-resistant strains has become available (Davidse, 1982a). Mutation of the resistance gene modifies the β-subunit of tubulin in such a way that, although binding of the fungicide to the protein is practically nullified, normal formation and function of tubulin microtubules is not affected.

High-level resistance to benzimidazoles has been observed in many fungi and

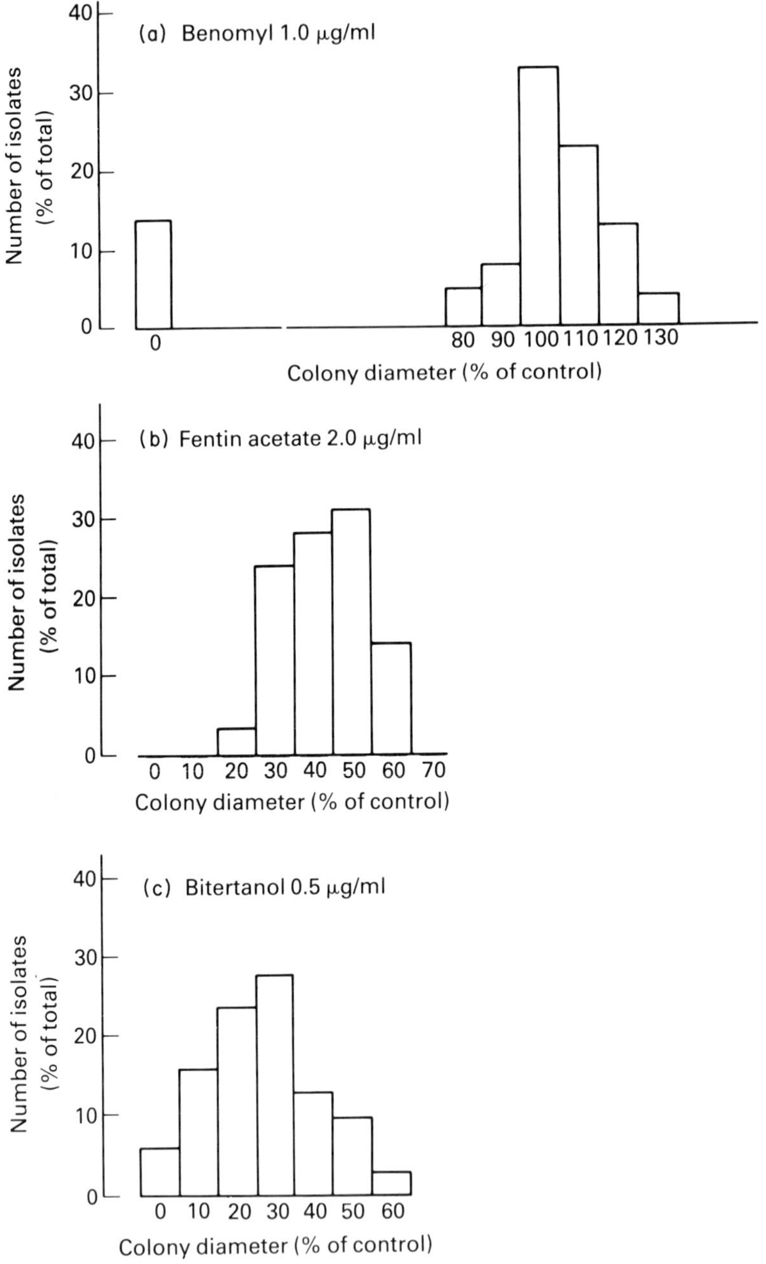

Figure 18.1. Frequency distribution with respect to sensitivity to three fungicides of 400 isolates of *Cercospora beticola* obtained in 1981, 1982, and 1983 from several areas of northern Greece. The degree of inhibition of each isolate is given by the diameter of colonies growing on medium containing the fungicide concentration indicated, expressed as percentage of the diameter of the same isolate on medium without fungicide. Sensitivities outside the distribution shown have never been observed. In the areas from which the isolates were obtained 10–15 spray applications of benomyl, approximately 100 applications of fentin, and 8–10 applications of EBIs had been made. It is evident that the distribution has become discontinuous with respect to benomyl sensitivity but has remained unimodal with respect to sensitivity to fentin and to bitertanol (an EBI).

complete loss of effectiveness has often been reported (Georgopoulos, 1977). A one-step change under simple genetic control and a small fitness deficit in the absence of the fungicide can be assumed in all cases.

There seem to be no great differences between sensitive fungi in the frequency of mutation to benzimidazole resistance (van Tuyl, 1977). The fact that some diseases may still be effectively controlled with benzimidazoles does not necessarily mean that appropriate genetic variability is absent. As calculated by Skylakakis (1983), even highly resistant and fit mutants will find it difficult to become predominant in diseases with a low-infection rate. The fact that resistance to benzimidazoles is now causing failures of control of eyespot of cereals (Brown *et al.*, 1984), a slow-spreading disease, indicates that it is only a matter of time for pathogens with a slow rate of increase to overcome the effect of benzimidazole treatments.

(b) Acylalanines

There can be little doubt that major genes are involved in resistance to the fungicides known as the acylalanines, which are active specifically against oomycetes. Data summarized by Staub & Sozzi (1981) and by Davidse (1982b) show that several and probably all of the acylalanine-sensitive fungi have the potential to develop, in a single step, high-level resistance to metalaxyl and other acylalanines. Resistant strains can grow and reproduce on plants treated with much higher doses than those effective against sensitive strains (Davidse, 1982b; Georgopoulos, 1984). Selection has resulted in complete loss of effectiveness within one growing season in some cases although the response has been slower in others (see Crute, this volume). Metalaxyl-resistant strains also seem to be highly effective in the absence of fungicide (Cohen *et al.*, 1983; Grigoriu & Georgopoulos, unpublished results). Unfortunately, the genetics and the biochemical mechanism(s) of resistance to acylalanines are not yet known.

(c) Polyoxins

Although polyoxins have been used against only a few diseases, resistance has caused complete loss of effectiveness against *Alternaria kikuchiana* and *A. mali* in Japan (Nishimura *et al.*, 1976; Uesugi, 1983). Genetic analyses have not been made with these organisms, but the differences in sensitivity and the stability of resistance in the absence of fungicide suggest involvement of major genes and a small coefficient of selection. Control failures were first noticed in 1971, only a few years after the first introduction of polyoxins, and resistant strains were still common at least 5 years after the use of these fungicides was stopped.

High-level resistance to polyoxins results from low activity of a dipeptide permease, responsible for polyoxin uptake (Hori *et al.*, 1977). Apparently the fungus does not need dipeptide influx under normal conditions, so that growth is not affected by the level of activity of the particular permease.

(d) Carboxanilides

In *Ustilago maydis* single-gene mutations decrease the sensitivity of the succinic dehydrogenase system to carboxin and to related fungicides (Georgopoulos & Ziogas, 1977). Resistant haploids grow normally in the absence of fungicide and combine to form plant parasitic dikaryons which do not seem to differ from dikaryons lacking carboxin resistance genes in growth and reproduction in untreated corn seedlings (Georgopoulos *et al.*, 1975). Such dikaryons should be selected by the treatments and, on this basis, carboxanilides should be classified as high-risk fungicides. Their limited use, mainly against smut diseases with low infection rates (Skylakakis, 1983), has apparently not permitted development of practical problems. However, when oxycarboxin is used against a fast-developing disease, chrysanthemum rust, resistance does cause failures of control (Abiko *et. al.*, 1977; Grouet *et al.*, 1981).

Serious effects on fitness: moderate-risk fungicides

Single-gene mutations causing single-step, large changes in sensitivity, comparable with those described above, have also been recognized with the antibiotic kasugamycin (Taga *et al.*, 1978) and with fungicides of the aromatic hydrocarbon group (Georgopoulos & Panopoulos, 1966), with which the dicarboximides must be included (Grindle, 1983). With these chemicals, however, resistance problems do not arise as quickly and with the same intensity as with the fungicides of the previous group because resistance mutations have, apparently invariably, serious adverse effects on fitness. Because of reduced competitiveness, resistant strains require sustained selection in order to become predominant. Loss of effectiveness may be complete, particularly in closed environments, but only after continuous selection for long periods and under favourable conditions. After a more or less short interval, however, the fungicide can be used again.

(a) Kasugamycin

Populations of *Pyricularia oryzae* from areas where effectiveness of kasugamycin against rice blast was declining, gave a bimodal frequency distribution with no intermediate types with respect to sensitivity (Miura *et al.*, 1976). When the problem appeared in one district of Japan in 1971, most isolates were resistant but their proportion decreased strikingly after the use of the antibiotic was stopped in 1972 (Uesugi, 1982). It seems that resistant strains are inferior to sensitive strains as pathogens because they require longer times for appressorium formation and infection (Ito & Yamaguchi, 1979).

Kasugamycin inhibits protein synthesis, and resistance mutations seem to decrease the affinity of the ribosomes for the inhibitor (Misato & Ko, 1975). It appears that this change in the ribosome affects the time required for the infection process.

(b) Aromatic hydrocarbon group

With some members of this group, e.g. diphenyl, mutation of one gene is sufficient for complete insensitivity to the maximum concentration of the fungicide which can be made available to the organism. Since the early 1960s the selection pressure for resistance has been very intense in some situations, leading to the failures of sodium orthophenylphenate and diphenyl against penicillium rots of citrus (Eckert, 1982).

Recently, selection of dicarboximide-resistant strains (which are cross-resistant to other members of the aromatic hydrocarbon group) has been reported repeatedly, particularly following the use of these fungicides against *Botrytis cinerea*. Failures of dicarboximides have been noticed in protected winter vegetable crops, where up to 100% of *B. cinerea* isolates have been found to be resistant (Katan, 1982; Panayotakou & Malathrakis, 1983). In the absence of selection, however, the proportion of resistant isolates falls rapidly (Hunter & Brent, 1983; Locher *et al.*, 1983), due to reduced sporulation and slower growth and colonization of host tissue which apparently result from the osmotic sensitivity (Grindle, 1983) of the mutants.

Polygenic control of sensitivity differences

With other fungicides, sensitivity differences between isolates are apparently under more complex genetic control. Variation remains continuous even after long exposures of field populations to the fungicide, so that resistant and sensitive subpopulations cannot be distinguished (Skylakakis, this volume). The risk from resistance development is low or moderate, depending on the degree of selection.

Evidence of selection available: moderate-risk fungicides

This group should include dodine, the fentin derivatives, the 2-aminopyrimidines, the phosphorothiolates, and the ergosterol biosynthesis inhibitors (EBIs). These fungicides allow for a variety of responses, but in most cases a number of steps in the development of resistance can be recognized in the laboratory. For dodine (Kappas & Georgopoulos, 1970) and the EBIs (van Tuyl, 1977), genes which have small effects on sensitivity, but which interact positively in recombinants to give higher levels of resistance, have been identified. Modifiers have been recognized and apparently they also contribute to variation in sensitivity in nature.

There is good evidence of selection responses to each member of the group, though they may not be large. With some of these fungicides, this may be due to the low levels of resistance attainable. With others, sensitivity differences may be great, but resistance appears to carry a physiological burden. It is possible that the burden increases with the number of mutations required for a given change in sensitivity to each fungicide. The result may be a gradually less satisfactory performance of the chemical, but complete loss of effectiveness appears unlikely, except under rare circumstances.

(a) Dodine

This fungicide is used against apple and pear scab for the control of which numerous applications per growing season are often required; resistance to it has been reviewed recently (Gilpatrick, 1982). The most resistant isolates of *Venturia inaequalis* found in orchards are only 4 times less sensitive to dodine than wild type isolates. Resistance became a problem only after continued and exclusive use of the fungicide for 9–10 years, during which it was applied about 80 times. The distribution of sensitivity is unimodal and exposure to the fungicide causes the mean sensitivity to decrease. It appears that effectiveness is not completely lost, however, and with increasing fungicide dosage some control is achieved (Szkolnik & Gilpatrick, 1973). Dodine resistance, at least of the level found in the orchards, does not appear to affect the pathogen seriously, since resistant strains were isolated from the problem areas 10 years after the use of dodine was discontinued.

(b) Fentins

There has been only one report (Giannopolitis, 1978) of poor performance of fentin acetate and fentin hydroxide caused by resistance. After many years of intensive use against *C. beticola* in Greece, decreased effectiveness of the fungicide was noticed in 1976. The distribution of sensitivity in populations from a location with a long history of fentin applications was unimodal, but the mean sensitivity was lower than in a location where fentins had never been used. Following this observation, fentin usage has been reduced to only two or three applications per season. The distribution of sensitivity of the fungus population continues to be unimodal (Figure 18.1B) and narrow, although fentins must have been used more than 100 times in most areas during the last 20 years. In mixed inoculations in the absence of fungicides the less sensitive isolates are less competitive (Giannopolitis & Chrysayi-Tokusbalidis, 1980). Because of this and the low degree of resistance there has been only partial loss of effectiveness.

(c) 2-aminopyrimidines

Resistance caused failures of dimethirimol against the powdery mildew of cucurbits, *Sphaerotheca fuliginea*, in glasshouses the second year after the introduction of this fungicide in the Netherlands. This happened following year-round, almost universal use of the fungicide, resulting in near-perfect control in the isolated glasshouse environment. In other areas, where selection has not been so intense, no resistance has been reported after about 10 years of use (Brent, 1982). Withdrawal of the fungicide in the Netherlands was followed by a gradual decline in the proportion of resistant isolates, which permitted re-use of dimethirimol.

Sensitivity to ethirimol within populations of *Erysiphe graminis* f. sp. *hordei* (powdery mildew of barley) varies greatly (Skylakakis, this volume; Wolfe, this volume). Populations from treated crops show a unimodal frequency distribution, and differ from those from untreated crops in the absence of only the most sensitive strains (Hollomon, 1980). The response to ethirimol, therefore, seems weak, and although less sensitive forms tend to be selected on treated plants, under field conditions it was the strains of intermediate sensitivity which became predominant (Brent, 1982). No correlation between selection and control failures seems to have been generally established. The use of ethirimol, however, was considerably less than universal and certainly the situation did not parallel that of dimethirimol on cucurbits. Ethirimol sensitivity differences between *E. graminis* f. sp. *hordei* isolates could not be attributed to specific genes (Hollomon, 1981). It appears that interaction of genes of minor effect may be required for high resistance to ethirimol and that such an interaction hinders the pathogen in the absence of the fungicide.

(d) Phosphorothiolates

Work on resistance to the organophosphorous fungicides IBP and edifenfos in *P. oryzae* has been reviewed by Uesugi (1982). A major gene responsible for the high resistance obtained in the laboratory has been identified. Resistance found in the field, however, is intermediate and has not been assigned to any specific genes. Phosphorothiolate fungicides were in use for 10 years before strains with lower sensitivity were obtained from the field. Field populations cannot be divided into sensitive and resistant fractions and, although some selection seems to have taken place, the unimodal distribution has been maintained. Effectiveness, however, is diminishing because the fungicides are applied as granules to submerged paddy fields and the small amounts which reach the sensitive tissues are apparently insufficient to inhibit even moderately resistant strains. No evidence of differences in absolute fitness between strains seems to have been obtained.

(e) Ergosterol biosynthesis inhibitors (EBIs)

The genetics of resistance to this group of systemic fungicides has been studied only in the non-pathogen *Aspergillus nidulans* (van Tuyl, 1977) where at least eight chromosome loci are involved in resistance to EBIs. The frequency of mutation for imazalil resistance was 70 times higher than that for resistance to benomyl. Single-gene mutations result in small decreases in sensitivity, but interaction between resistance genes at different loci and also modifying genes may result in a resistance factor of up to 100.

In *Cladosporium cucumerinum* the degree of resistance to triarimol and to triforine was inversely correlated with pathogenicity in a test on untreated cucumber seedlings (Fuchs *et al.*, 1977). Generally, in competition experiments with mixed inocula of *Penicillium italicum* in the absence of fungicide the percentage of resistant conidia decreased (De Waard *et al.*, 1982). Resistant strains of the latter pathogen were effectively controlled by imazalil at the recommended rate, but control was not complete at half this dosage.

Although some of the EBIs have been in use for 10 years or more, concern about their performance in the field has been expressed so far only in the control of the powdery mildews of barley and cucumber. In the former fungus, the use of triazoles in England from 1981 to 1983 increased the percentage of field isolates which could infect plants from seed treated with a low dose of triadimenol (Wolfe *et al.*, 1983). In central and north-western Europe, high percentages of isolates with lower sensitivity were also obtained from areas where EBIs had been extensively used (Limpert &

Schwarzbach, 1981). Even in such areas, however, the sensitivity distribution remained continuous and the dose recommended by the manufacturer continued to give satisfactory control. Whether further selection may render recommended rates of EBIs ineffective is not yet known (Skylakakis, this volume; Wolfe, this Volume).

In the case of *S. fuliginea*, lower sensitivities have been associated with a history of intensive usage of fenarimol and other EBIs over several years (Huggenberger *et al.*, 1984). Under conditions of high disease pressure, inoculum of low sensitivity could cause control failures. Even in such cases, however, acceptable disease control can be achieved by increasing the fungicide rate or by shortening the spray interval.

In recent years EBIs have been used successfully against several diseases caused by non-obligate parasites. Field isolates of significantly lower sensitivity have not been reported in any of these fungi. In Greece, for example, nuarimol and bitertanol have been used against *C. beticola* (two to three applications per season) and the sensitivity of field isolates to bitertanol has remained within the original range shown in Figure 18.1C.

No response to selection: low-risk fungicides

This group includes most of the protectant fungicides, i.e. sulphur, coppers, dithiocarbamates, phthalimides, quinones, dinocap, and chlorothalonil. Genetically controlled differences in sensitivity to any of these chemicals have never been obtained, at least in filamentous fungi. In addition, selection of less sensitive forms with the rates which are applied in practice has not been demonstrated.

These fungicides are not persistent, they do not enter the plant to be protected from environmental erosion, and are less effective than many systemics. With their application, therefore, the degree and duration of selection (Skylakakis, this volume) are lower than with many systemics. On existing evidence, however, it is questionable whether appropriate variability is available to target fungi to overcome the toxicity of these fungicides. Firstly, induction of stable, genetic resistance to these chemicals has not been possible in the laboratory, in spite of several attempts which have been recorded and several others which have not. Secondly, there have been no reports of loss of effectiveness even under conditions of high disease pressure and with very frequent applications.

A particularly good example of the continuing effectiveness of these fungicides is provided by the control of downy mildew of cucurbits (N.E. Malathrakis, personal communication). In the Greek island of Crete this disease, due to *Pseudoperonospora cubensis*, was probably unknown until 1965 when growers started to use walk-in plastic houses to grow protected winter crops. Even today, the dry climate during spring and summer does not permit the development of the disease on cucurbits grown in the open, although inoculum is plentiful. All of the *P. cubensis* population on the island is, therefore, produced in the plastic houses where dithiocarbamates have been in general use for more than 15 years. In these plastic houses, poor aeration and moderate winter temperatures are conducive to serious downy mildew epidemics. Growers routinely remove and destroy infected foliage and spray with dithiocarbamates every week. If this spraying is done regularly and carefully, downy mildew does not establish at all. However, once established in a plastic house, the disease is very difficult to control and growers are forced to reduce the spray interval from 7 to 5 or even 4 days.

Better results were obtained by some growers who used metalaxyl during the 1978–1979 growing season, but failures of this fungicide because of the selection of resistant strains were noticed a few months later. In contrast, the performance of the dithiocarbamates does not seem to have been affected, in spite of so many years with conditions extremely favourable for adaptation. For example, in an experiment in 1980–81 only 20% of foliage became infected in plots treated with mancozeb, in spite of abundant inoculum from adjacent plots where the foliage was 78% and 76% diseased for the control and the metalaxyl treatment respectively (N.E. Malathrakis, personal communication). If all plants in a plastic house are regularly and carefully sprayed with dithiocarbamates, the disease may not even establish at all, indicating high effectiveness to this date.

It is conceivable that alleles for resistance to a particular type of toxicant may not be present in a sensitive species. Some species of higher plants, for example, seem incapable of evolutionary adaptation to metal-contaminated environments to which other species adapt readily (Bradshaw, 1984). What is peculiar about the chemical control of fungal diseases is that several of the fungicides available appear not to involve a risk of resistance, independent of the target species. On the other hand, with insecticides and antibacterial drugs (Datta, 1984), it is very exceptional that a compound continues to be effective after many years of use. This difference is very important for our strategies to delay or counteract fungicide resistance. While, for example, in entomology the use of a mixture may cause both insecticides to lose effectiveness (Brown, 1976), in fungicide mixtures one of the partners can be a low-risk compound, which may delay resistance to a high- or moderate-risk chemical and also provide some control of the target pathogen when such resistance eventually develops (see also Skylakakis, this volume).

The multiplicity of sites of action of the protectant fungicides, which makes it improbable for resistance to be acquired by accumulation of the many mutant genes required for site modification, was thought to be responsible for the apparent inability of fungi to develop resistance to these chemicals. However, resistance to a multisite inhibitor can develop by a single mechanism, for example detoxification. According to a recent report (Barak & Edgington, 1984) such detoxification of captan is possible in some isolates of *B. cinerea* due to a larger pool of glutathione-like compounds. Why such mechanisms have not become common over the many years of use of protectant fungicides is not known.

Acknowledgements

The author is grateful to the Greek Ministry of Research and Technology which provides financial support (Contract No. 80006) for his research on fungicides.

References

Abiko K., Kishi K. & Yoshioka A. (1977) Occurrence of oxycarboxin-tolerant isolates of *Puccinia horiana* P. Henn. in Japan. *Annals of the Phytopathological Society of Japan* **43**, 145–50.

Barak E. & Edgington L.V. (1984) Glutathione synthesis in response to captan: a possible mechanism for resistance of *Botrytis cinerea* to the fungicide. *Pesticide Biochemistry and Physiology* **21**, 412–16.

Borck K. & Braymer H.D. (1974) The genetic analysis of resistance to benomyl in *Neurospora crassa. Journal of General Microbiology* **85**, 51–6.

Bradshaw A.D. (1984) Adaptation of plants to soils containing toxic metals – a test for conceit. In: *Origins and Development of Adaptation.* (Ciba Foundation Symposium **102**), pp. 4–19. Pitman Books.

Brent K.J. (1982) Case study 4: Powdery mildews of barley and cucumber. In: *Fungicide Resistance in Crop Protection* (Ed. by J. Dekker & S.G. Georgopoulos), pp. 219–30. Pudoc, Wageningen.

Brown A.W.A. (1976) How have entomologists dealt with resistance? *Proceedings of the American Phytopathological Society* **3**, 67–74.

Brown M.C., Taylor G.S. & Epton H.A.S. (1984) Carbendazim resistance in the eyespot pathogen *Pseudocercosporella herpotrichoides. Plant Pathology* **33**, 101–11.

Cohen Y., Reuveni M. & Samoucha Y. (1983) Competition between metalaxyl-resistant and -sensitive strains of *Pseudoperonospora cubensis* on cucumber plants. *Phytopathology* **73**, 1516–20.

Datta N. (1984) Bacterial resistance to antibiotics. In: *Origins and Development of Adaptation.* (Ciba Foundation Symposium 102), pp. 204–18. Pitman Books.

Davidse L.C. (1982a) Benzimidazole compounds: selectivity and resistance. In: *Fungicide Resistance in Crop Protection* (Ed. by J. Dekker & S.G. Georgopoulos), pp. 60–70. Pudoc, Wageningen.

Davidse L.C. (1982b) Acylalanines: Resistance in downy mildews, *Pythium* spp. and *Phytophthora* spp. In: *Fungicide Resistance in Crop Protection* (Ed. by J. Dekker & S.G. Georgopoulos), pp. 118–27. Pudoc, Wageningen.

Eckert J.W. (1982) Case study 5: Penicillium decay of citrus fruits. In: *Fungicide Resistance in Crop Protection* (Ed. by J. Dekker & S.G. Georgopoulos), pp. 231–50. Pudoc, Wageningen.

Fuchs A., de Ruig S.P., van Tuyl J.M. & de Vries F.W. (1977) Resistance to triforine: a nonexistent problem? *Netherlands Journal of Plant Pathology* **83** (Suppl. 1), 189–205.

Georgopoulos S.G. (1977) Development of fungal resistance to fungicides. In: *Antifungal Compounds* (Ed. by M.R. Siegel & H.D. Sisler), Vol. 2, pp. 439–95. Marcel Dekker, New York.

Georgopoulos S.G. (1984) Adaptation of fungi to fungitoxic compounds. In: *Origins and Development of Adaptation.* (Ciba Foundation Symposium 102), pp. 190–203. Pitman Books.

Georgopoulos S.G., Chrysayi M. & White G.A. (1975) Carboxin resistance in the haploid, the heterozygous diploid, and the plant parasitic dicaryotic phase of *Ustilago maydis. Pesticide Biochemistry and Physiology* **5**, 543–51.

Georgopoulos S.G. & Panopoulos N.J. (1966) The relative mutability of the *cnb* loci in *Hypomyces. Canadian Journal of Genetics and Cytology* **8**, 347–9.

Georgopoulos S.G. & Ziogas B.N. (1977) A new class of carboxin resistant mutants of *Ustilago maydis. Netherlands Journal of Plant Pathology* **83** (Suppl. 1), 235–42.

Giannopolitis C.N. (1978) Occurrence of strains of *Cercospora beticola* resistant to triphenyltin fungicides in Greece. *Plant Disease Reporter* **62**, 205–8.

Giannopolitis C.N. & Chrysayi-Tokusbalidis M. (1980) Biology of triphenyltin-resistant strains of *Cercospora beticola* from sugar beet. *Plant Disease Reporter* **64**, 940–3.

Gilpatrick J.D. (1982) Case study 2: *Venturia* of pome fruits and *Monilinia* of stone fruits. In: *Fungicide Resistance in Crop Protection* (Ed. by J. Dekker & S.G. Georgopoulos), pp. 195–206. Pudoc, Wageningen.

Grindle M. (1983) Effects of synthetic media on the growth of *Neurospora crassa* isolates carrying genes for benomyl resistance and vinclozolin resistance. *Pesticide Science* **14**, 481–91.

Grouet D., Montfort F. & Leroux P. (1981) Mise en evidence, en France, d'une souche de *Puccinia horiana* resistante a l'oxycarboxine. *Phytiatrie-Phytopharmacie* **30**, 3–12.

Hollomon D.W. (1980) Resistance of barley powdery mildew to fungicides. *ADAS Quarterly Review* **39**, 226–233.

Hollomon D.W. (1981) Genetic control of ethirimol resistance in a natural population of *Erysiphe graminis* f. sp. *hordei. Phytopathology* **71**, 536–40.

Hori M., Kakiki K. & Misato T. (1977) Antagonistic effect of dipeptides on the uptake of polyoxin A by *Alternaria kikuchiana. Journal of Pesticide Science* **2**, 139–49.

Huggenberger F., Collins M.A. & Skylakakis G. (1984) Decreased sensitivity of *Sphaerotheca fuliginea* to fenarimol and other ergosterol-biosynthesis inhibitors. *Crop Protection* **3**, 137–49.

Hunter T. & Brent K.J. (1983) Effects of different spray regimes on dicarboximide resistance in *Botrytis cinerea* on strawberries. *Proceedings of the 10th International Congress of Plant Protection*, Vol. 2, p. 631 (abstract).

Ito I. & Yamaguchi T. (1979) Competition between sensitive and resistant strains of *Pyricularia oryzae* Cav. against kasugamycin. *Annals of the Phytopathological Society of Japan* **45**, 40–6.

Kappas A. & Georgopoulos S.G. (1970) Genetic analysis of dodine resistance in *Nectria haematococca. Genetics* **66**, 617–22.

Katan T. (1982) Resistance to 3,5-dichlorophenyl-N-cyclic imide (dicarboximide) fungicides in the grey mould pathogen *Botrytis cinerea* on protected crops. *Plant Pathology* **31**, 133–41.

Katan T., Shabi E. & Gilpatrick J.D. (1983) Genetics of resistance to benomyl in *Venturia inaequalis* from Israel and New York. *Phytopathology* **73**, 600–3.

Limpert E. & Schwarzbach E. (1981) Virulence analysis of powdery mildew of barley in different European regions in 1979 and 1980. *Barley Genetics* **4**, 458–65.

Locher F., Lorenz G. & Beetz K.J. (1983) Influence of vinclozolin mixtures on the development of resistance and on disease control in *Botrytis cinerea* Pers. of grapes. *Proceedings of the 10th International Congress of Plant Protection*, Vol. 2, p. 627 (abstract).

Misato T. & Ko K. (1975) The development of resistance to agricultural antibiotics. In: *Environmental Quality & Safety*, Suppl. 3 (Ed by F. Coulston & F. Korte), pp. 437–440. Georg Thieme, Stuttgart.

Miura H., Katagiri M., Yamaguchi T., Uesugi Y. & Ito H. (1976) Mode of occurrence of kasugamycin resistant rice blast fungus. *Annals of the Phytopathological Society of Japan* **42**, 117–23.

Nishimura S., Kohmoto K. & Udagawa H. (1976) Tolerance to polyoxin in *Alternaria kikuchiana* Tanaka, causing black spot disease of japanese pear. *Review of Plant Protection Research* **9**, 47–57.

Panayotakou M. & Malathrakis N.E. (1983) Resistance of *Botrytis cinerea* to dicarboximide fungicides in protected crops. *Annals of Applied Biology* **102**, 293–9.

Shabi E. & Katan T. (1979) Genetics, pathogenicity and stability of carbendazim-resistant isolates of *Venturia pirina. Phytopathology* **69**, 267–9.

Skylakakis G. (1983) Theory and strategy of chemical control. *Annual Review of Phytopathology* **21**, 117–35.

Staub T. & Sozzi D. (1981) First practical experiences with metalaxyl resistance. *Netherlands Journal of Plant Pathology* **87**, 245 (abstract).

Szkolnik M. & Gilpatrick J.D. (1973) Tolerance of *Venturia inaequalis* to dodine in relation to the history of dodine usage in apple orchards. *Plant Disease Reporter* **57**, 817–21.

Taga M., Nakagawa H., Tsuda M. & Ueyama A. (1978) Ascospore analysis of kasugamycin resistance in the perfect stage of *Pyricularia oryzae. Phytopathology* **68**, 815–17.

Tuyl J.M. van (1977) Genetics of fungal resistance to systemic fungicides. *Mededelingen Landbouwhogeschool, Wageningen* **77–2**, 236 pp.

Uesugi Y. (1982) Case study 3: *Pyricularia oryzae* of rice, In: *Fungicide Resistance in Crop Protection* (Ed. by J. Dekker & S.G. Georgopoulos), pp. 207–18. Pudoc, Wageningen.

Uesugi Y. (1983) Fungicide resistance: Problem with modern fungicides in Japan. *Tropical Agriculture Research Series* No. 16, pp. 105–11.

Waard M.A. de, Groeneweg H. & Nistelrooy J.G.M. van (1982) Laboratory resistance to fungicides which inhibit ergosterol biosynthesis in *Penicillium italicum. Netherlands Journal of Plant Pathology* **88**, 99–112.

Wolfe M.S. (1982) Dynamics of the pathogen population in relation to fungicide resistance. In: *Fungicide Resistance in Crop Protection* (Ed. by J. Dekker & S.G. Georgopoulos), pp. 139–48. Pudoc, Wageningen.

Wolfe M.S., Slater S.E. & Minchin P.N. (1983) Fungicide insensitivity and host pathogenicity in barley mildew. *Proceedings of the 10th International Congress of Plant Protection*, Vol. 2, p. 645 (abstract).

Note added in proof: The review of the literature pertaining to this contribution was completed in June 1984.

19 Trying to understand and control powdery mildew*

M.S. WOLFE
Plant Breeding Institute, Trumpington, Cambridge CB2 2LQ, UK

Introduction

Erysiphe graminis DC. f. sp. *hordei* Marchal is a highly successful pathogen of barley, particularly in the U.K. The physical environment is close to the optimum for most of its life processes and now, more than ever before, there are large amounts of host tissue available the whole year round and over the whole country. Perhaps the most important factor, however, in its success is the life strategy of the pathogen. For example, during the late summer, the one period in the year when there is little or no living host tissue available, the pathogen passes through its resting stage as cleistothecia. The cleistothecia germinate in the autumn, nicely timed to take advantage of newly emerging volunteers and autumn-sown crops and with a high chance of success since the ascospores carry the maximum release of genetic variability of the organism. For the remainder of the year the pathogen reproduces asexually, growing superficially on its host in an ideal position for the release of copious numbers of conidiospores into the atmosphere.

This strategy evolved in the wild, to deal with the hot, dry conditions of over-summering in the Middle East (Koltin & Kenneth, 1970) but it also provides an ideal system for evolution on the agricultural crop. Past attempts to control the pathogen have usually been doomed to failure because they have taken little or no account of the organism's life strategy. However, lessons can be learnt from past attempts and more rational and practicable control strategies developed. In this paper I try to summarize some of the observations and interpretations that we have made at the Plant Breeding Institute and to persuade the reader of the value of the improved methods of control that are being developed.

Pathogen response to disease control

The influence of individual host resistance on the pathogen population

If the immense haploid population of the pathogen is not disturbed by the introduction of resistant varieties and fungicides, the pathogenicity genes are sorted into combinations whose frequencies depend on the individual gene frequencies. This

* Delivered as the 1983 British Society for Plant Pathology Presidential Address.

Wolfe M.S. & Caten C.E. (1987) *Populations of Plant Pathogens: their Dynamics and Genetics.* Blackwell Scientific Publications, Oxford.

arrangement follows the product rule (Wolfe *et al*., 1976; Wolfe & Knott, 1982) and is most clearly evident from the data of Leijerstam (1965), re-analysed on the basis of genotype frequencies (Wolfe & Schwarzbach, 1975).

If an introduced host gene for resistance is initially effective, it is assumed that the apparent low frequency of the matching pathogenicity character is due to previous selection against it. Of course, the low frequency may be due to drift or other stochastic effects (Gale, this volume); if so, then the host resistance might be suddenly and rapidly overcome, leading perhaps, to fixation of the pathogenicity gene. In each of the examples of large-scale introduction of major resistance genes in barley, the change in pathogenicity was progressive and it is likely that the matching pathogenicity was being held at a low frequency, less than 1%, by a mutation–selection equilibrium (Barrett, this volume). Parallel examples with the rust fungi and other diploid pathogens are more difficult to interpret because there is little information on the scale of the reservoir of recessive pathogenicity genes in heterozygotes (Johnson, this volume).

The first example of a consciously introduced major resistance gene was that of *Mlg* (more strictly, *Mlg* plus *Ml-CP*, the latter being a gene of small effect that often accompanies *Mlg* in European barley varieties). The *Mlg* gene was effective during the 1930s and 1940s in Germany and remained so, at least while the area under cultivation remained small. From the late 1940s, the area increased rapidly, followed by the frequency of the matching pathogenicity gene, *Vg*, until the latter became sufficiently common to cause the *Mlg* varieties to lose their mildew resistance (Wolfe & Schwarzbach, 1978b).

Because *Mlg* became widely used in European breeding programmes, the matching pathogenicity, *Vg*, also became widespread, partly by selection and partly by immigration from areas where races carrying it were already common. It is only in the last year or so, half a century after the introduction of *Mlg*, that we have begun to see a decline in the frequency of *Vg*. For most of this time its high frequency has been maintained by the large proportion of spring barley crops possessing *Mlg*. The recent decline in *Vg* indicates that the gene did not become fixed and that, in the absence of *Mlg*, selection against it still occurs. Indeed, there is no reason to suppose that such selection is different now from the period before the increased use of *Mlg*; this principle has been supported by Leonard & Czochor (1980) (see also Leonard, this volume). It is important to add, however, that selection against *Vg* may vary on different hosts (see below). Also, there has been ample opportunity for selection of improved genetic backgrounds for *Vg*, which may help to slow its decline in frequency in the absence of *Mlg* (Gale, this volume).

We obtained evidence for selection within a matching pathogen population in a field of the variety Keg (*Mlk* + *Mla 7*) in 1980 (Wolfe *et al*., 1981). Two populations of isolates were collected, one from the lower, older leaves, and one from the younger, upper leaves. From the age of the crop we estimated that the population on the younger leaves was established about a month later than that on the older leaves and had therefore been subjected to about four more generations of selection on

Keg. The mean pathogenicity for Keg did not differ significantly between the populations but the distribution was changed; in the population from the younger leaves there were fewer isolates with very low, or high, levels of pathogenicity for the variety. So, by the end of the season stabilizing selection had fitted the pathogen population better to growing on Keg. It is probable that selection was acting on the range of genetic backgrounds containing *Vk* + *Va7*; however, we cannot exclude the possibility that alternative forms of the major genes were also being selected.

In essence, the story of *Mlg* can be repeated for most of the major gene introductions in barley in Europe. The example of *Mla12*, however, takes the process a stage further. *Mla12*, first introduced in the variety Sultan, did not become as widely used as *Mlg* before it started to decline in popularity. It was superseded by other varieties with different resistance genes and *Va12* declined in frequency. During this period breeders were still using the resistance gene but the programmes containing it were being selected under natural infection with populations still with high frequencies of *Va12*. Some new varieties thus emerged which contained *Mla12* but either in resistant backgrounds or combined with other major resistance genes. By the time these varieties became widely grown, *Va12* had decreased in frequency and so, for a period, the value of *Mla12* was regained until these newer varieties were eventually overcome. This sequence is now in a third cycle.

By plotting the frequency of *Mla12* against the frequency of the matching pathogenicity gene (Figure 19.1), a repeating cyclical pattern is generated which mimics the frequency dependence or limit cycle invoked by Clarke (1976) for

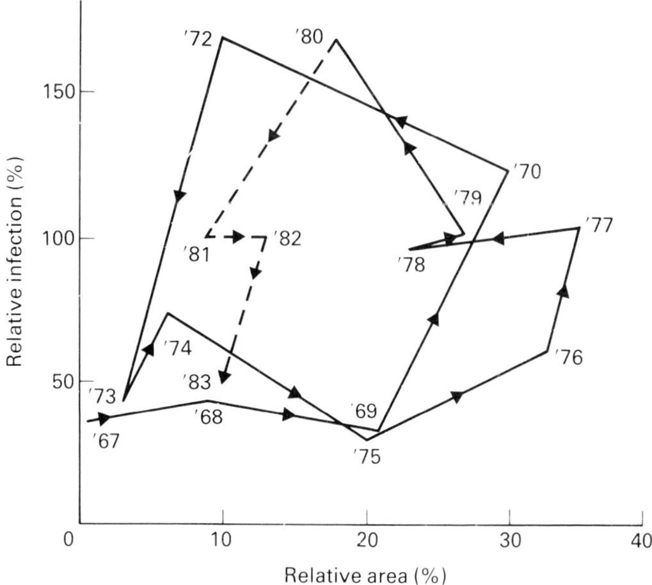

Figure 19.1. Relative area of the national spring barley crop occupied by varieties with *Mlal2* mildew resistance plotted against the mildew infection on those varieties relative to the remainder, from 1967 to 1983. Data derived mainly from the national foliar disease surveys of the ADAS Harpenden Laboratory.

natural host–pathogen systems. The analogy is not exact because in the natural system, the increasing pathogen frequency is directly responsible for the decline in the matching host frequency. In agriculture the influence of the pathogen is indirect, mediated by the farmer's perception of the loss in performance of the matching host. However, it is likely that, in both the agricultural and natural systems, the amplitude of the limit cycles may be influenced by changing genetic backgrounds in host and pathogen in each revolution.

The exceptional host resistance gene is *mlo*; it has been widely introduced in Europe in several varieties and there are no reports of matching pathogen isolates in the field (to December 1983). However, Schwarzbach (1979), was able to select for increased pathogenicity on *mlo* in the laboratory, which suggests that the effective life of *mlo* will be limited even though it has so far proved to be more durable than other simply inherited resistance characters (see Dinoor & Eshed, this volume).

The influence of individual fungicides on the pathogen population

The first major systemic fungicides to be used for control of barley mildew were ethirimol, used largely as a seed treatment, and tridemorph, used solely as a foliar spray. We detected an increased frequency of mutants with reduced sensitivity to ethirimol shortly after its introduction (Wolfe & Dinoor, 1973) and suggested to the manufacturer that the use of the product on winter barley should be discouraged. This advice was based on observations in East Anglia that the frequency of mutants insensitive to ethirimol decreased rapidly on untreated winter barley crops. At the time winter barley was a minor crop so that the pathogen population was forced through a bottleneck for survival between successive spring crops. Further, the fungicide was never used on more than one-quarter of the spring barley area. Under these conditions insensitivity increased to a level that caused some loss of effectiveness of the chemical, though insufficient to be considered economically serious.

Later in the 1970s Brent (1982) found that forms with extreme sensitivity or insensitivity had been lost from the population leading to an apparent equilibrium at an intermediate level of insensitivity. Using a simple model, Wolfe (1982) argued that on the untreated crop area, mutants with reduced sensitivity to ethirimol were at a selective disadvantage compared with the wild type. The *effective* treated area of crop was a function of the area sown with treated seed and the persistence of the fungicide in the developing crop, which was probably for less than half of the growing period. During this early and limited time on the treated crop, insensitive mutants would have been at a selective advantage compared with the wild type, allowing them to reproduce to a greater extent and thus to compensate to some degree for their relatively poor reproduction on the larger, untreated, crop area. Evidently, the effective treated area was insufficient to allow complete compensation for deficiencies in the insensitive mutants. They were unable to predominate and were probably maintained in a dynamic, unstable equilibrium by continued use of the fungicide (see also Skylakakis, this volume).

Mutants insensitive to tridemorph were also found during the 1970s (Walmsley-Woodward *et al.*, 1979) following the commercial introduction of this fungicide, but they did not increase to any great extent. This was partly because the effective treated area was less than for ethirimol. Further, because the material was used as a foliar spray, selection for insensitivity was imposed on an established sensitive population not all of which was killed. With ethirimol seed treatment emerging seedlings selected the initial airborne inoculum directly for insensitive mutants that had survived the winter. Indeed, other things being equal, seed treatments generally may be expected to have a greater impact than foliar sprays because of their earlier influence on establishment of the pathogen population.

Because of their performance and persistence the triazole fungicides quickly became popular and reached an effective treated area much greater than had been experienced with ethirimol. For this reason we expected the pathogen to respond, and it did so. Insensitivity was first detected among isolates from northern Britain, shortly after the introduction of the fungicide (Fletcher & Wolfe, 1981). At this stage, the most insensitive isolates had ED_{50} values about 10–20 times higher than those of the wild type, but they were not common. More recently they have become much more common and other forms with ED_{50}s of 50–60 times those of the wild type are now not difficult to find. There can be no doubt that selection has been accelerated by the recent increase in cultivation of winter barley and the use of the fungicides on both the winter and spring crops.

The triazole fungicides are less effective against powdery mildews than they were 5 years ago but they are still able to provide satisfactory economic returns in most cases. However, the continued extensive use of these fungicides, creating almost year-round selection on the pathogen, seems an unfortunate policy because of the likelihood of further pathogen adaptation and the consequent risk of large-scale and more severe loss of effectiveness of the compounds in controlling mildew and other pathogens (see also Skylakakis, this volume).

The interaction of pathogenicity genes with non-corresponding host genes

So far, pathogen responses to host varieties have been considered individually. In practice, several forms of resistance are used simultaneously, imposing disruptive selection and an evolutionary dilemma for the pathogen. Spores that do not move far from their origin will be selected in each pathogen generation for adaptation to the particular host on which they were produced. Spores that are carried some distance from their origin will be selected for versatility, or the ability to grow on several common hosts. Since adaptation to one host does not cause adaptation to all others, it is clear that these two major directions of selection are not wholly compatible. A compromise results and the principal questions are first, whether or not that compromise is acceptable to growers in terms of crop damage, and second, if not, can we modify the compromise to a more acceptable outcome?

Wolfe & Schwarzbach (1978a) pointed out the value of versatility to the pathogen at the beginning of the epidemic when selection is exerted by the host range on the immigrant air spora landing on newly emerged crops. They also pointed out that the effects of selection within the crop, i.e. adaptation, become predominant as soon as infections are established within the crop and begin to reproduce. This view was elaborated by Wolfe *et al.* (1983a) who suggested that within-crop selection was considerably more effective than between-crop selection. In other words, the advantages of versatility that are gained are usually insufficient to allow the highly versatile phenotypes to compete successfully within a field against the phenotypes selected only for maximum reproduction in that field.

An insight into this problem has been gained from continuous monitoring of the pathogen population over a number of years for the UK Cereal Pathogen Virulence Survey (and previously, the UK Physiologic Race Survey). The first objective of the Survey is to provide early warning of a pathogen response to hitherto effective resistant varieties or fungicides. The data produced also reveal the fate of each monitored pathogenicity character on the range of non-corresponding hosts.

Each year, many isolates are collected and classified according to the group of varieties with a particular resistance phenotype (BMR, or Barley Mildew Resistance group) from which they were collected. They are tested for their ability to produce colonies on detached leaf segments of seedling leaves of test varieties, including a susceptible control, Golden Promise. The test varieties are representatives of the same groups as the source varieties. This means that the tests provide data on the pathogenicity of the population collected from each BMR group on all other BMR groups. Conversely, they also reveal the pathogenicity or frequency of each BMV (Barley Mildew Virulence) character on all BMR groups. Generally, varieties in the same BMR group behave similarly, as do isolates with the same pathogenicity character, so that the interactions observed largely reflect the behaviour of the major genes involved.

Table 19.1 gives the average for the last 14 years of the relative numbers of colonies produced by each pathogenicity character from all sources. There are, of course, many fluctuations concealed in mean values derived from such a long period. However, in relation to the principal deviations in Table 19.1, described below, the fluctuations are small. Virtually all isolates are pathogenic on BMR 1, the winter barley group, since this is almost essential for winter survival. There is also increasing evidence to show that pathogen populations from spring barley (BMR 2–6) migrate back on to winter barley during the period of rapid increase of the epidemic on the spring crop. Population samples from winter barley taken either in the summer or the early autumn should therefore reflect the overall relative frequencies of the major BMV characters. If there are no interactions between the non-corresponding pathogenicity characters and each BMR group source, the values obtained within each column of Table 19.1 should be equal, and equal to the value obtained from the population from BMR 1 and, of course, the mean for the column.

In Table 19.1, the mean values for the non-corresponding pathogenicity characters in the samples obtained from the spring BMR groups are correlated with those

Table 19.1. Numbers of colonies of *Erysiphe graminis* f. sp. *hordei* on leaf segments of test varieties relative to those on segments of the susceptible control variety, Golden Promise, averaged over 14 years, 1970–83. The samples obtained were classified according to source variety; the source, and test, varieties were assigned to BMR (Barley Mildew Resistance) groups according to the specific resistance genes that each carries

	Test variety:				
Source	BMR 2	BMR 3	BMR 4	BMR 5	BMR 6
BMR 1[a]	83	48	48	37	27
BMR 2	–	56	47	30	23
BMR 3	93	–	37	26	17
BMR 4	79	29	–	29	5
BMR 5	85	13	29	–	5
BMR 6	103	28	23	22	–
Mean	90	32	34	27	13

[a] BMR 1 varieties possess either, or both, *Ml(37/136)* or *Ml(41/145)*, BMR 2 varieties have *Mlg*, BMR 3 have *Mla6*, BMR 4 have *Mlv*, BMR 5 have *Mla 12* and BMR 6 have *Mlk* + *Mla7*.

from the winter barley group. This correlation largely reflects the usage of each corresponding BMR group which, over the past 14 years in the UK, follows approximately in numerical order, BMR 2 having been used the most and BMR 6 the least, until recently. However, there are obvious differences between the rows in the Table. Some of the non-corresponding values are higher, and most lower, than would be expected from the winter barley values.

It appears then, that pathogenicity for a particular host group can be influenced by a *non-corresponding* host resistance. This concept was introduced by Wolfe *et al.* (1983a or b) and is somewhat contrary to the generally accepted view of the gene-for-gene theory, which assumes an all-or-nothing relationship between a host resistance gene and the corresponding gene for pathogenicity and no relationship between that host gene and other, non-corresponding, genes for pathogenicity (Christ *et al.*, this volume).

In Table 19.1, the values in the BMR 1 row represent, in turn, the frequencies of the non-corresponding pathogenicity characters, BMV 2–6, in the populations sampled on BMR 1. Since all of the isolates involved are pathogenic on BMR 1, these values also represent, respectively, the frequencies of the non-corresponding characters combined with the corresponding character; for example, the frequency of BMV 2 and 1 + 2 is 83.

The expected value for the frequency of a pair of *non*-corresponding characters on BMR 1, for example, BMV 2 + 3 is normally calculated as the product of the total frequencies of BMV 2 and 3, i.e. $(83 \times 48)/100 = 40$. If follows, therefore, that among the isolates with BMV 2, the proportion that possesses BMV 3 is $(40/83)100 = 48$.

Thus the expected frequency of BMV 3 among the successful immigrant spores filtered out of the atmosphere by fields of BMR 2 varieties, which all carry the corresponding character BMV 2, is 48, equal to the frequency of BMV 3 on BMR 1. In a similar way, the frequency of BMV 4 on BMR 1 provides an expected value for BMV 2 + 4 on BMR 2. The same argument can be applied to each other row in Table 19.1, assuming that there are no interactions between the host resistances and the non-corresponding pathogenicity characters.

For examining the data in detail, the diagonals from bottom left to top right in Table 19.1 provide comparisons of reciprocal pairs of corresponding and non-corresponding pathogenicity characters against expectation. For example, the first diagonal shows that in the samples from BMR 2, non-corresponding pathogenicity for BMR 3 (equivalent to BMV 2 + 3) had a value 56 which was greater than that determined from the samples from BMR 1 (48). Similarly, in the samples from BMR 3 non-corresponding pathogenicity for BMR 2 (also equivalent to BMV 2 + 3) had a value 93 which was also greater than that determined from BMR 1 (83). In other words, in the samples from both BMR 2 and BMR 3 the frequency of the pathogenicity combination BMV 2 + 3 was higher than expected from the samples from BMR 1 and, indeed, from BMR 4–6. This may relate to the domination of the pathogen population by this pathogenicity combination in the late 1960s and early 1970s (Wolfe & Barrett, 1976).

There is one other example, not illustrated in Table 19.1, in which a pathogenicity combination occurred more often than expected. Whenever pathogenicity for BMR 7 was detected in the surveys it occurred more on BMR 5 varieties than on any others. This suggests some positive interaction between the characters for pathogenicity against the two *Mla* alleles, *Mla1* (BMR 7) and *Mla12* (BMR 5). This phenomenon is in marked contrast with the apparent selection against the combination of characters for pathogenicity against the *Mla* alleles *Mla12* and *Mla7* (BMR 6), described below.

The reciprocal values for corresponding and non-corresponding pathogenicity on BMR 2 and 4, 2 and 5, and 2 and 6 were similar to the expected values obtained from the BMR 1 samples. In each case, single varieties carrying the corresponding combination of host resistance genes (e.g. BMR 2 + 4 = Georgie, Koru, Sundance, etc.) has been widely used in agriculture. Varieties with BMR 3 + 4 and 4 + 5, have also been widely used in recent years, producing a similar effect.

The values of the remaining pairs of pathogenicity characters were all more or less lower than expectation, particularly BMV 4 and 6, and BMV 5 and 6, which produced low values consistently throughout the survey period. For example, in the samples from BMR 6, pathogenicity for BMR 4 had a value 23, less than half that expected from the BMR 1 samples (48). Similarly, in the samples from BMR 4, pathogenicity for BMR 6 (5) was considerably less than the value 27, obtained in the samples from BMR 1. In other words, in the samples from both BMR 4 and BMR 6, the value for the pathogenicity combination BMV 4 + 6 was lower than expected from the samples from BMR 1 and, indeed, from the other BMR groups. The symmetry of this and the other examples suggest that selection is acting against the combination of pathogenicity genes. Unfortunately, the data do not prove whether selection is acting against the combination, or against each non-corresponding character on each host.

Varieties with the resistance combination BMR 4 + 6 were not used until very recently, which may account for the deficiency of the corresponding pathogenicity combination. The remaining three combinations, BMR 3 + 5, 3 + 6, and 5 + 6, have not been used at all since the genes involved are allelic or closely linked at or near the *Mla* locus on chromosome 5. Nevertheless, the opportunity for selection of the appropriate pathogenicity combinations has been considerable since varieties from each of the host resistance groups have been used separately on a large scale and simultaneously for a number of years.

In the example of BMV 5 + 6, it has always been easy to find population samples pathogenic for either BMR 5 or 6 but virtually non-pathogenic on the other host, despite the fact that for several years each host occupied about a quarter of the spring barley area. However, Limpert & Schwarzbach (1981) noted that in Austria, during the same period (1979–81), the combination BMV 5 + 6 increased rapidly. There were, however, two major differences between the regions. In Austria, the area separately occupied by BMR 5 and 6 increased to about 90% of the total spring barley area so that there was a much greater selective advantage for the pathogenic combination. Further, there is some evidence to suggest that the barley area in Austria is on the receiving end of a 'Mildew Path' of spores being blown into the area on prevailing westerly winds. This would have the effect of increasing the relative efficiency of the between-field phase of selection.

The difference between the fate of BMV 5 + 6 in the UK compared with in Austria indicates that the size of the interaction between these characters and their non-corresponding hosts cannot be great. Nevertheless, the simultaneous exploitation of a number of different *Mla* alleles could have a significant effect on pathogen spread. For example, in all of the comparisons in Table 19.1, with the exception of BMV 2, the pathogenicity of each population was mostly less than 50% for each non-corresponding variety. Had there been more resistances, then the values would have been still lower. Under those conditions, one could expect varietal diversification to be more effective, since relatively few of the spores from any one variety would be pathogenic on any other.

The particular advantage of the *Mla* locus in this respect is that it limits breeders from combining any of the alleles. Individually, the alleles are easy to incorporate into breeding programmes and are highly effective in terms of disease control. Greater concentration on exploitation of the *Mla* locus could have the added benefit of increasing the isolation of subpopulations of the pathogen, thus limiting overall epidemic spread following the emergence of a corresponding pathogenicity character. This advantage has also been used to great effect in variety mixtures (see below) using components with different *Mla* alleles, particularly those which have been shown to select for pathogenicity combinations that occur at lower than expected frequencies. Isolation could be further reinforced if the group of *Mla* alleles used in winter barley varieties was different from that used in spring barleys.

The accumulation of *Mla* alleles generally in breeding programmes also allows for diversification in time. For example, *Mla12* is now in a third cycle of use in the UK because of its re-introduction with different sets of background and other resistance genes. If a relatively large number of alleles was in constant use, each would appear in new varietal backgrounds at different times and thus continue to provide the benefit of maintaining some isolation of subpopulations of the pathogen.

The value of enforced diversification provided by the *Mla* locus raises the question of the evolution of the complex locus and its possible, similar advantage under natural selection. Complex loci occur in host resistance in other post-pathogen systems, which may also serve a similar function. Haldane (1948) suggested a different advantage of such systems in that they involve genes that can provide great diversity of host resistance without any severe impact on the remainder of the host genotype and its other functions.

In summary, the frequency of a pathogenicity character is determined largely by the frequency of the corresponding resistance and by the direction and degree of selection acting on it in the populations on other, non-corresponding varieties. Vanderplank (1968) suggested that in the *absence* of the matching host, the frequency of a particular pathogenicity character would depend on whether that host was 'strong' or 'weak'. It is more difficult for the pathogen to overcome some hosts than others, but this characteristic need not necessarily correlate with the ability of the pathogen to survive on hosts carrying resistance genes other than the one in question (Chin & Wolfe, 1984). For example, in Table 19.1, BMR 6 appears 'strong' in terms of the low frequency of BMV 6 on BMR 4, but less so in relation to the pathogenicity of BMV 6 on BMR 2. There may be some biological truth in the 'strong-weak' hypothesis but it is untestable in practice, and many exceptions and more direct explanations can be found (Johnson, this volume).

In relation to fungicides, however, the arguments of Georgopoulos (1984; this volume) imply an analogy with the 'strong—weak' gene theory. For example, benomyl may be regarded as 'weak' in the sense that insensitive forms decline in frequency very slowly in the absence of the fungicide. Ethirimol and triadimenol can be regarded as 'stronger' in that highly insensitive forms decrease relatively quickly on untreated host tissue. Of course, with fungicides there are no characters involved other than the direct action of the chemical on the pathogen, so that it is perhaps not surprising that such relationships are evident.

The interaction of insensitivity genes with non-corresponding fungicides

Recombination and selection among fungicide insensitivity genes presumably follow a pattern similar to that observed for pathogenicity genes. Again, the advantage of versatility is likely to be offset by the physiological problems of carrying two or more insensitivity characters. For example, isolates with some insensitivity to ethirimol and to the triazole fungicides were not as insensitive to either fungicide as isolates adapted to only one or other of these compounds (unpubl.). This suggests that their increase may be restricted relative to the simpler forms, though we have no information on their fitness on untreated host tissue. As with pathogenicity gene combinations, we cannot exclude the possibility that under continued selection, the genetic backgrounds for combinations of required genes might compensate for their disadvantages, allowing them to become predominant.

A further complication in fungicide insensitivity compared with host pathogenicity is that there is more evidence of different genes controlling the same mechanism and of different mechanisms for insensitivity to a particular fungicide, for example, with the triazole fungicides (van Tuyl, 1977). This may explain the variation in insensitivity to triazoles that we have observed in *E. graminis* f. sp. *hordei* (Wolfe *et al.*, 1984b). There is no analogous evidence for variation in the genetic control of pathogenicity in this organism, possibly because it has not been sought. The occurrence of alternative mechanisms or genes obviously provides a greater potential for the recombination of appropriate insensitivity and pathogenicity characters.

The interaction of pathogenicity and fungicide insensitivity

Selection for fungicide insensitivity will tend to occur first on varieties that are susceptible at the time of introduction of the fungicide because this is where the chemical will be used first and these varieties support large pathogen populations. Moreover, if the varieties involved possess a common, known resistance gene, it is likely that a linkage disequilibrium may be formed in the pathogen population between the corresponding pathogenicity gene and a gene for fungicide insensitivity.

This may have occurred on a large scale twice in recent years. The first occasion followed the introduction of ethirimol which was frequently used at the beginning of the 1970s to control mildew on Sultan (*Mla12*), a spring variety that had recently become highly susceptible. From that time, pathogen isolates insensitive to ethirimol and pathogenic on Sultan occurred more often than expected from their separate frequencies. However, this combination was found to be widespread throughout the U.K. and highly persistent, even after the decline in use of both the fungicide and the *Mla12* host. It was also detected in northern Germany (Nicklahs, personal communication), even though ethirimol had not been used in that region at the time. The possibility still remains, therefore, that the association between these two characters may have involved other than a simple 'hitch-hiking' effect.

More recently, following the introduction of the triazole fungicides, we quickly found that pathogenicity for the variety Midas (*Mla6*) was commonly associated with the first isolates of the pathogen showing reduced sensitivity to triadimefon and triadimenol. The probable explanation was that the fungicides were used intensively first in Scotland and northern England. The mildew populations in these areas had a preponderance of pathogenicity for Midas, which was the only variety with a recognized gene used there on a large scale at the time. The disequilibrium became easily detectable in the south of England, encouraged by the much increased use of triazole fungicides to the extent that, in 1983, pathogenicity for Midas occurred at a high frequency over the whole of England even though the matching resistance was being used on only a small area. In other words, there appeared to be a founder effect in England caused by south-ward drift of pathogen genotypes with insensitivity to triazoles combined with pathogenicity for Midas. Intensive use of the fungicide in the south selected for the insensitivity and jeopardized the resistance of the *Mla6* varieties being grown in the region.

Unfortunately, the linkage disequilibrium led to some loss of the separation in the pathogen population caused by allelism at the *Mla* locus of the resistances in BMR 3 and 6. During the early 1980s and the widespread intensification of triazole application, there was a large and rapid increase in the use of the spring barley, Triumph, which was BMR 6. Monoculture of Triumph quickly led to a decline in its resistance and an increased use of fungicide on the variety. This caused selection for pathogen genotypes with BMV 6 and triazole-insensitivity combined. These came initially from the population with a high frequency of the combination of fungicide insensitivity and BMV 3, which increased the probability of selecting genotypes with the combination (BMV 3 + 6 + triazole insensitivity).

The linkage disequilibrium of BMV 3 with triazole-insensitivity provided a logical reason for advising growers to use fungicides with a different mode of action on BMR 3 varieties. Field trials at the PBI in 1982–3 supported this view: a greater response to chemical treatment of BMR 3 varieties was obtained with ethirimol than with triadimenol (Wolfe *et al.*, 1983). However, because there is evidence of selection against BMV 3 on varieties that do not possess the corresponding resistance, it is likely that the linkage disequilibrium will be transient, particularly if selection for fungicide insensitivity is continued on a large scale on a range of varieties.

This leads to the speculation that, if it were possible in practice, diversification between fungicides could be integrated with diversification between varieties, to the benefit of both. For example, if the use of fungicide A was both limited and restricted to use on varieties that carried one *Mla* allele, whilst the use of fungicide B was similarly limited and restricted to varieties with a different *Mla* allele, then the isolation of the selected sub-populations of the pathogen would be reinforced. Similar arguments could be advanced for the separation of resistance genes and fungicides between winter and spring sown crops. Such systems would undoubtedly delay the rate of pathogen evolution relative to current practice which is, effectively, the random exploitation of host resistance and fungicides.

Maintaining disease control

Maintaining host resistance

The observations discussed so far have resulted from the application of the 'replacement strategy' (Duvick, 1977) both for resistant varieties and for fungicides: varieties or fungicides are replaced by others as soon as their massive exploitation has led to a noticeable loss of effectiveness, which frequently occurs, sooner or later. This happens because, following the introduction, for example, of a new disease resistant variety, the rapidly expanding area occupied by pure stands of the variety act as immensely effective spore traps for filtering out of the atmosphere any immigrant spore that happens to contain an initially rare, corresponding pathogenicity character. If the spore infects and reproduces, any daughter spore that is produced is in an ideal situation for further reproduction because of the high probability that it will land on a compatible host surface, either on the plant on which it was produced or on any other nearby plant. In this way, the absolute population size of the corresponding pathogen genotype increases at a maximum rate (Gale, this volume; Barrett, this volume). Indeed, it would be difficult to design a system that would lead more rapidly to a loss of effectiveness. Unfortunately, with the single possible exception of the *mlo* resistance, there has been no evidence of long-term durability among the resistant varieties that have been used, which has prevented exploitation of the durable resistance strategy, proposed by Johnson (1981).

Following the argument that within-field selection is more efficient than between fields, it is likely that if the numbers of different resistant varieties and fungicides could be increased and maintained in roughly equal proportions, the greater diversification of hosts and fungicides would slow down the required rate of replacement. Unfortunately, such a system is impossible to organize both in terms of development and introduction of the required numbers of resistant hosts and fungicides, and in their distribution and use. One possibility, however, energetically supported by Dr Erik Schwarzbach, is to try to ensure that different resistance genes are used in the winter and spring crops.

An alternative to diversification is to build a number of known resistance genes into a single variety that should thus select only for the exceedingly rare pathogen recombinant possessing all of the matching pathogenicity genes. Three practical difficulties militate against this approach for barley mildew in Europe. First, many of the desired genes are alleles at, or closely linked to, the *Mla* locus and so cannot be easily combined. Second, within the small epidemiological zone of Europe, it would be impossible to prevent prior use of the component resistance genes alone or in simple combinations, allowing a degree of pre-selection within the European pathogen population before the introduction of the complex variety (Wolfe & Barrett, 1976). Third, if such a complex variety could be produced, its commercial introduction and success could not be guaranteed. Each of these problems is of less importance in the programmes for breeding wheat varieties for resistance to the less variable stem rust pathogen (*Puccinia graminis* f. sp. *tritici*) in Australia and the United States. In regions of these countries, the pathogen population is monitored to predict the resistance gene combinations that should be introduced in new wheat varieties; this procedure has been refined and used successfully for a number of years (e.g. McIntosh, 1976).

Considerations such as these led John Barrett and me to look more closely at the use of variety mixtures for disease control (Wolfe & Barrett, 1979). The use of variety or species mixtures in this way has been considered many times, the first known occasion being that recorded by Tozzetti (1767). More recently, a branch of this thinking led to the development of the multiline concept (Browning & Frey, 1969), though this has proved to be of limited value in practice (Wolfe & Barrett, 1980).

By using appropriately designed trials in which the plots are protected as far as possible from local external influences, it has been possible to show repeatedly with many different mixtures of only three well-chosen component varieties, that diseases can be reduced usually by more than one-half and sometimes up to 80%, relative to the mean of the components grown as pure stands. These reductions are correlated with yield increases of 6–10% above the mean of the components.

The principle of using variety mixtures is simple; by growing an intimate mixture of plants with different resistances, the spread of the pathogen selected on any one plant will be restricted. There is, of course, selection for the pathogen genotype able to reproduce on all of the host components in the mixture. Since such genotypes are less well able to compete than simpler genotypes on individual plants, they may not necessarily predominate. Further, the outcome of the competition between simple and complex genotypes is also dependent on the distribution of spores within and between plants. For example, if a large proportion of spores is distributed between plants, this will tend to favour the more complex pathogen genotypes. However, this proportion varies at different stages of plant development, which adds further to the unpredictability of the final outcome (see Barrett, 1980; Barrett & Wolfe, 1980; Chin & Wolfe, 1984).

In field trials in which we have followed the dynamics of the pathogen on pure stands and mixtures, we have so far not detected any instances in which complex races have predominated at the end of the epidemic (Barrett & Wolfe, 1980). In another example involving varieties that did not differ greatly in their resistance genotypes, we carried out six field trials between 1978 and 1983 on the three winter barley varieties Athene, Igri and Sonja, and the mixture of equal parts of the three varieties. All three varieties have a resistance gene in common, though Athene has a second gene and was more resistant than the other two. Because of their similarity, it was not expected that a mixture of the three varieties would greatly reduce mildew infection; on average the mixture had one-third less infection than the mean of the components, and in four of the six trials yielded more than the expected value.

To assess the dynamics of the pathogen population in the mixture, we exposed seedlings of each of the three varieties in plots of each variety and of the mixture, at intervals during each season. After exposure, the seedlings were incubated and the numbers of colonies of the pathogen that developed on each were counted. For convenience, in Figure 19.2, the values for the colony numbers on seedlings of Athene, Igri and Sonja exposed in pure stands of those varieties, are set to 100 for each exposure date. In the majority of the comparisons, the equivalent values observed in the mixture plots started equal to or higher than those in the pure

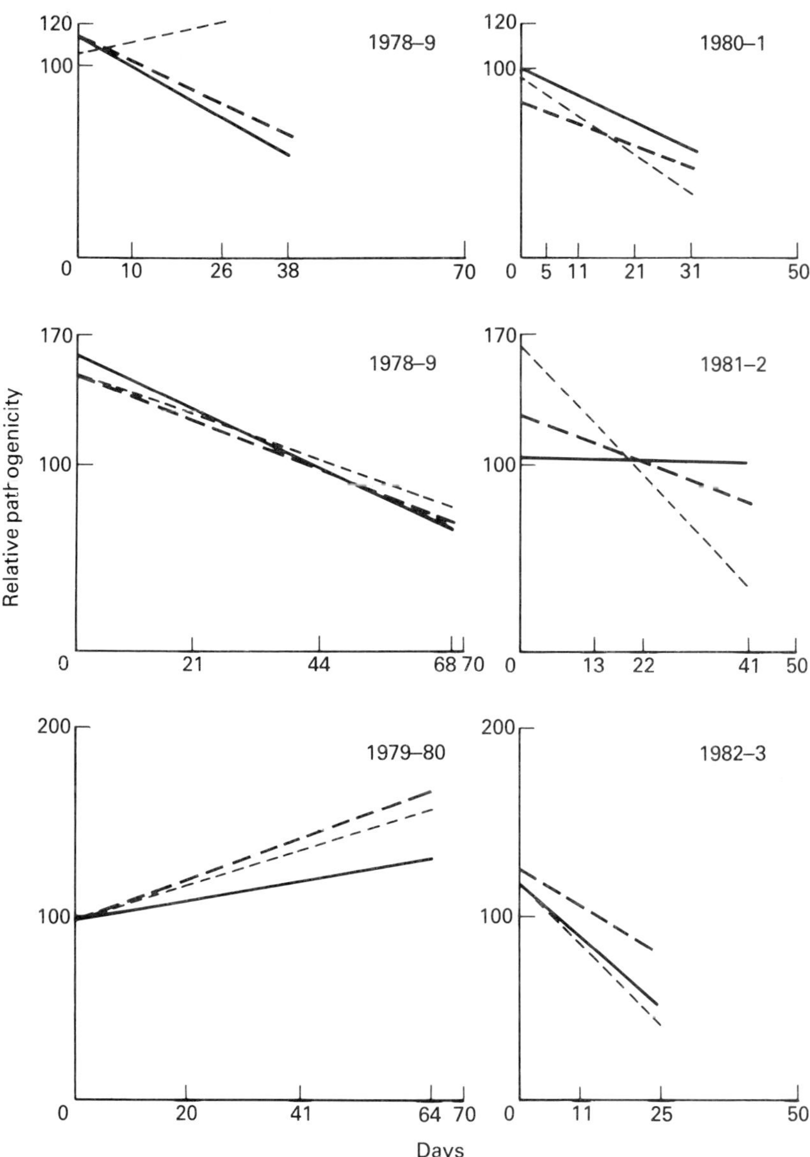

Figure 19.2. Relative numbers of mildew colonies on seedlings of the winter barley varieties, Athene, Igri and Sonja in field plots of the same varieties grown as pure stands or as a mixture of equal parts of those varieties. The seedlings were exposed on several occasions over a period of about 40 days in each of six trials. The means of the pure stand values on each test variety are adjusted to 100 for each exposure date. ——— = Athene, ------ = Igri, · · · · · = Sonja.

stands, but subsequently decreased. The exception was the trial in 1979–80, for which we have no explanation.

The relative decrease in pathogenicity in the mixed plots, after starting at a high level, is not inconsistent with other observations (Barrett & Wolfe, 1980) or the model proposed by Barrett (1980). The likely explanation is that, early in the epidemic, there is an advantage for immigrant spores pathogenic on all three varieties since they have a higher probability of initial survival; pathogenicity values in the mixed plots would therefore be high. Later, if pathogen genotypes specifically adapted to each host begin to predominate, then the average pathogenicity for each host in the mixture would be expected to fall.

During the 5-year period there was no evidence of an increase in the frequency of a genotype of the pathogen able to grow on all three varieties, despite the fact that the winter barley acreage was dominated by Igri and Sonja, and that Athene was also exploited commercially.

However, the mixture was grown on only a very small area and we cannot predict the outcome if it had been more widely grown. Indeed, because of the impossibility of making such predictions, the use of variety mixtures has been proposed as a two-part system, first, that appropriate varieties should be grown in mixtures, and second, that *the mixtures should be diversified as far as possible, both in space and time.*

Given that a system of using variety mixtures works in terms of disease control and restraining pathogen evolution, the key question is whether or not the system is practicable, and more so than the other alternatives. I believe it is, and this is borne out by commercial development of the system, particularly in the UK and Denmark. Indeed, there was little difficulty in persuading the variety rights legislators in the EEC to accept mixing of cereal varieties within member states from 1979. There is now wide interest in the potential of the system for the control of other diseases in other crops.

The variety mixture system is complementary to other facets of the development and commercialization of varieties, in that mixing is the final process carried out on the seed before sowing. Mixing can be included in conventional variety diversification and in systems involving fungicide treatments (see below). It can be applied to varieties with durable resistance, to varieties with combinations of resistance genes, to hybrid varieties and to species mixtures which may or may not incorporate variety mixtures.

So far as cereals are concerned, appropriate mixtures are likely to reduce the level of any foliar disease, so that mixtures can be designed to deal with particular local problems in addition to a principal target disease. Control may be insufficient to eliminate the need for fungicide, but the amount required will be reduced and the timing of spray application made less critical since epidemic development will be slow.

Whenever foliar disease occurs, there is likely to be some control and thus some yield benefit; in the absence of disease there is no yield penalty. Because of the nature and flexibility of the system the yield advantage can be exploited in two ways. First, by continually changing the mixture composition to take advantage of the improved yield of new varieties, the yield potential of mixtures can be maintained close to the maximum possible. Second, variety mixtures are higher yielding and more predictable in yield performance than most single varieties (Wolfe *et al.*, 1984a). Their stability appears to be similar to that of the mean of their components, with the bonus of a yield increase, particularly if disease occurs (Wolfe *et al.*, to be published).

The remaining principal advantage to the farmer, is that the mixture system costs little to implement in terms of seed cost and management. This may make the system even more advantageous in cereal-producing areas outside western Europe, where crop inputs are relatively more expensive.

The main disadvantages that have emerged are first, that the system only works well for foliar diseases that depend more on within-crop spread over several pathogen generations than on immigration of inoculum with few generations of within-crop increase. This also assumes that diversification of disease resistance among components is available; if there is no diversification, then, of course, the crops involved may be at considerable risk.

The second disadvantage is that there is considerable reluctance from maltsters, particularly in the UK, to accept mixtures of known composition. Unfortunately, this applies even to Maris Tricorn, a three variety mixture that we produced for the quality market, and which was accepted as at least adequate for malting. Indeed, as a pathologist I would argue that the malting industry places too great a constraint on barley growers in forcing them into massive monoculture of the one or two varieties that it happens to regard as suitable at a particular time. The result is inevitable, as we are now seeing with the high-yielding malting variety Triumph, eagerly sought by the trade. Because of over-exposure, the field performance of Triumph is now declining rapidly and can be maintained only by increasing amounts of fungicide application. The approach to quality in wheat is much more satisfactory in that the industrial users have improved and altered their technology so that they are now able to handle a higher proportion of home-grown wheat; they have modified their methods to ease the problems faced by the breeders.

A third disadvantage has been the slow response of the seed trade to involvement in the system. The technology required is simple and available but requires some investment. It seems that the investment will not be made more widely, however, unless the demand from farmers becomes more obvious. This in turn depends on cereal advisers projecting the value of low input variety mixtures, particularly for feed use, to the farming public. Here I must confess to some disappointment; the advantages of the system seem obvious and outweigh the disadvantages but there seems often to be a reluctance to push forward with a scheme which departs from the recently established and hazardous tradition of monoculture of uniform varieties.

Maintaining fungicide activity

Currently, fungicide use also follows a 'replacement strategy'. The drive towards maximizing the use of a single fungicide, the equivalent of variety monoculture, is even greater than with host varieties because of the higher costs of developing, launching and maintaining a fungicide. Fortunately, at least under cereal cultivation, problems due to fungicide insensitivity are generally slow to develop because the period of selection imposed by a fungicide on a crop is short relative to the life of the crop; this contrasts with resistance genes which are often functional throughout the life of the host. Nevertheless, problems have arisen and so there is now increasing interest in various diversification strategies for fungicide use (Skylakakis, this volume), although one of the simplest and most effective strategies is to avoid prophylactic treatment.

In the absence of tested models, and due to commercial constraints, the major strategy emerging from the manufacturers at present is the use of fungicide mixtures, even though they are expensive and there is little evidence for their ability to usefully delay the increase of insensitivity. A major disadvantage, indeed, of tank-mix formulations, is that the components are applied jointly to each plant in the crop. If one material is more persistent than the others then at some stage after application there will be uniform exposure to the pathogen of a single compound on all plants. A better system may be to arrange that individual plants or groups of plants receive a different fungicide. The effect on the pathogen population would then be more akin to the use of mixtures of resistant varieties (Wolfe, 1981).

To diversify in time by alternating between treatments with different fungicides usually requires recommendation by independent advisers, since manufacturers are often constrained by interests in very few different compounds. There are some indications of the value of alternating between different materials from barley mildew control in Scotland. Following the development of insensitivity to the triazoles shortly after their introduction for use on the widely grown, highly susceptible variety, Golden Promise, Scottish farmers followed the advice to diversify between the triazoles, morpholines and pyrimidines. As a result, triazole insensitivity is now more common in England than in Scotland, since English farmers have continued to depend much more havily on the triazoles.

Integrating strategies for using resistant varieties and fungicides

For the farmer, the most desirable strategy would be to grow durably resistant varieties that required no further treatment for disease. This goal seems unreachable in the face of the variability and mobility of *E. graminis* f. sp. *hordei*. The next simplest strategy is to diversify among the available varieties and fungicides. In barley, there is little diversification in practice among varieties, partly because of the constraints imposed by the malting industry and partly because there is little diversity of resistance among the available winter barleys. Diversification among fungicides only seems to gain popularity when a major fungicide is already losing effectiveness.

The studies of population dynamics suggest, for example, that if we could separate the resistance genes used in winter and spring barley and simultaneously limit the use of specific fungicides to each crop, there could be a considerable isolation of the pathogen populations on the two sorts of barley which would be beneficial in limiting overall population increase. In a similar way, the potential for isolation offered by the *Mla* alleles could be reinforced if particular fungicides were restricted to use on varieties with particular alleles. Continuous monitoring of the pathogen populations could also be used to develop annual recommendations for which fungicides are likely to be most effective on which varieties. Our field trial evidence from 1982–3 showed that, at least for those years, triazoles would have been more effective on *Mla12* than on *Mla6* varieties, and that ethirimol would have been more effective on *Mla6* and *Mlk* plus *Mla7* varieties than on *Mla12* varieties (Wolfe *et al.*, 1983, 1984a).

The limiting factor for these recommendations is that the biological diversification may involve commercial constraints that are not practically feasible. A more practicable alternative is to integrate the use of variety mixtures and fungicides (Wolfe, 1981; Wolfe & Riggs, 1983). The simplest method is to treat the seed of only one component of a mixture, usually the most susceptible, with a single mildew or broad spectrum fungicide. This has the obvious benefit for the farmer of limiting the cost of fungicide treatment and is another feature of Maris Tricorn: the commercial mixture involves treatment of the component variety Porter with imazalil and thiabendazole. Disease control is effective but we do not yet have information on the dynamics of fungicide insensitivity in the mixture. The hypothesis is that any spores carrying fungicide insensitivity would have a limited probability of landing on fungicide-treated tissue, considerably less than in a conventionally treated crop. They would be more likely to land on untreated tissue, on which they would be at a selective disadvantage and thus tend to decrease in relative frequency. Pathogen genotypes from untreated host tissue would be limited in their spread by the barrier of treated plants. Because of the diversity within each crop of Maris Tricorn, it is likely that pathogen evolution towards improved performance on the crop as a whole would be slow.

Ideally, in future seasons, the varietal composition would be altered and the treatment modified either by using different fungicides or by using the same material on different components. Such additional diversification in time would provide further protection of the component varieties and fungicides, all at minimal cost to the farmer.

Unfortunately, and for reasons not related directly to the concept, it is likely that Maris Tricorn will not be a commercial success. Nevertheless, the introduction of the concept has advanced sufficiently far to show that it is a practical proposition and that commercial seed production is perfectly feasible. Perhaps the steadily increasing interest in what might now be termed 'conventional' variety mixtures, will help to pave the way for later introductions of more successful fungicide-integrated mixtures, and for mixtures of species or genera, or, indeed, involving different levels, and designed for low input production of stable, high yields of feed crops that prevent, or at least severely delay, pathogen evolution. Eventually, we may follow Darwin's (1872) fundamental view of divergence, when he pointed out that in the comparison of 'one variety and several mixed varieties' then, in the latter case, 'a greater number of individual plants ... would succeed in living on the same piece of ground'.

Acknowledgements

So many people have contributed to the development of these studies over more than 20 years, to all of whom I am sincerely grateful, that it is impossible to mention them all individually. Nevertheless, I feel I must mention John Barrett, Roy Johnson, Peter Minchin, Peter Scott and Susan Slater for their long-term help, influence and support.

References

Barrett J.A. (1980) Pathogen evolution in multilines and variety mixtures. *Zeitschrift für Pflanzenkrankheit und Pflanzenschutz* **87**, 383–96.

Barrett J.A. & Wolfe M.S. (1980) Pathogen response to host resistance and its implication in breeding programmes. *EPPO Bulletin* **10**, 341–7.

Brent K.J. (1982) Case study 4: powdery mildews of barley and cucumber. In: *Fungicide Resistance in Crop Protection* (Ed. by J. Dekker & S.G. Georgopoulos), pp. 219–30. Centre for Agricultural Publishing and Documentation, Wageningen.

Browning J.A. & Frey K.J. (1969) Multiline cultivars as a means of disease control. *Annual Review of Phytopathology* **7**, 355–82.

Chin K.M. & Wolfe M.S. (1984) Selection on *Erysiphe graminis* in pure and mixed stands of barley. *Plant Pathology* **33**, 535–46.

Clarke B. (1976) The ecological genetics of host–parasite relationships. *Symposia of the British Society for Parasitology* **14**, 87–103.

Darwin C.R. (1872) *On the Origin of Species by Means of Natural Selection*, (6th edn). Murray, London.

Duvick D.N. (1977) Major United States crops in 1976. *Annals of the New York Academy of Sciences* **287**, 86–96.

Fletcher J.T. & Wolfe M.S. (1981) Insensitivity of *Erysiphe graminis* f. sp. *hordei* to triadimefon, triadimenol and other fungicides. *Proceedings of the British Crop Protection Conference–Pests and Diseases 1981*, pp. 633–40.

Georgopoulos S. G. (1984) Adaptation of fungi to fungitoxic compounds. *CIBA Foundation Symposium* **102**, 190–203.

Haldane J.B.S. (1948) Disease and evolution. *La Ricerca Scientifica Supplement* **19**, 68–76.

Johnson R. (1981) Durable resistance, definition of, genetic control, and attainment in plant breeding. *Phytopathology* **71**, 567–8.

Koltin Y. & Kenneth R. (1970) The role of the sexual stage in the oversummering of *Erysiphe graminis* DC. f. sp. *hordei* Marchal under semi-arid conditions. *Annals of Applied Biology* **65**, 263–8.

Leijerstam B. (1965) Studies on powdery mildew on wheat in Sweden. II. Physiological races in Scandinavia in 1962 and 1963 and the resistance in a number of wheats to Scandinavian races. *National Swedish Institute for Plant Protection Contribution 13* **103**, 171–83.

Leonard K.J. & Czochor R.J. (1980) Theory of genetic interactions among populations of plants and their pathogens. *Annual Review of Phytopathology* **18**, 237–58.

Limpert E. & Schwarzbach E. (1981) Virulence analysis of powdery mildew of barley in different European regions in 1979 and 1980. *Proceedings of the 4th International Barley Genetics Symposium*, pp. 458–65.

McIntosh R.A. (1976) Genetics of wheat and wheat rusts since Farrer. *Journal of the Australian Institute of Agricultural Science* **42**, 203–16.

Schwarzbach E. (1979) Responses to selection for virulence against the mlo-based mildew resistance in barley. *Barley Genetics Newsletter* **9**, 85–8.

Tozzetti G.T. (1767) (True nature, cause and sad effects of the rust, the bunt, the smut and other maladies of wheat and of oats in the field) *Phytopathological Classics No. 9* (1952), 159 pp.

Tuyl J.M. van (1977) Genetics of fungal resistance to systemic fungicides. *Mededelingen Landbouwhogeschool Wageningen* **77–2**, 1–136.

Vanderplank J.E. (1968) *Disease Resistance in Plants.* Academic Press, London.

Walmsley-Woodward D.J., Laws F.A. & Whittington W.J. (1979) The characteristics of isolates of *Erysiphe graminis* f. sp. *hordei* varying in response to tridemorph and ethirimol. *Annals of Applied Biology* **92**, 211–19.

Wolfe M.S. (1981) Integrated use of fungicides and host resistance for stable disease control. *Philosophical Transactions of the Royal Society of London, Series B,* **295**, 175–84.

Wolfe M.S. (1982) Dynamics of the pathogen population in relation to fungicide resistance. In: *Fungicide Resistance in Crop Protection* (Ed. by J. Dekker & S. G. Georgopoulos), pp. 219–30. Centre for Agricultural Publishing and Documentation, Wageningen.

Wolfe M.S. & Barrett J.A. (1976) The influence and management of host resistance on control of powdery mildew on barley. *Proceedings of the 3rd International Barley Genetics Symposium, Garching*, pp. 433–9.

Wolfe M.S. & Barrett J.A. (1979) Disease in crops: controlling the evolution of plant pathogens. *Journal of the Royal Society of Arts* **127**, 321–33.

Wolfe M.S. & Barrett J.A. (1980) Can we lead the pathogen astray? *Plant Disease* **64**, 148–55.

Wolfe M.S. & Dinoor A. (1973) The problem of fungicide tolerance in the field. *Proceedings of the 7th British Insecticide and Fungicide Conference*, pp. 11–19.

Wolfe M.S. & Knott D.R. (1982) Populations of plant pathogens: some constraints on analysis of variation in pathogenicity. *Plant Pathology* **31**, 79–90.

Wolfe M.S. & Riggs T.J. (1983) Fungicide integrated into host mixtures for disease control. *Proceedings of the 10th International Congress of Plant Protection,* 834 pp.

Wolfe M.S. & Schwarzbach E. (1975) The use of virulence analysis in cereal mildews. *Phytopathologische Zeitschrift* **82**, 297–307.

Wolfe M.S. & Schwarzbach E. (1978a) Patterns of race changes in powdery mildews. *Annual Review of Phytopathology* **16**, 159–80.

Wolfe M.S. & Schwarzbach E. (1978b) The recent history of the evolution of barley powdery mildew in Europe. In: *The Powdery Mildews* (Ed. by D.M. Spencer), pp. 129–57. Academic Press, London.

Wolfe M.S., Barrett J.A. & Slater S.E. (1983a) Pathogen fitness in cereal mildews. In: *Durable Resistance in Crops* (Ed. by F. Lamberti, J. M. Waller & N. A. Van der Graaff), pp. 81–100. Plenum Press, New York.

Wolfe M.S., Minchin P.N. & Slater S.E. (1981) *Annual Report of the Plant Breeding Institute for 1980*, pp. 88–9.

Wolfe M.S., Minchin P.N. & Slater S.E. (1984a) *Annual Report of the Plant Breeding Institute for 1983*, pp. 87–91.

Wolfe M.S., Minchin P.N. & Slater S.E. (1984b) Dynamics of triazole sensitivity in barley mildew, nationally and locally. *Proceedings of the British Crop Protection Conference–Pests and Diseases 1984*, 465–470.

Wolfe M.S., Slater S.E. & Minchin P.N. (1983b) Fungicide insensitivity and host pathogenicity in barley mildew. *Proceedings of the 10th International Congress of Plant Protection*, p. 645.

Wolfe M.S., Barrett J.A., Shattock R.C., Shaw D.S. & Whitbread R. (1976) Phenotype–phenotype analysis: field applications of the gene-for-gene hypothesis in host–pathogen relations. *Annals of Applied Biology* **82**, 369–74.

Index